Probability
and
Statistics
for
Engineers

2nd edition

Probability and Statistics for Engineers

IRWIN MILLER
Vice President,
Arthur D. Little, Inc.

JOHN E. FREUND
Arizona State University

Prentice-Hall, Inc., *Englewood Cliffs, New Jersey   07632*

Library of Congress Cataloging in Publication Data

MILLER, IRWIN
   Probability and statistics for engineers.

   Bibliography: p.
   Includes index.
   1. Engineering—Statistical methods. 2. Probabili-
ties. I. Freund, John E., joint author. II. Title.
TA340.M5 1977   519.2′023′62   76–14351
ISBN 0–13–711945–3

10 9 8 7 6

Printed in the United States of America

Prentice-Hall International, Inc., *London*
Prentice-Hall of Australia Pty. Limited, *Sydney*
Prentice-Hall of Canada, Ltd., *Toronto*
Prentice-Hall of India Private Limited, *New Delhi*
Prentice-Hall of Japan, Inc., *Tokyo*
Prentice-Hall of Southeast Asia Pte. Ltd., *Singapore*
Whitehall Books Limited, *Wellington, New Zealand*

Contents

v

Preface

This book has been written for an introductory course in probability and statistics for students of engineering and the physical sciences. In the Second Edition, the authors have revised and updated the examples in the text, and set them apart from the explanatory material to make the book easier for students to use. All exercise material has been revised and additional exercises have been added where needed. Applications to operations research have been woven into the chapters on distribution theory (Chapters 3 and 4) rather than being treated separately as in the First Edition. The treatment of Operating Characteristic Curves has been expanded (Chapter 7) and a section on Analysis of Covariance has been added (Section 12.6). Throughout, the text has been revised with the aim of improving clarity and bringing the applications up to date. It has been tested extensively in courses for university students as well as through in-plant training of practicing engineers. The authors have found that the material in the book can be covered in a two-semester or three-quarter course consisting of three lectures a week. However, through a choice of topics the book also lends itself as a text for shorter courses with emphasis upon either theory or applications.

Chapters 2, 3, 4, and 6 provide a brief, though rigorous, introduction to the theory of statistics, and together with some of the material in Chapter 15, they are suitable for an introductory semester (or quarter) course on the mathematics of probability and statistics. Chapters 5, 7, 8, 9, and 10 contain conventional material on the elementary methods of statistical inference, including a treatment of nonparametric methods. Chapters 11, 12, and 13 comprise an introduction to some of the standard, though more advanced, topics of experimental statistics, while Chapters 14 and 15 deal with special applications which have become increasingly important in recent years.

The mathematical background expected of the reader is a year course in calculus; actually, calculus is required mainly for Chapters 4 and 6 dealing with basic distribution theory in the continuous case, for Chapter 15 dealing with applications to reliability theory, and for the least-squares methods of Chapters 11 and 12. The treatment of probability in Chapter 2 is modern in the sense that it is based on the elementary theory of sets.

The authors would like to express their appreciation and indebtedness to the D. Van Nostrand Company for permission to reproduce the material in Table 2; to the late Sir Ronald A. Fisher, F.R.S., Cambridge, and to Messrs. Oliver and Boyd, Ltd., Edinburgh, for permission to reprint Table 4 from their book, *Statistical Methods for Research Workers*; to Professor E.S. Pearson and the *Biometrika* trustees for permission to reproduce the material in Tables 5, 6, and 9; to A. H. Bowker and G. J. Lieberman and Prentice-Hall, Inc., for permission to reproduce Table 8, to Donald B. Owen and Addison-Wesley, Inc., for permission to reproduce part of the Table of Random Numbers from their *Handbook of Statistical Tables*; to Frank J. Massey, Jr., and the *Journal of the American Statistical Association* for permission to reproduce the material in Table 10; to D. B. Duncan, H. L. Harter, and *Biometrics* for permission to reproduce Table 12; to the American Society for Testing and Materials for permission to reproduce Table 13; and to the McGraw-Hill Book Company for permission to reproduce the material in Table 14.

The authors would also like to express their appreciation to the editorial staff of Prentice-Hall, Inc., for their courteous cooperation in the production of this book, to the various secretaries who helped in the typing of the manuscript, and above all to their wives for not complaining too much about the demands made on their husbands throughout the writing of this book.

IRWIN MILLER AND JOHN E. FREUND

1

Introduction

1.1 MODERN STATISTICS

The origin of statistics can be traced to two areas of interest which, on the surface, have little in common: *games of chance and what is now called political science.* Mid-eighteenth century studies in probability (motivated largely by interest in games of chance) led to the mathematical treatment of errors of measurement and the theory which now forms the foundation of statistics. In the same century, interest in the numerical description of political units (cities, provinces, counties, etc.) led to what is now called *descriptive statistics.* At first, descriptive statistics consisted merely of the presentation of data in tables and charts; nowadays, it includes also the summarization of data by means of numerical descriptions.

In recent decades, the growth of statistics has made itself felt in almost every major phase of human activity, and the most important feature of

1

its growth has been the shift in emphasis from descriptive statistics to the methods of *statistical inference,* or *inductive statistics.* Statistical inference concerns generalizations based on sample data; it applies to such problems as *estimating* an engine's average emission of pollutants on the basis of some trial runs, *testing* a manufacturer's claim on the basis of measurements performed on samples of his product, and *predicting* the fidelity of an audio system on the basis of sample data pertaining to the performance of its components.

When one makes a statistical inference, namely, an inference which goes beyond the information contained in a set of data, one must always proceed with caution. One must decide carefully how far one can go in generalizing from a given set of data, whether such generalizations are at all reasonable or justifiable, whether it might be wise to wait until there are more data, and so forth. Indeed, some of the most important problems of statistical inference concern the appraisal of the risks and the consequences to which one might be exposed by making generalizations from sample data. This includes an appraisal of the probabilities of making wrong decisions, the chances of making incorrect predictions, and the possibility of obtaining estimates which do not lie within permissible limits.

In recent years, attempts have been made to treat all these problems within the framework of a unified theory called *decision theory.* Although this theory has many conceptual as well as theoretical advantages, it poses some problems of application that are difficult to overcome. To understand these problems, one must realize that *no matter how objectively an experiment or an investigation is planned, it is impossible to eliminate all elements of subjectivity.* It is at least partly a subjective decision whether to base an experiment (say, the determination of a specific heat) on 5 measurements, on 12 measurements, or on 25 or more. Also, subjective factors invariably enter the design of equipment, the hiring of personnel, and even one's deciding *how* to formulate a hypothesis and the alternative against which it is to be tested. (This important problem will be discussed in some detail in Chapter 7.) An element of subjectivity enters even when we define such terms as "good" or "best" in connection with the description of decision criteria—for instance, in Chapter 11 we shall ask for a straight line which "best" fits a given set of paired data. Above all, subjective judgments are virtually unavoidable when one is asked to put "cash values" on the various risks to which one is exposed. In other words, it is impossible to be completely objective in specifying "rewards" for being right (or close) and "penalties" for being wrong (or not close enough). After all, if a scientist is asked to judge the safety of a piece of equipment, how can he possibly put a cash value on the possibility that he might make an error, when such an error may lead to the loss of human lives?

Whether or not statistical inference is viewed within the broader framework of decision theory, it depends heavily on the theory of probability. This *is* a mathematical theory, but the question of objectivity vs. subjectivity arises in its application and in its interpretation. As we shall see in Chapter 2, statements such as "there is a fifty-fifty chance that a given fabric will ignite when exposed to a cigarette ash" or "the probability is 0.05 that a defective part will not be caught in the final inspection" can be looked upon as objective or subjective evaluations of the corresponding uncertainties.

We shall approach the subject of statistics as a science, developing each statistical idea insofar as possible from its probabilistic foundation, and applying each idea to problems of physical or engineering science as soon as it has been developed. The great majority of the methods we shall use in stating and solving these problems belong to the *classical approach*, because they do not *formally* take into account the various subjective factors mentioned above; in selected applications we present also the *Bayesian approach* (see Sections 7.3 and 9.2) which accounts formally for at least some of these subjective factors. In any case, we shall endeavor continually to make the reader aware that the subjective factors do exist, and to indicate whenever possible what role they might play in making the final decision. Subjectivity plays an important role in the choice among statistical methods or formulas to be used in a given situation, in deciding on the size of a sample, in specifying the probabilities with which we are willing to risk errors, and so forth. This "bread-and-butter" approach to statistics presents the subject in the form in which it has so successfully contributed to engineering science, as well as to the natural and social sciences, in the last twenty years.

1.2 STATISTICS AND ENGINEERING

There are few areas where the impact of the recent growth of statistics has been felt more strongly than in engineering and industrial management. Indeed, it would be difficult to overestimate the contributions statistics has made to problems of production, to the effective use of materials and manpower, to basic research, and to the development of new products. As in the other sciences, statistics has become a vital tool to the engineer; in fact, some knowledge of statistics has become a necessity without which he cannot possibly appreciate, understand, or apply much of the work done in his field.

In this text, our attention will be directed largely towards engineering applications, but we shall not hesitate to refer also to other areas, to impress upon the reader the great generality of most statistical techniques.

Thus, the reader will find that the statistical method which is used to estimate the coefficient of thermal expansion of a metal serves also to estimate the average time it takes a secretary to perform a given task, the average thickness of a pelican egg, or the average I.Q. of an immigrant arriving in the United States. Similarly, the statistical method which is used to compare the strength of two alloys serves also to compare the effectiveness of two teaching methods, the merits of two insect sprays, or the performance of men and women in a current-events test.

In spite of the generality of most statistical techniques, there are also instances where the requirements of different fields have led to the development of special techniques. Thus, problems of economic forecasting led to special methods used in the analysis of series of business data; the pharmacological problem of determining critical dosages led to what is called probit analysis; problems in psychological testing led to factor analysis, and so forth. So far as engineering is concerned, we shall introduce the reader to three areas which have required the development of special techniques. In Chapter 14 we shall present some of the special methods used in problems of *quality assurance*, including the problem of controlling (establishing and maintaining) quality in mass production, the problem of establishing tolerance limits, and problems of sampling inspection. Then, in Chapter 15 we shall present special techniques which have been developed to meet the reliability needs of the highly complex products of space-age technology. The third area, *operations research*, is a new technology which is characterized by the application of scientific techniques (including probability theory and statistics) to problems involving the operations of a "system" looked upon as a whole. Thus, it applies to the conduct of a war, the management of a firm, the manufacture of a product, and so forth. This new technology will not be presented in a special chapter, but some of its methods will be illustrated throughout the text in the exercises and examples.

2

Probability

2.1 SAMPLE SPACES

In statistics, a set of all the possible outcomes of an experiment is called a *sample space*, and it owes its name to the fact that in most instances it consists of all the things that can happen when one takes a sample. Sample spaces are usually denoted by the letter S. To avoid misunderstandings about the words "experiment" and "outcome" as they were used above, it should be understood that statisticians use these terms in a very wide sense. An *experiment* may consist of the simple process of noting whether a switch is turned on or off; it may consist of determining what proportion of a sample of patients shows an adverse reaction to a new medication; or it may consist of the very complicated process of finding the mass of an electron. Correspondingly, an *outcome* of an experiment may be a simple choice between two alternatives; it may be the result of a measurement or count; or it may be an answer obtained after extensive measurements and calculations.

When dealing with problems in which uncertainties are connected with the various outcomes, it is convenient to think of the outcomes of an experiment, the *elements* of the sample space, as points in a space of one or more dimensions.

> **Example.** Suppose that a government agency must decide where to locate two new atomic power plants, and that (for a certain purpose) it is of interest only how many of these plants will be located in Texas. The possible outcomes are 0, 1, and 2, and the sample space may be pictured as in Figure 2.1. If it were of interest to indicate also how many of the new power plants will be located in California, the possible outcomes could be denoted (0, 0), (1, 0), (0, 1), (2, 0), (1, 1), and (0, 2), where the first coordinate pertains to the number of new power plants to be located in Texas and the second coordinate pertains to the number of new power plants to be located in California. The corresponding sample space may be pictured as in Figure 2.2.

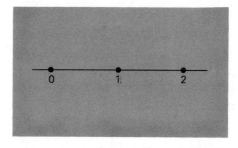

Figure 2.1. Sample space for the number of new atomic power plants to be located in Texas.

In the second part of this example, the six possible outcomes could have been represented by means of six points on a line, or six points *in any kind of pattern*, but the use of coordinates and a corresponding number of dimensions has the advantage that it makes it easy to identify each outcome with the corresponding point.

Generally, sample spaces are classified according to the *number of elements* (points) which they contain. Thus, the sample spaces of Figures 2.1 and 2.2 have 3 and 6 elements, respectively, and they are both referred to as *finite*. Other examples of finite sample spaces are the one used to represent the various ways in which a president and a vice-president can be elected from among the 25 members of a union local (it has 600 ele-

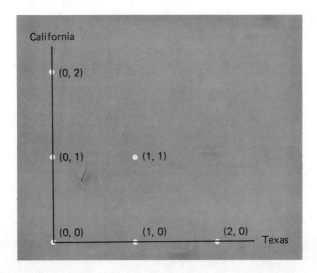

Figure 2.2.   Sample space for the number of new atomic power plants to be located in Texas and in California.

ments); and the one used to represent the various ways in which a student can answer the 12 questions on a true-false test (the answer is $2^{12} = 4,096$). The following are examples of sample spaces that are *not finite*:

Example. Suppose that a person checks the nitrogen-oxide emission of automobiles and that we are interested in the number of cars he has to inspect before he observes the first one that does not meet government regulations. It could be the first, the second, ... , the fiftieth, ... , and for all we know he may have to check a million or more cars before he finds one that does not meet government regulations. Not knowing how far we might have to go, it is appropriate in an example like this to consider the sample space to be the whole set of natural numbers, of which there is a *countable infinity*.

To go one step further, if we were interested in the nitrogen-oxide emission of a given car in grams per mile, the sample space would have to consist of all the points on a continuous scale (a certain interval on the line of real numbers), of which there is a *continuum*.

In general, a sample space is said to be *discrete* if it has finitely many or a countable infinity of elements. If the elements (points) of a sample space constitute a continuum, for example, all the points on a line, all the points on a line segment, all the points in a plane, and so forth, the sample space is said to be *continuous*. In the remainder of this chapter we shall consider only discrete and mainly finite sample spaces.

At times it can be quite difficult, or at least tedious, to determine the number of elements in a finite sample space by direct enumeration. The following illustrates a method that can often be used to simplify this task:

Example. A consumer testing service rates lawn mowers according to their ease of operation, retail price, and average cost of repairs. If E_1, E_2, and E_3 represent a lawn mower's being easy, average, or difficult to operate, P_1 and P_2 represent its being expensive or inexpensive, and C_1, C_2, and C_3 represent its average cost of repairs being high, average, or low, the various ways in which a lawn mower can thus be rated may be visualized by means of the *tree diagram* of Figure 2.3. Following a given path (from left to right) along the branches

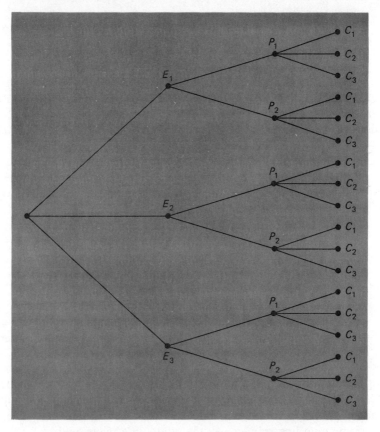

Figure 2.3. Tree diagram for ratings of lawn mowers.

of the "tree," we obtain a particular rating, namely, a particular element of the sample space, and it can be seen that altogether there are 18 possibilities.

This result could also have been obtained by observing that there are three "E-branches," that each E-branch forks into two "P-branches," and that each P-branch forks into three "C-branches." Thus, there are $3 \cdot 2 \cdot 3 = 18$ combinations of branches, or paths. This result is generalized by the following theorem:

THEOREM 2.1. *If sets A_1, A_2, ..., A_k contain, respectively n_1, n_2, ..., n_k elements, there are $n_1 \cdot n_2 \cdot \ldots \cdot n_k$ ways of selecting first an element from A_1, then an element from A_2, ..., and finally an element from A_k.*

This theorem can be verified by constructing a tree diagram similar to that of Figure 2.3, where we had $n_1 = 3$, $n_2 = 2$, $n_3 = 3$, and, hence, $3 \cdot 2 \cdot 3 = 18$ possibilities.

Example. To illustrate the use of this theorem, consider again the two finite sample spaces referred to on pages 6 and 7. A union local with a membership of 25 can elect a president in 25 ways; subsequently it can elect a vice-president in 24 ways, and the whole selection can be made in $25 \cdot 24 = 600$ ways. So far as the true-false test with 12 questions is concerned, each question can be answered in two ways so that there are altogether $2 \cdot 2 \cdot 2 \cdot 2 \cdot 2 \cdot 2 \cdot 2 \cdot 2 \cdot 2 \cdot 2 \cdot 2 \cdot 2 = 2^{12} = 4,096$ ways in which a student can answer the 12 questions of the test.

2.2 EVENTS

Probabilities are always associated with the occurrence or the nonoccurrence of *events*, such as the event that 8 of 20 road building jobs are finished on time, the event that a vacuum cleaner will not exceed a guaranteed maximum level of sound, the event that we get 0 heads in 4 flips of a coin, the event that in one week there will be at least four industrial accidents in a given plant, and so forth. Thus, in connection with probabilities, we always refer to an individual outcome or to a set of outcomes of an experiment as an event. (The event of getting 0 heads in 4 flips of a coin pertained to an individual outcome while the event of there being at least four industrial accidents pertains to a set of outcomes.)

In other words, *we shall think of an event as a subset of an appropriate sample space.*

Example. To illustrate this in more detail, suppose that the laboratory of an engineering firm has to check the suitability of 4 solid crystal lasers and 3 carbon dioxide lasers for a given task. If we are interested only in how many of each type are suitable (and not which ones in particular), the number of elements in the sample space is $5 \cdot 4 = 20$ and the corresponding two-dimensional sample space may be pictured as in Figure 2.4.

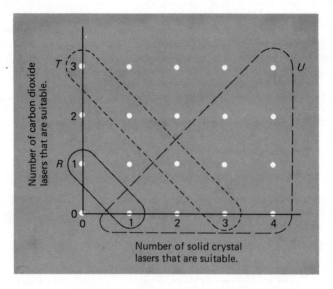

Figure 2.4. Events in a sample space.

Now suppose we let R stand for the event that only 1 of all the lasers is suitable for the given task, we let T stand for the event that exactly 3 of the lasers are suitable for the given task, and we let U stand for the event that more of the solid crystal lasers than the carbon dioxide lasers are suitable for the given task. The elements of the sample space which correspond to these events are indicated, respectively, by means of the solid, dotted, and dashed curves of Figure 2.4. Note that events R and T have no elements in common—they are *mutually exclusive events.*

In many problems of probability we are interested in events which are actually combinations of two or more events, formed by taking *unions*, *intersections*, and *complements*. Although the reader must surely be familiar with these terms, let us review briefly that if A and B are any two sets in a sample space S, their *union* $A \cup B$ is the subset of S which contains all elements that are either in A, in B, or in both; their *intersection* $A \cap B$ is the subset of S which contains all elements that are in both A and B; and the *complement* A' of A is the subset of S which contains all the elements of S that are not in A.

Continuation of Example. From Figure 2.4 it can be seen that $R \cup T$ contains the elements $(0, 1)$, $(0, 3)$, $(1, 0)$, $(1, 2)$, $(2, 1)$, and $(3, 0)$, while $R \cap T$ has no elements at all. Following the practice of denoting the *empty set*, or *null set*, by the symbol $\varnothing$, we can write $R \cap T = \varnothing$, and this is the mathematical way of saying that R and T are mutually exclusive. Finally, U' contains the elements $(0, 0)$, $(0, 1)$, $(1, 1)$, $(0, 2)$, $(1, 2)$, $(2, 2)$, $(0, 3)$, $(1, 3)$, $(2, 3)$, and $(3, 3)$, where in each case the number of solid crystal lasers that are suitable for the given task is less than or equal to the corresponding number of carbon dioxide lasers. Observe also that the complement of S with respect to itself is the null set, that is $S' = \varnothing$.

Sample spaces and events, particularly relationships among events, are often depicted by means of *Venn diagrams* like those of Figures 2.5

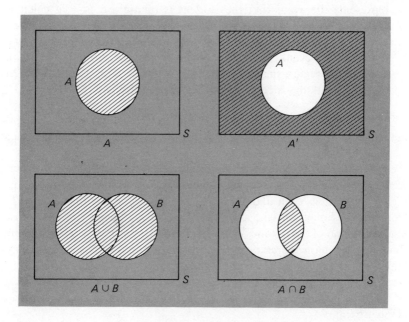

Figure 2.5. Venn diagrams.

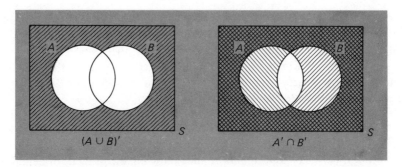

Figure 2.6. Use of Venn diagrams to verify that $(A \cup B)' = A' \cap B'$.

and 2.6. In each case the sample space is represented by a rectangle, while events are represented by regions within the rectangle, usually by circles or parts of circles. Thus, the shaded regions of the four Venn diagrams of Figure 2.5 represent, respectively, the event A, the complement of A, the union of A and B, and the intersection of A and B.

> **Example.** If A is the event that a certain student is taking a course in calculus and B is the event that she is taking a course in applied mechanics, then the region shaded in the first diagram of Figure 2.5 represents the event that the student is taking a course in calculus, the region shaded in the second diagram represents the event that she is *not* taking a course in calculus, the region shaded in the third diagram represents the event that she is taking calculus and/or applied mechanics, and the region shaded in the fourth diagram represents the event that she is taking calculus as well as applied mechanics.

Venn diagrams are often used to verify relationships among sets, thus making it unnecessary to give formal proofs based on the algebra of sets. To illustrate, let us demonstrate that $(A \cup B)' = A' \cap B'$, which expresses the fact that the complement of the union of two sets equals the intersection of their respective complements. To begin, note that the shaded region of the first Venn diagram of Figure 2.6 represents the set $(A \cup B)'$ (compare this diagram with the third diagram of Figure 2.5). The crosshatched region of the second Venn diagram of Figure 2.6 was obtained by shading the region representing A' with lines going in one direction and that representing B' with lines going in another direction. Thus, the crosshatched region represents the intersection of A' and B', and it can be seen that it is identical with the shaded region of the first Venn diagram of Figure 2.6.

EXERCISES

1. A building inspector has to check the wiring in a new apartment building either on Monday, Tuesday, Wednesday, or Thursday, and at 8 A.M., 1 A.M., or 2 P.M. Draw a tree diagram which shows the various ways in which he can schedule the inspection of the wiring of the new apartment building.

2. If the five finalists in the Miss Universe contest are Miss Spain, Miss U.S.A., Miss Uruguay, Miss Portugal, and Miss Japan, draw a tree diagram which shows the various ways in which the judges can choose the winner and the first runner-up.

3. A student can study either 0, 1, or 2 hours for a test in computer programming on any given night. Construct a tree diagram to show that there are 10 different ways in which he can study altogether 6 hours for the test on four consecutive nights.

4. If the tone quality of five audio systems is to be rated superior, average, or inferior and we are interested only in how many of the systems get each of these ratings, draw a tree diagram which shows the 21 different possibilities.

5. Suppose that in a baseball World Series (in which the winner is the first team to win four games) the National League champion leads the American League champion two games to one. Construct a tree diagram which shows the number of ways in which these teams can win or lose the remaining games.

6. In an optics kit there are six concave lenses, four convex lenses, and three prisms. In how many ways can one choose one of the concave lenses, one of the convex lenses, and one of the prisms?

7. A questionnaire sent through the mail as part of a market study consists of eight questions, each of which can be answered in three different ways. In how many different ways can a person answer the eight questions on this questionnaire?

8. In each of the following experiments decide whether it would be appropriate to use a sample space which is finite, countably infinite, or continuous:
   (a) The amount of cosmic radiation to which passengers are exposed during a transcontinental jet flight is measured by means of a suitable counter.
   (b) Five of the members of a professional society with 12,600 members are chosen to serve on a nominating committee.
   (c) An experiment is conducted to measure the heat of vaporization of water.
   (d) A study is made to determine in how many of 450 airplane accidents the main cause is pilot error.

(e) Measurements are made to determine the uranium content of a certain ore.

(f) In a torture test, a watch is dropped from a tall building until it stops running.

9. Two professors and three graduate assistants are responsible for the supervision of a physics lab, and at least one professor and one graduate student has to be present at all times.

(a) Using two coordinates so that (1, 2), for example, represents the event that one of the professors and two of the graduate assistants are present, and (2, 3) represents the event that both professors and all three graduate assistants are present, draw a diagram (similar to that of Figure 2.4) showing the points of the corresponding sample space.

(b) Describe *in words* the event which is represented by each of the following sets of points of the sample space: the event B which consists of the points (1, 3) and (2, 3), the event C which consists of the points (1, 1) and (2, 2), and the event D which consists of the points (1, 2) and (2, 1).

(c) With reference to part (b), list the points of the sample space which represent the event $C \cup D$, and describe this event *in words*.

(d) With reference to part (b) are events B and D mutually exclusive?

10. Mr. Jones, a salesman of industrial chemicals, has four customers in Sacramento, whom he may or may not be able to visit on a two-day visit to this city. He will not visit any of these customers more than once.

(a) Using two coordinates so that (2, 1), for example, represents the event that he will visit two of his customers on the first day and one on the second day, and (0, 2) represents the event that he will not visit any of his customers on the first day but two on the second day, draw a diagram (similar to that of Figure 2.4) showing the points of the corresponding sample space.

(b) List the points of the sample space of part (a) which constitute the following events: event X that he will visit all four of his customers, event Y that he will visit more of his customers on the first day than on the second day, and event Z that he will visit at least three of his customers on the second day.

(c) With reference to part (b), describe each of the following events in words and list the points which they contain: X', $X \cup Y$, $X \cap Z$, and $X' \cap Y$.

(d) With reference to part (b), check whether any two of the three events X, Y, and Z are mutually exclusive.

11. Suppose that a group of efficiency experts is visiting a large firm and that A is the event that they inspect the accounting department, Q is the event that they inspect the quality control department, and R is the event that they inspect the research department. With reference to the Venn diagram of Figure 2.7 list (by numbers) the regions or combinations of regions which represent the following events:

(a) The event that they will visit the accounting and quality control departments, but not the research department.

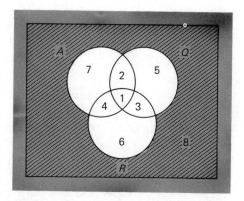

Figure 2.7. Venn diagram for Exercises 11 and 12.

(b) The event that they will visit the accounting department, but neither the quality control nor the research department.

(c) The event that they will visit the accounting and research departments.

(d) The event that they will visit the quality control department.

(e) The event that they will visit none of the three departments.

12. With reference to Exercise 11 and the Venn diagram of Figure 2.7, explain *in words* what events are represented by:
   (a) Region 1;   (b) Region 5;
   (c) Regions 1 and 2 together;   (d) Regions 3 and 5 together;
   (e) Regions 3, 5, and 6 together; (f) Regions 1, 2, 4, and 7 together.

13. Use Venn diagrams to verify that
   (a) $(A \cap B)' = A' \cup B'$;   (b) $A \cup (A \cap B) = A$;
   (c) $(A \cap B) \cup (A \cap B') = A$;
   (d) $A \cup B = (A \cap B) \cup (A \cap B') \cup (A' \cap B)$;
   (e) $A \cup (B \cap C) = (A \cup B) \cap (A \cup C)$.

2.3 PROBABILITY

In this section we shall define probabilities as the values of *set functions*, *additive set functions* to be exact. Since the reader is probably most familiar with functions for which the elements of the domain and the range are all numbers, let us first consider a very simple example where the elements of the domain are sets while the elements of the range are real numbers. Specifically, we shall study a function, a correspondence, which assigns real numbers to the subsets of a given set, namely, to the subsets of a sample space.

Example. The set function which we shall consider is the one which assigns to each subset *A* of a given finite set *S* the *number of elements in A*, written *N(A)*. Suppose, then, that 500 machine parts are inspected before they are being shipped, that *I* denotes that a machine part is improperly assembled, and *D* denotes that it contains one or more defective components. The distribution of these 500 machine parts among the various categories is as shown in the Venn diagram of Figure 2.8 and, using this diagram, we can now determine the value

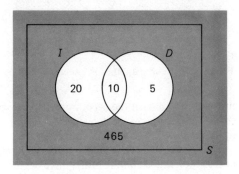

Figure 2.8. Classification of the 500 machine parts.

N(A) for any one of the 16 subsets of the sample space. (To see that there are 16 subsets, consider the four regions into which the circles of Figure 2.8 divide the sample space. We must decide whether or not each region is to be included in any given subset; hence, there are $2 \cdot 2 \cdot 2 \cdot 2 = 16$ possibilities, including the set *S* of all 500 machine parts and the empty set $\varnothing$.) The numbers in Figure 2.8 are the number of machine parts that are improperly assembled but do not have any defective components, the number of machine parts that are improperly assembled and have one or more defective components, the number of machine parts that are properly assembled but have one or more defective components, and the number of machine parts that are properly assembled and do not have any defective components. Symbolically,

$$N(I \cap D') = 20, \qquad N(I \cap D) = 10, \qquad N(I' \cap D) = 5,$$
$$\text{and} \quad N(I' \cap D') = 465$$

To find the total number of machine parts that are improperly assembled, we have only to add the number of improperly assembled machine parts that have one or more defective components to the number of improperly assembled machine parts which do not have any defective components, and we obtain

$$N(I) = N(I \cap D) + N(I \cap D') = 10 + 20 = 30$$

Similarly, we find that the total number of machine parts with one or more defective components is

$$N(D) = N(I \cap D) + N(I' \cap D) = 10 + 5 = 15$$

Finally, since $N(S) = 500$, where S is the set of all 500 machine parts, we obtain by subtraction $N(I') = N(S) - N(I) = 500 - 30 = 470$ and $N(D') = N(S) - N(D) = 500 - 15 = 485$.

The set function which we have introduced in this example is said to be *additive*, meaning that the number which it assigns to the union of two subsets which have no elements in common equals the *sum* of the numbers assigned to the individual subsets.* It was this property which enabled us to calculate $N(I)$ by adding $N(I \cap D)$ and $N(I \cap D')$, $N(D)$ by adding $N(I \cap D)$ and $N(I' \cap D)$, $N(I')$ by subtracting $N(I)$ from $N(S)$, and $N(D')$ by subtracting $N(D)$ from $N(S)$. Note that this property applies only to subsets which have no elements in common, namely, to subsets which are *disjoint*; as we indicated on page 10, events that correspond to disjoint subsets are called *mutually exclusive events*.

When dealing with two subsets A and B which may have some elements in common, we must use the more general *addition formula*

$$N(A \cup B) = N(A) + N(B) - N(A \cap B)$$

where we subtracted $N(A \cap B)$ because these elements are counted once in $N(A)$ and once in $N(B)$.

Continuation of Example. For our illustration we thus get

$$N(I \cup D) = N(I) + N(D) - N(I \cap D)$$
$$= 30 + 15 - 10$$
$$= 35$$

for the number of machine parts that are improperly assembled, have one or more defective components, or both.

---

*A simple example of a set function which is *not* additive is the one which assigns to every subset the square of the number of elements which it contains.

Using the concept of an additive set function, let us now explain what we mean by the *probability of an event*. Given a finite sample space S and an event A in S, we define $P(A)$, the probability of A, to be a value of an additive set function **P**, which must satisfy the following three conditions:

AXIOM 1. $0 \leq P(A) \leq 1$ *for each event A in S.*

AXIOM 2. $P(S) = 1$.

AXIOM 3. *If A and B are any mutually exclusive events in S, then* $P(A \cup B) = P(A) + P(B)$.

We refer to such a function P as a probability function, and it should be noted that the first axiom states that a probability function assigns to every event A in S some real number from 0 to 1, inclusive. The second axiom states that the sample space as a whole is assigned the number 1 and it expresses the idea that the probability of a certain event, an event which must happen, is equal to 1. The third axiom states that the probability function must be additive, and with the use of mathematical induction it can be extended to include any finite number of mutually exclusive events. In other words, it can be shown that

---

$$P(A_1 \cup A_2 \cup \ldots \cup A_n) = P(A_1) + P(A_2) + \ldots + P(A_n)$$

---

where $A_1, A_2, \ldots, A_n$ are mutually exclusive events in S. (In Section 3.3 we shall explain how the third axiom will have to be modified so that we can consider also sample spaces which are not finite.)

Axioms for a mathematical theory require no proof, but if such a theory is to be applied to the physical world, we must show somehow that the axioms are "realistic"; that is, we must show that they yield reasonable results. To do so with regard to the axioms of probability, let us say a few words about two of the most popular concepts of probability: the *frequency interpretation* and *subjective probabilities*. According to the widely-held frequency interpretation, a probability is looked upon as a *relative frequency, or proportion, in the long run.*

Example. If we say "the probability that a jet from New York to Los Angeles will arrive on time is 0.82," we mean that if present conditions prevail, 82 percent of such flights will arrive on time. Also, if we say "the probability is 0.98 that a ceramic insulator will be able to withstand a certain thermal shock," we mean that 98 percent of the insulators should withstand such a thermal shock. Note, however, that if we say "the probability of getting heads with a balanced coin is 0.50," this does *not* mean that we must necessarily get 10 heads and 10 tails in 20 flips of the coin or 50 heads and 50 tails in 100 flips; it means that if the coin is flipped a large number of times we should get close to 50 percent heads and 50 percent tails.

To demonstrate that the three axioms of probability are consistent with the frequency interpretation, we need only observe that the proportion of the time an event occurs cannot be negative or exceed 1 and that one outcome or another must occur 100 percent of the time, that is, with a probability of 1. Also, if A and B are mutually exclusive events, the proportion of the time that either one or the other occurs is the *sum* of the proportions of the time that they occur.

Example. If the proportion of voters favoring a piece of legislation is 0.42 and the proportion undecided is 0.19, then $0.42 + 0.19 = 0.61$ is the proportion of the voters who are either for the legislation or undecided.

A point of view, which is currently gaining favor, is to interpret probabilities as *personal* or *subjective* evaluations. Such probabilities express the *strength of one's belief* with regard to the uncertainties that are involved, and they apply especially when there is *very little direct evidence*, so that there may really be no choice but to consider collateral (indirect) information, educated guesses, and perhaps intuition and other subjective factors. Subjective probabilities are best determined by referring to a *risk-taking situation* and putting the issue on a "put up or shut up" basis.

Example. For instance, to determine how a businessman may feel about the success of a new venture (say, the marketing of some new power tools), we ask him for the *odds* at which he would be willing to bet (or consider it *fair* to bet) that the new venture will be a success. If he "feels" that the proper odds are, say, 3 to 2, this means that it would be fair to bet $300 against $200, or perhaps $1,500 against $1,000, that the new venture will be a success.

To convert such odds into the corresponding probability, we make use of the rule that *if somebody considers it fair to bet a dollars against b dollars* *that a given event will occur, he or she is, in fact, assigning the event the probability $\frac{a}{a+b}$.*

Continuation of Example. Thus, the businessman who considers it fair to bet $300 against $200 that the new power tools will be a commercial success, is actually assigning this event the subjective probability $\frac{300}{300 + 200} = 0.60$.

So far as the three axioms of probability are concerned, the first one is consistent with the subjective approach since $\frac{a}{a+b}$ cannot be negative or exceed 1 so long as a and b are positive amounts bet for and against an event. With regard to the second axiom we can say that we should be willing to give "better" and "better" odds when we become more and more certain that an event will occur—say, 100 to 1, 1,000 to 1, or perhaps even 1,000,000 to 1. The corresponding probabilities are $\frac{100}{100 + 1}$, $\frac{1,000}{1,000 + 1}$, and $\frac{1,000,000}{1,000,000 + 1}$ (or approximately 0.99, 0.999, and 0.999999). Thus, *the more certain we are that an event will occur, the closer its probability will be to 1.* The third postulate of probability is not necessarily satisfied by subjective probabilities, as is illustrated by the following example:

Example. Asked about the chances that a mining operation will be a great success or a modest success, a petroleum engineer "feels" that the odds are 1 to 5 that the operation will be a great success (or 5 to 1 that it will *not* be a great success), 2 to 1 that it will be a modest success, and 3 to 1 that it will be either a modest success or a great success. The corresponding probabilities are $\frac{1}{1 + 5} = \frac{1}{6}$, $\frac{2}{2 + 1} = \frac{2}{3}$, and $\frac{3}{3 + 1} = \frac{3}{4}$, and, as can easily be verified, $\frac{1}{6} + \frac{2}{3}$ does *not* equal $\frac{3}{4}$.

Thus, proponents of the subjective approach to probability impose the third axiom of probability as a *consistency criterion*. This means that if a person's subjective probabilities "behave" in accordance with the third axiom of probability, he is said to be *consistent*; otherwise, he is said to be *inconsistent* and his probability judgments must be taken with a grain of salt.

Before we give some examples of useful probability functions, it is important to stress the point that *the three axioms do not tell us how to assign probabilities to the various outcomes of an experiment, they merely restrict the ways in which it can be done.* In actual practice, probabilities are assigned either on the basis of estimates obtained from past experience, on the basis of a careful analysis of conditions underlying the experiment, on the basis of assumptions—say, the common assumption that various outcomes are equiprobable, or on the basis of subjective evaluation.

Example. The following are three examples of *permissible* ways of assigning probabilities in an experiment where there are three possible and mutually exclusive outcomes A, B, and C:

(1) $P(A) = \frac{1}{3}$, $P(B) = \frac{1}{3}$, $P(C) = \frac{1}{3}$

(2) $P(A) = 0.57$, $P(B) = 0.24$, $P(C) = 0.19$

(3) $P(A) = \frac{24}{27}$, $P(B) = \frac{2}{27}$, $P(C) = \frac{1}{27}$

However,

(4) $P(A) = 0.64$, $P(B) = 0.38$, $P(C) = -0.02$

and

(5) $P(A) = 0.35$, $P(B) = 0.52$, $P(C) = 0.26$

are *not* permissible because (4) violates Axiom 1 and (5) violates Axiom 2.

As it can be shown that a sample space of n points (outcomes) has 2^n subsets, it would seem that the problem of specifying a probability function (namely, a probability for each subset or event) can easily become very tedious. Indeed, for $n = 20$ there are already more than a million possible events. Fortunately, this task can be simplified considerably by the use of the following theorem:

THEOREM 2.2. *If A is an event in the finite sample space S, then $P(A)$ equals the sum of the probabilities of the individual outcomes comprising A.*

To prove this theorem, let $E_1, E_2, \ldots, E_n$ be the n outcomes comprising A, so that we can write $A = E_1 \cup E_2 \cup \ldots \cup E_n$. Since the E's are individual outcomes they are mutually exclusive, and by the extension of Axiom 3 on page 18 we have

$$P(A) = P(E_1 \cup E_2 \cup \ldots \cup E_n)$$
$$= P(E_1) + P(E_2) + \ldots + P(E_n)$$

which completes the proof.

Example. To illustrate the use of this theorem, let us return to the rating of lawn mowers in the example on page 8, and let us suppose that the probabilities of the 18 different ratings are as shown in Figure 2.9 (which, except for the probabilities, is identical with Figure

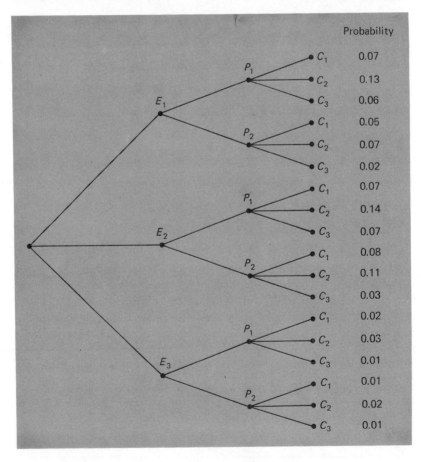

Figure 2.9.   Ratings of lawn mowers and their probabilities.

2.3 on page 8). From this figure, we determine $P(E_1)$, the probability that a lawn mower will be rated easy to operate by *adding* the probabilities of the *six* outcomes comprising the event E_1, getting

$$P(E_1) = 0.07 + 0.13 + 0.06 + 0.05 + 0.07 + 0.02 = 0.40$$

Similarly, we find that

$$P(P_1) = 0.07 + 0.13 + 0.06 + 0.07 + 0.14$$
$$+ 0.07 + 0.02 + 0.03 + 0.01 = 0.60$$
$$P(C_1) = 0.07 + 0.05 + 0.07 + 0.08 + 0.02 + 0.01 = 0.30$$
$$P(E_1 \cap P_1) = 0.07 + 0.13 + 0.06 = 0.26$$

and

$$P(E_1 \cap C_1) = 0.07 + 0.05 = 0.12$$

where the last value is the probability that a lawn mower will be rated easy to operate but having a high average cost of repairs.

Example. To consider another illustration, let us refer again to the example on page 16, which dealt with the 500 machine parts, and let us suppose that one of the machine parts is to be chosen by lot for inspection. Assuming that each machine part, thus, has a probability of $\frac{1}{500}$ of being selected, it can easily be verified by using Theorem 2.2 that the probability that the machine part is improperly assembled is $P(I) = \frac{3}{50}$, the probability that it contains one or more defective components is $P(D) = \frac{3}{100}$, and the probability that it is improperly assembled, contains one or more defective components, or both is $P(I \cup D) = \frac{7}{100}$. (Using Theorem 2.2, each of these probabilities was obtained by adding $\frac{1}{500}$ as many times as there are individual outcomes in the respective events.)

In this last example we could have used to advantage the following theorem which applies to experiments where the individual outcomes are all *equiprobable*:

THEOREM 2.3. *If an experiment has n possible outcomes which are equiprobable and if s of these are labeled "success," then the probability of a "success" is s/n.*

This theorem follows immediately from Theorem 2.2 and the generalization of the third axiom of probability given on page 18.

Continuation of Example. We could, thus, have found directly that the probability of selecting a machine part which is improperly assembled is

$$P(I) = \frac{N(I)}{N(S)} = \frac{30}{500} = \frac{3}{50}$$

and that the probability of selecting a machine part which is improperly assembled and also has one or more defective components is

$$P(I \cap D) = \frac{N(I \cap D)}{N(S)} = \frac{10}{500} = \frac{1}{50}$$

Theorem 2.3 is particularly useful in problems dealing with games of chance, where it is assumed that if a deck of cards is properly shuffled each card has the same chance of being selected, if a coin is properly flipped heads is as likely as tails, if a die is properly rolled each face is as likely to come up as any other, and so on. Thus, Theorem 2.3 applies also when selections are made by means of lots or other gambling devices to decide, say, which items coming off an assembly line are to be inspected, which test cars are to be assigned to each of four drivers, which plots of farm land are to be used to try a special variety of wheat, which coal-fired generating plants in a geographical area are to be used to experiment with antipollution devices, and so forth.

2.4 SOME ELEMENTARY THEOREMS

Using the axioms of probability, it is possible to derive many theorems which play an important role in applications. First let us show that

---

THEOREM 2.4. *If A is any event in S, then* $P(A') = 1 - P(A)$.

---

which, in accordance with the frequency interpretation, expresses the fact that if 34 percent of the shipments from a vendor arrive on time, then 66 percent of the shipments do not arrive on time, or if 16 percent of all applicants fail a certain civil service test, then 84 percent do not fail the test. So far as subjective probabilities are concerned, Theorem 2.4 expresses the fact that if the odds for the occurrence of an event are 8 to 1, then the odds against its occurrence are 1 to 8; the corresponding probabilities are $\frac{8}{8+1} = \frac{8}{9}$ and $\frac{1}{1+8} = \frac{1}{9}$, and $\frac{1}{9} = 1 - \frac{8}{9}$. To prove

Theorem 2.4, observe that A and A' are mutually exclusive by definition, and that $A \cup A' = S$ (that is, among them, A and A' contain all of the elements of S). Hence, we have

$$P(A \cup A') = P(A) + P(A')$$

according to Axiom 3,

$$P(A \cup A') = P(S) = 1$$

according to Axiom 2, and it follows that

$$P(A) + P(A') = 1$$

which completes the proof of Theorem 2.4. As a special case we find that $P(\varnothing) = 1 - P(S) = 1 - 1 = 0$, since the empty set $\varnothing$ is the complement of S.

Example. Referring again to the lawn mowers and the results on page 23, we thus find that the probability that a lawn mower will *not* be rated easy to operate is $1 - P(E_1) = 1 - 0.40 = 0.60$, and that the probability that a lawn mower will be rated as either not being easy to operate or not having a high average cost of repairs is $1 - P(E_1 \cap C_1) = 1 - 0.12 = 0.88$.

On page 18 we observed that Axiom 3 can be extended to include more than two mutually exclusive events. Another useful and important extension of this axiom allows us to find the probability of the union of *any two* events in S regardless of whether they are mutually exclusive. To motivate the theorem which follows, let us consider again the example on page 17, the one dealing with the 500 machine parts, where we obtained the number of machine parts that are improperly assembled, have one or more defective components, or both, by writing

$$N(I \cup D) = N(I) + N(D) - N(I \cap D)$$
$$= 30 + 15 - 10$$
$$= 35$$

As we pointed out in connection with the general formula for $N(A \cup B)$ on page 17, we had to subtract $N(I \cap D)$, the number of machine parts that are improperly assembled *and also* contain one or more defective components, because they would otherwise be counted twice, once in $N(I)$ and once in $N(D)$. If we divide each of the above figures by 500, the total number of machine parts, we could similarly have argued that the *proportion* of machine parts that are improperly assembled, have one or more defective components, or both, equals the *proportion* of machine parts that are improperly assembled *plus* the *proportion* of machine parts that contain one or more defective components *minus* the *proportion* of machine parts that fit both of these descriptions. In line with this intuitive justification, let us now prove the following theorem, which is usually called the *general law of addition*:

THEOREM 2.5. *If A and B are any events in S, then*

$$P(A \cup B) = P(A) + P(B) - P(A \cap B)$$

To prove this theorem, note first from the Venn diagram of Figure 2.10 that

$$A \cup B = (A \cap B) \cup (A \cap B') \cup (A' \cap B)$$

and also that

$$A = (A \cap B) \cup (A \cap B')$$
$$B = (A \cap B) \cup (A' \cap B)$$

(The reader was asked to verify these important relations in parts (c) and (d) of Exercise 13 on page 15.) Since $A \cap B$, $A \cap B'$, and $A' \cap B$ are evidently mutually exclusive, the extension of Axiom 3 on page 18 yields

$$P(A \cup B) = P(A \cap B) + P(A \cap B') + P(A' \cap B)$$

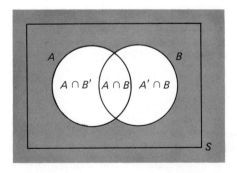

Figure 2.10. Diagram for proof of Theorem 2.5.

and after adding and subtracting $P(A \cap B)$ we obtain

$$P(A \cup B) = [P(A \cap B) + P(A \cap B')] + [P(A \cap B)$$
$$+ P(A' \cap B)] - P(A \cap B)$$
$$= P(A) + P(B) - P(A \cap B)$$

Note that when A and B are mutually exclusive, Theorem 2.5 reduces to Axiom 3, since in that case $P(A \cap B) = 0$. For this reason, we often refer to Axiom 3 as the *special law of addition*, whereas Theorem 2.5 is called the *general law of addition*.

Example. To illustrate the use of this theorem, let us refer again to the lawn mower example and let us determine the probability that a lawn mower will be rated easy to operate, having a high average cost of repairs, or both. Making use of the results obtained on page 23, namely, $P(E_1) = 0.40$, $P(C_1) = 0.30$, and $P(E_1 \cap C_1) = 0.12$, we can immediately write the answer as

$$P(E_1 \cup C_1) = 0.40 + 0.30 - 0.12$$
$$= 0.58$$

Example. To give another example, suppose that the probabilities are, respectively, 0.28, 0.15, and 0.09 that a person traveling through the state of Arizona will visit the Grand Canyon, the Petrified Forest, or both. Substituting these values into the formula of Theorem 2.5, we find that the probability is

$$0.28 + 0.15 - 0.09 = 0.34$$

that a person traveling through Arizona will visit *at least one* of these scenic attractions.

EXERCISES

1. With reference to the example on page 16 and Figure 2.8, list the 16 different subsets that can be formed and the number of elements (machine parts) contained in each.

2. In a group of 160 graduate engineering students, 92 are enrolled in an advanced course in statistics, 63 are enrolled in a course in operations

research, and 40 are enrolled in both. How many of these students are not enrolled in either course?

3. The personnel manager of a manufacturing plant claims that among the 400 employees 312 got a raise in 1975, 248 got increased pension benefits, 173 got both, and 43 got neither. Draw a Venn diagram and fill in the figures which correspond to the various regions to check whether this claim should be questioned.

4. Among 150 persons interviewed as part of an urban mass transportation study, some live more than three miles from the center of the city (*A*), some now regularly drive their own car to work (*B*), and some would gladly switch to public mass transportation if it were available (*C*). Use the information given in Figure 2.11 to find

(a) $N(A)$; (b) $N(B)$;
(c) $N(C)$; (d) $N(A \cap B)$;
(e) $N(A \cap C)$; (f) $N(A \cap B \cap C)$;
(g) $N(A \cup B)$; (h) $N(B \cup C)$;
(i) $N(A' \cup B' \cup C)$; (j) $N[B \cap (A \cup C)]$.

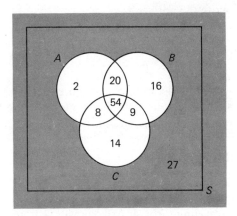

Figure 2.11. Diagram for Exercise 4.

5. An experiment has the four possible mutually exclusive outcomes *A*, *B*, *C*, and *D*. Check whether the following assignments of probability are permissible:

(a) $P(A) = 0.38$, $P(B) = 0.16$, $P(C) = 0.11$, $P(D) = 0.35$;
(b) $P(A) = 0.31$, $P(B) = 0.27$, $P(C) = 0.28$, $P(D) = 0.16$;
(c) $P(A) = 0.32$, $P(B) = 0.27$, $P(C) = -0.06$, $P(D) = 0.47$;
(d) $P(A) = \frac{1}{2}$, $P(B) = \frac{1}{4}$, $P(C) = \frac{1}{8}$, $P(D) = \frac{1}{16}$;
(e) $P(A) = \frac{5}{18}$, $P(B) = \frac{1}{6}$, $P(C) = \frac{1}{3}$, $P(D) = \frac{2}{9}$.

6. Referring to the example on page 10 and the sample space of Figure 2.4, suppose that the points (0, 0), (1, 0), (2, 0), (3, 0), (4, 0), (0, 1), (1, 1), (2, 1),

(3, 1), (4, 1), (0, 2), (1, 2), (2, 2), (3, 2), (4, 2), (0, 3), (1, 3), (2, 3), (3, 3), and (4, 3) of the sample space are assigned the respective probabilities 0.01, 0.01, 0.01, 0.03, 0.04, 0.01, 0.01, 0.01, 0.03, 0.04, 0.03, 0.03, 0.03, 0.09, 0.09, 0.05, 0.06, 0.07, 0.14, and 0.21.
   (a) Verify that this assignment of probabilities is permissible.
   (b) Calculate the probabilities of events R, T, and U described on page 10.
   (c) Calculate the probabilities that, respectively, 0, 1, 2, or 3 of the carbon dioxide lasers will be suitable for the given task.

7. Referring to the example on page 8 and Figure 2.9, find
   (a) $P(E_2)$;   (b) $P(P_2)$;
   (c) $P(C_2)$;   (d) $P(E_2 \cap P_2)$;
   (e) $P(E_2 \cap C_2)$;   (f) $P(P_2 \cap C_2)$.

8. With reference to Exercise 9 on page 14 suppose that each point (i, j) of sample space (for $i = 1$ or 2, and $j = 1, 2,$ or 3) is assigned the probability
$$\frac{\frac{15}{28}}{i + j}.$$
   (a) Verify that this assignment of probabilities is permissible.
   (b) Calculate the probabilities of events B, C, and D described in that exercise.
   (c) Calculate the probabilities that, respectively, 1, 2, or 3 of the graduate students will be supervising the physics lab.

9. Explain how one might conceivably assign a probability to the success of an off-shore oil drilling operation, using
   (a) the frequency concept of probability;
   (b) subjective probabilities.

10. Discuss whether it is possible for two persons to arrive at correct, yet different, probabilities regarding the commercial success of a new latex paint, if they both use
   (a) the frequency concept of probability;
   (b) subjective probabilities.

11. If a student is willing to bet $3 against $1, but not $4 against $1 that he will get a passing grade in a certain course, what does this tell us about the subjective probability he assigns to his getting a passing grade in the course?

12. The supplier of delicate optical equipment feels that the odds are 7 to 5 against a shipment arriving late and 11 to 1 against it not arriving at all. Furthermore, he feels that there is a fifty-fifty chance (the odds are 1 to 1) that such a shipment will either arrive late or not at all. Are the corresponding probabilities *consistent*?

13. There are two Ferraris in a race, and an expert feels that the odds against their winning are, respectively, 2 to 1 and 3 to 1. Furthermore, she claims that there is a less-than-even chance that either of the two Ferraris will win. Discuss the *consistency* of these claims.

14. When rolling a pair of balanced dice, what are the probabilities of getting
    (a) 7;   (b) 11;
    (c) 7 or 11;   (d) 2 or 12;
    (e) 3;   (f) 2 or 3 or 12?

15. If a card is drawn from a well-shuffled deck of 52 playing cards, find the probabilities of drawing
    (a) a red king;   (b) a black card;
    (c) a 3, 4, 5, or 6;   (d) a red ace or a black 7.

16. A lottery sells tickets numbered from 00001 through 50000. What is the probability that the number drawn will be divisible by 200?

17. Referring to Exercise 4 and Figure 2.11, suppose that the questionnaire filled in by one of the 150 persons interviewed is to be double-checked. Assuming the questionnaire is chosen by lot, find the probabilities that
    (a) the person lives more than three miles from the center of the city;
    (b) the person now regularly drives his or her car to work;
    (c) the person does not live more than three miles from the center of the city and would not want to switch to public mass transportation if it were available;
    (d) the person now regularly drives his or her car to work but would gladly switch to public mass transportation if it were available.

18. If A and B are mutually exclusive events, $P(A) = 0.23$, and $P(B) = 0.51$, find
    (a) $P(A')$;   (b) $P(A \cup B)$;
    (c) $P(A \cap B)$;   (d) $P(A' \cap B')$.

19. Given $P(A) = 0.35$, $P(B) = 0.73$, and $P(A \cap B) = 0.14$, find
    (a) $P(A \cup B)$;   (b) $P(A \cap B')$;
    (c) $P(A' \cup B')$;   (d) $P(A' \cap B)$.

20. The probability that a person stopping at a gas station will ask to have his tires checked is 0.12, the probability that he will ask to have his oil checked is 0.29, and the probability that he will ask to have them both checked is 0.07.
    (a) What is the probability that a person stopping at this gas station will have either his tires or his oil checked?
    (b) What is the probability that a person stopping at this gas station will have neither his tires nor his oil checked?

21. The probability that a new airport will get an award for its design is 0.16, the probability that it will get an award for the efficient use of materials is 0.24, and the probability that it will get both awards is 0.11.
    (a) What is the probability that it will get at least one of the two awards?
    (b) What is the probability that it will get only one of the two awards?

22. The probability that at least one of three events A, B, and C will occur is given by

$$P(A \cup B \cup C) = P(A) + P(B) + P(C) - P(A \cap B) - P(A \cap C)$$
$$- P(B \cap C) + P(A \cap B \cap C)$$

Verify this formula with the probabilities shown in Figure 2.12.

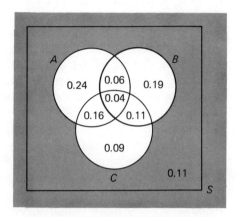

Figure 2.12. Diagram for Exercise 22.

23. Suppose that in the maintenance of a large medical-records file for insurance purposes the probability of an error in processing is 0.0010, the probability of an error in filing is 0.0009, the probability of an error in retrieving is 0.0012, the probability of an error in processing as well as filing is 0.0002, the probability of an error in processing as well as retrieving is 0.0003, the probability of an error in filing as well as retrieving is 0.0003, and the probability of an error in processing, filing, as well as retrieving is 0.0001. What is the probability of making at least one of these errors?

24. Referring to parts (c) and (d) of Exercise 13 on page 15, show that
    (a) $P(A) \geq P(A \cap B)$;   (b) $P(A) \leq P(A \cup B)$.

25. Explain why there must be a mistake in each of the following statements:
    (a) If the probability that an ore contains uranium is 0.28, the probability that it does not contain uranium is 0.62.
    (b) A company is working on the construction of two shopping centers; the probability that the larger of the two shopping centers will be completed on time is 0.35, and the probability that both shopping centers will be completed on time is 0.42.
    (c) The probability that a student will get an A in a geology course is 0.32, and the probability that he will get either an A or a B is 0.27.
    (d) In a certain city, the probability that a family has at least one color television set is 0.82, the probability that a family has at least one black-and-white television set is 0.64, and the probability that a family has at least one of each kind is 0.41.

2.5 CONDITIONAL PROBABILITY

As we have defined probability, it is meaningful to ask for the probability of an event only if we refer to a given sample space S. To ask for the probability that an engineer makes at least $20,000 a year is meaningless unless we specify whether we are referring to all engineers in the Western hemisphere, all engineers in the United States, all those in a particular industry, all those who are affiliated with a university, and so forth. Thus, when we use the symbol $P(A)$ for the probability of A, we really mean the probability of A *given some sample space* S. Since the choice of S is by no means always evident, and since there are problems in which we are interested in the probabilities of A with respect to more sample spaces than one, the notation $P(A \mid S)$ is used to make it clear that we are referring to the particular sample space S. We read $P(A \mid S)$ as "the *conditional probability* of A relative to S" and every probability is, thus, a conditional probability. Of course, we use the simplified notation $P(A)$ whenever the choice of S is clearly understood.

Example. To illustrate some of the ideas connected with conditional probabilities, let us consider again the 500 machine parts of which some are improperly assembled and some contain one or more defective components as shown in Figure 2.8. Assuming equal probabilities in the selection of one of the machine parts for inspection, we saw on page 23 that the probability of getting one with one or more defective components is $P(D) = \frac{3}{100}$. Let us now check whether this probability changes if the choice is restricted to the machine parts that are improperly assembled. To find $P(D \mid I)$ we have only to look at the reduced sample space of Figure 2.13, and to assume that each of the 30 improperly assembled machine parts, of which 10 contain one or more defective components, has the same probability of being

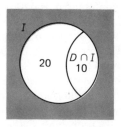

Figure 2.13. Reduced sample space.

selected. Using Theorem 2.3, we get

$$P(D \mid I) = \frac{N(D \cap I)}{N(I)} = \frac{10}{30} = \frac{1}{3}$$

Thus, the probability of getting a machine part with one or more defective components increases markedly on the reduced sample space I. Note that if we divide the numerator and the denominator of the last expression by $N(S)$, we get

$$P(D \mid I) = \frac{\dfrac{N(D \cap I)}{N(S)}}{\dfrac{N(I)}{N(S)}} = \frac{P(D \cap I)}{P(I)}$$

and it can be seen that the desired probability is the *ratio* of the probability that the machine part selected contains one or more defective components and belongs to the reduced sample space I to the probability that it belongs to I.

Looking at this example in another way, note that with respect to the whole sample space S we have

$$P(D \cap I) = \frac{N(D \cap I)}{500} = \frac{1}{50}$$

$$P(D' \cap I) = \frac{N(D' \cap I)}{500} = \frac{2}{50}$$

assuming, as before, that each of the 500 machine parts has the same probability of being selected for inspection. Thus, the probabilities that the machine part selected does or does not contain one or more defective components, given that it is improperly assembled, should be in the ratio $1 : 2$. Since all probabilities in the reduced sample space must add up to 1, it follows that

$$P(D \mid I) = \tfrac{1}{3} \quad \text{and} \quad P(D' \mid I) = \tfrac{2}{3}$$

which agress with the result obtained before. This also explains why we had to divide by $P(I)$ when we wrote

$$P(D \mid I) = \frac{P(D \cap I)}{P(I)}$$

above. Division by $P(I)$, or multiplication by $\dfrac{1}{P(I)}$, takes care of the *proportionality factor* which makes the sum of the probabilities over the reduced sample space equal to 1.

Following these observations, let us now make the following formal definition:

If A and B are any events in S and P(B) ≠ 0, the conditional probability of A relative to B is given by

$$P(A|B) = \frac{P(A \cap B)}{P(B)}$$

The following are some immediate applications of this formula:

Example. If the probability that a communication system will have high selectivity is $P(A) = 0.54$, the probability that it will have high fidelity is $P(B) = 0.81$, and the probability that it will have both is $P(A \cap B) = 0.18$, then the probability that a system with high fidelity will also have high selectivity is

$$P(A|B) = \frac{P(A \cap B)}{P(B)} = \frac{0.18}{0.81} = \frac{2}{9}$$

and the probability that a system with high selectivity will also have high fidelity is

$$P(B|A) = \frac{P(B \cap A)}{P(A)} = \frac{0.18}{0.54} = \frac{1}{3}$$

Example. If the probability that a research project will be well planned is 0.80 and the probability that it will be well planned and well executed is 0.72, then the probability that a research project which is well planned will also be well executed is $\frac{0.72}{0.80} = 0.90$.

Example. Returning to the lawn mower example, for which the probabilities of the individual outcomes are given in Figure 2.9 on page 22, let us now determine the probability $P(E_1|C_1)$ that a lawn mower will be easy to operate *given that it has a high average cost of repairs*. Using the results obtained on page 23, we get

$$P(E_1|C_1) = \frac{P(E_1 \cap C_1)}{P(C_1)} = \frac{0.12}{0.30} = 0.40$$

and it is of interest to note that this equals the value which we obtained earlier for $P(E_1)$. This means that the probability of a lawn mower's being rated easy to operate is the same regardless of whether or not it is given that it has a high average cost of repairs, and we say that E_1 is *independent* of C_1. As the reader will be asked to verify in Exercise 5 on page 41, it also follows from the results on page 23 that E_1 is *not independent* of P_1, namely, that a lawn mower's ease of operation *is* related to its retail price.

In general, if A and B are any two events in a sample space S, we say that *A is independent of B* if and only if $P(A \,|\, B) = P(A)$, and since it can be shown that *B is independent of A whenever A is independent of B*, namely, that $P(A \,|\, B) = P(A)$ implies $P(B \,|\, A) = P(B)$ provided $P(A) \neq 0$, it is customary to say simply that *A and B are independent.*

An immediate consequence of our definition of conditional probability is given by the following theorem, usually called the *general law of multiplication*:

THEOREM 2.6. *If A and B are any events in S, then*

$$P(A \cap B) = P(A) \cdot P(B \,|\, A) \qquad \text{if } P(A) \neq 0$$
$$= P(B) \cdot P(A \,|\, B) \qquad \text{if } P(B) \neq 0$$

The second of these relations follows from the definition of conditional probability, multiplying both sides by $P(B)$; the first is obtained by interchanging the letters A and B in the definition of conditional probability and then multiplying both sides by $P(A)$. Note that the definition of $P(B \,|\, A)$ and $P(A \,|\, B)$ assumes that $P(A) \neq 0$ and $P(B) \neq 0$, respectively.

Example. To illustrate the use of Theorem 2.6, suppose that the foreman of a group of 20 construction workers wants to get the opinion of two of them (to be selected at random) about certain new safety regulations. If 12 of them favor the new regulations while the other 8 are against them, let us find the probability that both of the workers chosen by the foreman will be against the new safety regulations. Assuming equal probabilities for each selection, the probability that the first construction worker selected will be against the new safety regulations is $\frac{8}{20}$, and the probability that the second construction worker selected will be against the new safety regulations *given that the first one is against them* is $\frac{7}{19}$. Substituting these values into the formula of Theorem 2.7, we find that the desired probability is $\frac{8}{20} \cdot \frac{7}{19} = \frac{14}{95}$. Similarly, the probability that both of the construction workers will be for the new safety regulations is $\frac{12}{20} \cdot \frac{11}{19} = \frac{33}{95}$, and, by subtraction, we find that the probability that one of them will be for the new safety regulations and one of them will be against them is $1 - \frac{14}{95} - \frac{33}{95} = \frac{48}{95}$.

In the special case where A and B are independent, Theorem 2.6 leads to the following, which is usually called the *special law of multiplication*:

THEOREM 2.7. *If A and B are independent events, then*

$$P(A \cap B) = P(A) \cdot P(B)$$

Example. For instance, the probability of getting two heads in two successive flips of a coin is $\frac{1}{2} \cdot \frac{1}{2} = \frac{1}{4}$, and the probability of drawing two aces in a row from a standard deck of 52 playing cards is $\frac{4}{52} \cdot \frac{4}{52} = \frac{1}{169}$ *provided the first card is replaced before the second is drawn.*

The special law of multiplication is readily extended to apply to more than two independent events: *if three or more events are independent, the probability that they will all occur is given by the product of their respective probabilities.*

Example. For instance, the probability of getting four heads in a row with a balanced coin is $\frac{1}{2} \cdot \frac{1}{2} \cdot \frac{1}{2} \cdot \frac{1}{2} = \frac{1}{16}$, and the probability of rolling a 3, a 1, a 6, and then some *other* number in four rolls of a balanced die is $\frac{1}{6} \cdot \frac{1}{6} \cdot \frac{1}{6} \cdot \frac{3}{6} = \frac{1}{432}$.

2.6 THE RULE OF BAYES

The general law of multiplication is useful in solving many problems in which the ultimate outcome of an experiment depends on the outcomes of various intermediate stages.

Example. Suppose, for instance, that an assembly plant receives its voltage regulators from two different suppliers, 75 percent from supplier B_1 and 25 percent from supplier B_2. In other words, the probability that any one voltage regulator received by the plant comes from supplier B_1 is $\frac{3}{4}$ and the probability that it comes from supplier B_2 is $\frac{1}{4}$. Suppose, furthermore, that 95 percent of the voltage regulators supplied by B_1 and 80 percent of those supplied by B_2 perform according to specifications. What we would like to know is the probability of the event A that *any one voltage regulator received by the plant will perform according to specifications.*

If we use the fact that $A = (A \cap B_1) \cup (A \cap B_2)$ and that $A \cap B_1$ and $A \cap B_2$ are mutually exclusive (see Figure 2.10 with B_1 and B_2 taking the place of B and B'), the special law of addition yields

$$P(A) = P(A \cap B_1) + P(A \cap B_2)$$

Then, if we apply the general law of multiplication to $P(A \cap B_1)$ and $P(A \cap B_2)$, we obtain

$$P(A) = P(B_1) \cdot P(A \mid B_1) + P(B_2) \cdot P(A \mid B_2)$$

and substituting the given probabilities $P(B_1) = \frac{3}{4}, P(B_2) = \frac{1}{4}, P(A \mid B_1)$ $= 0.95$, and $P(A \mid B_2) = 0.80$, we finally get

$$P(A) = \tfrac{3}{4}(0.95) + \tfrac{1}{4}(0.80) = 0.9125$$

for the desired probability that any one voltage regulator received by the plant will perform according to specifications.

In the preceding example there were only two alternatives at the intermediate stage, B_1 and B_2. In general, if there are n *mutually exclusive* alternatives $B_1, B_2, \ldots, B_n$ at the intermediate stage, the corresponding formula for the probability of the final outcome, event A, is given by

$$P(A) = \sum_{i=1}^{n} P(B_i) \cdot P(A \mid B_i)$$

sometimes called the *rule of elimination*. To visualize this result, we have only to construct a tree diagram like that of Figure 2.14, where the probability of the final outcome is given by the *sum* of the products of the probabilities corresponding to each individual branch.

Continuation of Example. To illustrate, suppose that there had been three suppliers in our example and that the required probabilities are as shown in Figure 2.15. Then, the probability that any one voltage

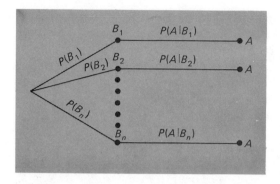

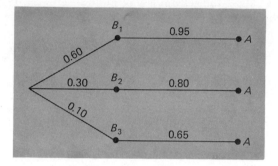

Figure 2.15. Tree diagram for example with three suppliers of voltage regulators.

regulator received by the plant will perform according to specifications is

$$P(A) = (0.60)(0.95) + (0.30)(0.80) + (0.10)(0.65)$$
$$= 0.875$$

To consider a problem which is closely related to the one we have just discussed, suppose we want to know the probability of the event that *a particular voltage regulator which is known to perform according to specifications came from supplier* B_3. Symbolically, we want to know the value of $P(B_3 \mid A)$. To solve this problem we write

$$P(B_3 \mid A) = \frac{P(A \cap B_3)}{P(A)}$$

and then substitute

$$P(A \cap B_3) = P(B_3) \cdot P(A \mid B_3)$$

and

$$P(A) = \sum_{i=1}^{3} P(B_i) \cdot P(A \mid B_i)$$

respectively, into the numerator and the denominator in accordance with the general law of multiplication and the rule of elimination. We thus obtain the formula

$$P(B_3 \mid A) = \frac{P(B_3) \cdot P(A \mid B_3)}{\sum\limits_{i=1}^{3} P(B_i) \cdot P(A \mid B_i)}$$

which expresses the required probability in terms of given probabilities. Substituting the values from Figure 2.15, we finally obtain

$$P(B_3 \mid A) = \frac{(0.10)(0.65)}{(0.60)(0.95) + (0.30)(0.80) + (0.10)(0.65)}$$

$$= 0.074$$

Note that the probability that a voltage regulator is supplied by B_3 decreases from 0.10 to 0.074 once it is known that it performs according to specifications.

The method used to solve the last example can easily be generalized to yield the following formula, called the *rule of Bayes*:

THEOREM 2.8. *If B_1, B_2, ..., B_n are mutually exclusive events of which one must occur, that is, $\sum\limits_{i=1}^{n} P(B_i) = 1$, then*

$$P(B_r \mid A) = \frac{P(B_r) \cdot P(A \mid B_r)}{\sum\limits_{i=1}^{n} P(B_i) \cdot P(A \mid B_i)}$$

for $r = 1, 2, \ldots,$ or n.

This rule provides a formula for finding the probability that the "effect" A was "caused" by the event B_r. For instance, in our example we found the probability that an acceptable voltage regulator was made by supplier B_3. The probabilities $P(B_i)$ are called the "prior" or "a priori" probabilities of the "causes" B_i, and in practice it is often difficult to assign them numerical values. For many years Bayes' rule was looked upon with suspicion because it was used with the often-erroneous assumption that the prior probabilities are all equal. A good deal of the controversy once surrounding the rule of Bayes has been cleared up with the realization that the probabilities $P(B_i)$ must be determined separately in each case from the nature of the problem, preferably on the basis of past experience. We shall return to this problem, as it relates to *Bayesian inference*, in Chapters 7 and 9.

> **Example.** To further illustrate the rule of Bayes, suppose that the four attendants at a gasoline service station are supposed to wash the windshield of each customer's car. Jack, who services 20 percent of all cars, fails to wash the windshiels 1 time in 20; Tom, who services 60 percent of all cars, fails to wash the windshield 1 time in 10; George, who services 15 percent of all cars, fails to wash the windshield 1 time in 10; and Peter, who services 5 percent of all cars, fails to wash the windshield 1 time in 20. If a customer complains later on that her windshield was not washed, the probability that her car was serviced by Jack may be obtained by substituting the corresponding probabilities into the formula for the rule of Bayes; thus, we get
>
> $$P(B_1 \mid A) = \frac{(0.20)(0.05)}{(0.20)(0.05) + (0.60)(0.10) + (0.15)(0.10) + (0.05)(0.05)}$$
> $$= 0.114$$
>
> and it is of interest to note that although Jack fails to wash the windshield of only 1 car in 20, namely, 5 percent, more than 11 percent of the windshields that are not washed at this service station are his responsibility.

EXERCISES

1. Referring to Figure 2.8, find $P(I \mid D)$ and $P(I \mid D')$; assume that originally each of the 500 machine parts has the same probability of being chosen for inspection.

2. If in the example on page 34 the probability is 0.75 that the research project will be well executed, what is the probability that the project was well planned *given that it is well executed?*

3. Referring to Exercise 4 on page 28, Figure 2.11, and assuming equal probabilities, find the probability that the person whose questionnaire is chosen to be double-checked

(a) lives more than three miles from the center of the city given that he or she would gladly switch to public mass transportation;

(b) regularly drives his or her car to work given that he or she lives more than three miles from the center of the city;

(c) would not want to switch to public mass transportation given that he or she does not regularly drive his or her car to work.

4. Referring to the probabilities of Figure 2.12, find

(a) $P(A|B)$;   (b) $P(B|C')$;

(c) $P(A \cap B|C)$;   (d) $P(B \cup C|A')$;

(e) $P(A|B \cup C)$;   (f) $P(A|B \cap C)$;

(g) $P(A \cap B \cap C|B \cap C)$;   (h) $P(A \cap B \cap C|B \cup C)$.

5. Referring to the lawn mower example and the probabilities given in Figure 2.9, find

(a) $P(E_1|P_1)$ and compare its value with that of $P(E_1)$;

(b) $P(C_2|P_2)$ and compare its value with that of $P(C_2)$;

(c) $P(E_1|P_1 \cap C_1)$ and compare its value with that of $P(E_1)$.

6. With reference to Exercise 21 on page 30, find

(a) the probability that the airport will get the design award given that it will get the award for the efficient use of materials;

(b) the probability that the airport will get the design award given that it will *not* get the award for the efficient use of materials.

7. Prove that $P(A|B) = P(A)$ implies $P(B|A) = P(B)$ provided $P(B) \neq 0$.

8. Among the 24 invoices prepared by a billing department, 4 contain errors while the others do not. If we randomly check two of these invoices, what are the probabilities that

(a) both will contain errors;

(b) neither will contain an error?

9. What is the probability of drawing two black cards from a standard deck of 52 playing cards if

(a) the first card is replaced before the second card is drawn;

(b) the first card is not replaced before the second card is drawn?

10. The probability that Mr. Jones will be late to work on any one day is 0.60. Assuming that his being late on any one day is independent of his being late on any other day, find the probability that in a given week he will be late on Monday, on time on Tuesday, Wednesday, and Thursday, and late again on Friday.

11. Referring to the lawn mower example, the probabilities of Figure 2.9, and the results given on page 23, check whether $P(E_1 \cap P_1 \cap C_1) = P(E_1) \cdot P(P_1) \cdot P(C_1)$. What can you conclude about the events E_1, P_1, and C_1?

12. Find the probabilities of getting

(a) six tails in a row with a balanced coin;

(b) five multiple-choice questions answered correctly, if for each question the probability of answering it correctly is $\frac{1}{3}$.

13. The problem of determining the probability that any number of events will occur becomes more complicated when the events are *not independent*. For three events A, B, and C, for example, the probability that they will all occur is obtained by multiplying the probability of A by the probability of B *given* A, and then multiplying the result by the probability of C *given* $A \cap B$. For instance, the probability of drawing (without replacement) 3 aces in a row from an ordinary deck of 52 playing cards is

$$\frac{4}{52} \cdot \frac{3}{51} \cdot \frac{2}{50} = \frac{1}{5,525}$$

Clearly, there are only 3 aces among the 51 cards which remain after the first ace has been drawn, and only 2 aces among the 50 cards which remain after the first two aces have been drawn.

(a) If six bullets, of which three are blanks, are randomly inserted into a gun, what is the probability that the first three bullets fired will all be blanks?

(b) In a certain city during the month of May, the probability that a rainy day will be followed by another rainy day is 0.80 and the probability that a sunny day will be followed by a rainy day is 0.60. Assuming that each day is classified as being either rainy or sunny and that the weather on any given day depends only on the weather the day before, find the probability that in the given city a rainy day in May is followed by two more rainy days, then a sunny day, and finally another rainy day.

(c) A department store which bills its charge-account customers once a month has found that if a customer pays promptly one month, the probability is 0.90 that he will also pay promptly the next month; however, if a customer does not pay promptly one month, the probability that he will pay promptly the next month is only 0.50. What is the probability that a customer who has paid promptly one month will not pay promptly the next three months?

14. The supervisor of an electronics plant knows from past experience that a new worker who has attended the company's training program has a probability of 0.85 of meeting his production quota, while a new worker who has not attended the training program has a probability of only 0.33. If 75 percent of all new workers attend the training program, what is the probability that a new worker will meet his production quota?

15. Use the result of Exercise 14 to find

(a) the probability that a new worker who meets his production quota will have attended the training program;

(b) the probability that a new worker who meets his production quota will not have attended the training program.

16. Members of a consulting firm rent cars from three rental agencies: 20 percent from agency D, 20 percent from agency E, and 60 percent from agency F. If 10 percent of the cars from rental agency D have bad tires, while the corresponding percentages for agencies E and F are, respectively, 12 and 4,

what is the probability that one of the consultants will get a rental car with bad tires?

17. Use the result of Exercise 16 to find the probability that a car with bad tires issued to one of the consultants came from rental agency F.

18. There is a fifty-fifty chance that Firm A will bid for the construction of a new city hall. Firm B submits a bid and the probability that it will get the job is $\frac{2}{3}$ provided Firm A does not bid; if Firm A submits a bid, the probability that Firm B will get the job is only $\frac{1}{3}$. If Firm B gets the job, what is the probability that Firm A did not bid?

19. With reference to the example on page 36, find the respective probabilities that a voltage regulator which performs according to specifications came
    (a) from supplier B_1;
    (b) from supplier B_2.

20. With reference to the example on page 40, find the probability that the car of the complaining customer was serviced by Tom.

21. An explosion in an LNG storage tank in the process of being repaired could have occurred as the result of static electricity, malfunctioning electrical equipment, an open flame in contact with the liner, or purposeful action (industrial sabotage). Interviews with engineers who were analyzing the risks involved led to estimates that such an explosion would occur with probability 0.25 as a result of static electricity, 0.20 as a result of malfunctioning electric equipment, 0.40 as a result of an open flame, and 0.75 as a result of purposeful action. These interviews also yielded subjective estimates of the prior probabilities of these four causes of 0.30, 0.40, 0.15, and 0.15, respectively. What was the most likely cause of the explosion?

2.7 MATHEMATICAL EXPECTATION AND DECISION MAKING

If an insurance agent tells us that in the United States a 40-year-old woman can expect to live 38 more years, this does not mean that anyone really expects a 40-year-old woman to live until her 78th birthday and then pass away the next day. Similarly, if we read that a radial tire can be expected to last 40,000 miles, that in the coal mining industry we can expect 41.5 nonfatal injuries per million man-hours, or that the U.S. Patent Office can expect to issue 27,000 certificates of trademarks each year, it must be understood that the word "expect" is not used in its colloquial sense. So far as the first statement is concerned, some 40-year-old women will live another 15 years, some will live another 30 years, some will live another 50 years, . . . , and the life expectance of "38 more years" will have to be interpreted as an average, namely, as a *mathematical expectation*.

Originally, the concept of a mathematical expectation arose in connection with games of chance, and in its simplest form it is *the product of the probability that a player will win and the amount he stands to win.*

Example. If we stand to receive $10 if a balanced coin comes up *tails* and nothing if it comes up *heads,* our mathematical expectation is $10 \cdot \frac{1}{2} = \$5$. Similarly, if we buy one of 1,000 raffle tickets issued for a grand prize of $500, the probability for each ticket is $\frac{1}{1,000}$ and our mathematical expectation is $500 \left(\frac{1}{1,000} \right) = \$0.50$. Note that 999 of the tickets will not pay anything at all, and only one of the tickets will pay $500. However, all the 1,000 tickets together pay $500, or on the average $0.50 per ticket—*this is the mathematical expectation for each ticket.*

In this example there was only a single "payoff," namely, a single prize or a single payment. To see how the concept of a mathematical expectation can be generalized, let us consider the following example:

Example. A distributor makes a profit of $20 on an item if it is shipped from the factory in perfect condition, but if it is not shipped in perfect condition, the profit is reduced to $12 because of the cost of repairs. This information in itself does not enable us to determine the distributor's average profit per item, so let us suppose that only 80 percent of the items are shipped from the factory in perfect condition. Then we can argue that for items shipped from the factory he should make a profit of $20 about 80 percent of the time, a profit of $12 about 20 percent of the time, and on the average he should, thus, make a profit of

$$20(0.80) + 12(0.20) = \$18.40$$

This result is the sum of the products obtained by multiplying each amount by the corresponding proportion or probability. Generalizing from this example, let us now make the following definition:

If the probabilities of obtaining the amounts $a_1, a_2, \ldots,$ or a_k are, respectively, $p_1, p_2, \ldots,$ and p_k, then the mathematical expectation is

$$a_1 p_1 + a_2 p_2 + \ldots + a_k p_k$$

So far as the a's are concerned, it is important to keep in mind that they are *positive* when they represent profits, winnings, or gains, and that they are *negative* when they represent losses, deficits, or penalties.

Example. To illustrate this definition, suppose that an engineering firm is faced with the task of preparing a proposal for a research contract; the cost of preparing the proposal is $5,000, and it is assumed to be known that the probabilities for potential gross profits of $50,000, $30,000, $10,000, or $0 are, respectively, 0.20, 0.50, 0.20, and 0.10, provided the proposal is accepted. Assessing the probability that the proposal will be accepted as 0.30, we find that there is a probability of $(0.30)(0.20) = 0.06$ of making a net profit of $45,000 ($50,000 minus the cost of the proposal). Similarly, the probabilities of making net profits of $25,000 or $5,000 are, respectively, $(0.50)(0.30) = 0.15$ and $(0.20)(0.30) = 0.06$, while the probability of a $5,000 loss is $(0.10)(0.30) + 0.70 = 0.73$, allowing for the eventuality that the proposal will not be accepted. Thus, we find that the *expected net profit* is

$$45,000(0.06) + 25,000(0.15) + 5,000(0.06) - 5,000(0.73) = \$3,100$$

Whether or not it is wise to risk $5,000 to make an expected profit of $3,100 is not a question of statistics. If a company has little capital and can make only few such proposals, the 0.73 chance of losing $5,000 may well make the venture unattractive. On the other hand, if a company is well capitalized and prepares many such proposals, the risk may be worth taking, for it promises an *expected return* of 62 percent on the original investment.

Example. The following is a typical inventory problem in which *expected values* are used to determine optimum conditions. Suppose it is known from past experience that the daily demand for a perishable product is as shown in the following table:

Number of orders	3	4	5	6	7	8	9
Probability	0.05	0.12	0.20	0.24	0.17	0.14	0.08

Each item costs $35 (including the cost of carrying it in stock), it is sold for $50 provided it is in stock, and an item remaining in stock at the end of a day represents a total loss. On the basis of this information we want to determine how many items should be stocked each day so as to *maximize the expected profit*.

For 3 units stocked, the profit in dollars is obviously $150 - 105 = 45$, as there is a probability of 1.00 that there will be a demand for 3 units or more. For 4 units stocked, there is a probability of 0.05 that exactly 3 units can be sold, a probability of 0.95 that there will be a demand for 4 or more, and there is, therefore, an expected dollar profit of

$$150(0.05) + 200(0.95) - 140 = 57.50$$

Similarly, for 5 units stocked there is a probability of 0.05 that exactly 3 units will be sold, a probability of 0.12 that exactly 4 units will be sold, a probability of 0.83 that there will be a demand for 5 or more, and there is, therefore, an expected dollar profit of

$$150(0.05) + 200(0.12) + 250(0.83) - 175 = 64.00$$

Continuing in this manner it can be shown that for 6 units stocked the expected profit is \$60.50, for 7 units stocked the expected profit is \$45.00, etc., and that the expected profit is a maximum when 5 units are stocked (see also Exercise 6 on page 47).

In recent years, mathematical expectations have been playing an increasingly important role in "scientific" decision making, as it is generally considered rational to select whichever alternative has the "most promising" mathematical expectation: the one which *maximizes expected profits, minimizes expected costs, maximizes tax advantages, minimizes expected losses*, and so forth. Although this approach to decision making has intuitive appeal and seems very logical, it is not without complications.

In applying the methods of decision making illustrated in this section, there are essentially two practical limitations: first, we must be able to assign "cash values" to the various outcomes, and second, we must have adequate information concerning all relevant probabilities. The problem of assigning cash values to the various outcomes can be much more difficult than it was in our examples. For instance, if a government agency were faced with the decision of whether to require crash barriers for interstate highway construction, it would be extremely difficult to assign a cash value to the savings in life and reductions in injuries that might result.

The specification of probabilities concerning the various outcomes can be equally difficult. In our examples, the probabilities were presumably based on past experience, and, hence, are really only estimates (or approximations). This may be questioned with respect to the last example (the one dealing with the preparation of the research proposal), where it would be extremely difficult even to estimate the probability that a given proposal will be accepted. This would have to depend not only on the quality of the proposal itself, but on such unknown factors as the number and quality of competing proposals as well as the policy of the sponsoring agency. The estimate of 0.30 in our example was purely subjective, being based on some executive's "feeling" about the situation measured against the background of his personal experience. While such estimates of probabilities are not without value, they do leave open the possibility that different individuals (acting in the same situation and having the same information) would arrive at different "optimum" decisions.

EXERCISES

1. If a charitable organization raises funds by selling 4,000 raffle tickets for a vacation trip worth $560, what is the mathematical expectation of a person who buys one of these tickets? What would be the mathematical expectation if there were also a second prize of a painting worth $320?

2. When the American League and National League champions are evenly matched, the probabilities that a World Series will end in 4, 5, 6, or 7 games are, respectively, $\frac{1}{8}$, $\frac{1}{4}$, $\frac{5}{16}$, and $\frac{5}{16}$. What is the expected length of a World Series when the two teams are evenly matched?

3. It costs $60 to test a certain component of a machine. If a defective component is installed, it costs $1,200 to repair the resulting damage to the machine. Is it more profitable to install the component without testing if it is known that
   (a) 3 percent of all components produced are defective;
   (b) 5 percent of all components produced are defective;
   (c) 8 percent of all components produced are defective?

4. The two finalists in a golf tournament play 18 holes, with the winner getting $20,000 and the runner-up getting $12,000. What are the two players' mathematical expectations if
   (a) they are evenly matched;
   (b) one of the two players should be favored by odds of 3 to 1?

5. If someone were to give us $6 each time we roll a 1 with a balanced die, how much should we have to pay him when we roll a 2, 3, 4, 5, or 6 to make the game equitable?

6. Referring to the example on page 45, verify that for 6 and 7 units stocked, the expected profits are $60.50 and $45.00, respectively.

7. To handle a liability suit, a lawyer has to decide whether to charge a straight fee of $600 or a contingent fee of $1,800 which he will get only if his client wins. What does the lawyer think about his client's chances if he feels that
   (a) taking the straight fee will give him a higher mathematical expectation;

(b) taking the contingent fee will give him a higher mathematical expectation?

8. The manufacturer of a new battery additive has to decide whether to sell his product for $0.80 a can, or for $1.20 with a "double-your-money-back-if-not-satisfied" guaranty. How does he feel about the chances that a person will ask for double his money back if
   (a) he decides to sell the product for $0.80;
   (b) he decides to sell the product for $1.20 with the guaranty;
   (c) he cannot make up his mind?

9. A contractor has to choose between two jobs. The first job promises a profit of $240,000 with a probability of 0.75 or a loss of $60,000 (due to strikes and other delays) with a probability of 0.25; the second job promises a profit of $360,000 with a probability of 0.50 or a loss of $90,000 with a probability of 0.50.
   (a) Which job should the contractor choose if he wants to maximize his expected profit?
   (b) Which job would the contractor probably choose if his business is in bad shape and he will go broke unless he can make a profit of at least $300,000 on his next job?

10. A merchant can buy an item for $2.10 and sell it for $4.50. The probabilities for a demand of 0, 1, 2, 3, 4, or "5 or more" items are, respectively, 0.05, 0.15, 0.30, 0.25, 0.15, and 0.10. Calculate the expected profits resulting from stocking 0, 1, 2, 3, 4, or 5 items and determine how many the merchant should stock so as to maximize his expected profit.

11. Mr. Brown and Mr. Jones are betting on repeated flips of a balanced coin. At the beginning Mr. Brown has m dollars, Mr. Jones has n dollars; at each flip the loser pays the winner one dollar, and they continue playing until one of them has lost all the money with which he began. To find p, the probability that Mr. Brown will win Mr. Jones' n dollars before he loses his m dollars, set up an equation in m, n, and p which makes use of the fact that in an *equitable game* each player's mathematical expectation is zero. What is the probability that Mr. Brown will win if he begins with $12 and Mr. Jones begins with $4?

12. The management of a mining company must decide whether to continue an operation at a certain location. If they continue and are successful, this will be worth $1,000,000; if they continue and are not successful, this will entail a loss of $600,000; if they do not continue but would have been successful, this will entail a loss of $400,000 (for competitive reasons); and if they do not continue and would not have been successful anyhow, this will be worth $100,000 to the company (because funds allocated to the operation remain unspent).
   (a) What decision would maximize the company's expected gain if the odds against success are 3 to 2?
   (b) Repeat part (a) with 4 to 1 odds against success.

(c) If the odds of part (a) are correct and it were possible to continue the operation for two months at a cost of $300,000, after which it would be known *for certain* whether the operation will be a success, would it be worthwhile to spend these additional funds?

(d) What could one expect the management of the mining company to decide if they did not know anything about the chances of success, but *in general* they are confirmed pessimists?

13. A power company is faced with the choice of whether to construct a light-water reactor (LWR) or a fossil-fuel power plant (FF). The LWR plant will cost $300 per kilowatt to construct, and the FF plant will cost $150 per kilowatt. Due to uncertainties concerning fuel availability and the impact of future air- and water-quality regulations, the useful operating life of each plant is unknown, but the following probabilities have been estimated:

Useful life (years)	10	20	30	40
Probabilities {LWR plant	0.05	0.25	0.50	0.20
{FF plant	0.10	0.50	0.30	0.10

(a) Which plant can be expected to cost less per year of useful life?

(b) What relative fuel cost per kilowatt would favor construction of the LWR plant?

(c) A further uncertainty concerning safety regulations makes it a 50-50 bet that $50 per kilowatt will be added to the actual construction cost of the LWR plant. What effect does this have on these results?

3

Probability Distributions

3.1 RANDOM VARIABLES

In most experiments we are interested only in a particular numerical description (or, perhaps, two or three particular descriptions) of the outcomes. For instance, in the inspection of manufactured products we may be interested only in the total number of defective items and not in the nature of the defects; in the study of an alloy we may want to know the percent of chromium present, without caring about the other elements; in the analysis of a road test we may be interested only in the average speed and the average fuel consumption, and not in any other performance characteristics; and in the study of the performance of a rotary stepping switch we may be interested in its lubrication, the current, and the humidity, but not in any other factors.

Example. To illustrate this idea more specifically, let us refer again to the rating of lawn mowers in the example on page 8 and the corresponding probabilities shown in Figure 2.9. Now let us suppose that we refer to E_1 (easy to operate), P_2 (inexpensive), and C_3 (low average cost of repairs) as preferred ratings, and that we are interested only in the *number of preferred ratings* which a lawn mower will get. To find the probabilities that a lawn mower will get 0, 1, 2, or 3 preferred ratings, let us refer to Figure 3.1, which is like Figure 2.9, except that we indicated for each outcome the corresponding number of preferred ratings. Adding the respective probabilities, we thus find

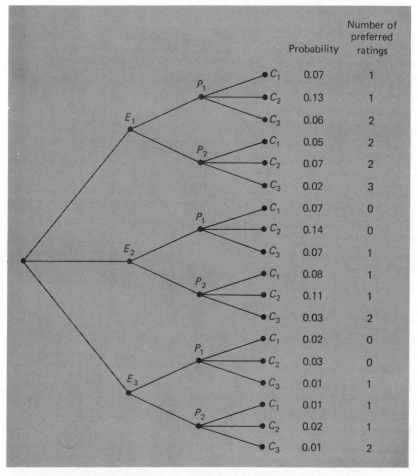

Figure 3.1. Ratings of lawn mowers and number of preferred ratings.

that the probability of a lawn mower getting zero preferred ratings is

$$P(0) = 0.07 + 0.14 + 0.02 + 0.03 = 0.26$$

Similarly, letting x represent the number of preferred ratings, we obtain the values shown in the following table.

Value of x	0	1	2	3
Probability	0.26	0.50	0.22	0.02

The numbers 0, 1, 2, and 3 in this table are values of the *random variable* describing the number of preferred ratings. Observe that to each outcome in the sample space there corresponds one and only one value x of this random variable, so that *a random variable may be thought of as a function defined over the elements of a sample space.* Note further that to find the probability that a random variable will take on any one value within its range, we have only to refer to the probabilities assigned to the individual outcomes and the extension of Axiom 3 given on page 18.* We thus define another function, which associates with each value of a random variable the probability that this value will be assumed. Such a function, an example of which is exhibited by the above table, is called a *probability distribution function*, or simply a *probability distribution*. To denote the values of a probability distribution we shall use such symbols as $f(x)$, $P(x)$, $\phi(z)$, $g(y)$, and so on.

Whenever possible, we try to define probability distributions by means of appropriate equations; otherwise we must give a table which actually exhibits the correspondence between the values assumed by the random variable and the associated probabilities.

Example. For instance,

$$f(x) = \tfrac{1}{6} \qquad \text{for } x = 1, 2, 3, 4, 5, 6$$

gives the probability distribution for the number of points we roll with a balanced die.

Of course, not every function defined for the values of a random variable can serve as a probability distribution. Since the values of a probability distribution are probabilities and since one value of a random variable must always occur, it follows that if $f(x)$ is a value of a probability

---

*It is assumed here that the sample space is finite; the changes that are required when a sample space is countably infinite or continuous will be discussed on pages 78 and 98.

distribution, then

$$f(x) \geq 0 \qquad \text{for all } x$$

and

$$\sum_{\text{all } x} f(x) = 1$$

Example. For instance

$$f(x) = \frac{x-2}{2} \qquad \text{for } x = 1, 2, 3, 4$$

cannot serve as a probability distribution because $f(1)$ is negative, and

$$h(x) = \frac{x^2}{25} \qquad \text{for } x = 0, 1, 2, 3, 4$$

cannot serve as a probability distribution because the sum of the five probabilities is $\frac{6}{5}$ instead of 1.

It is often helpful to visualize probability distributions by means of graphs like those of Figure 3.2. The one on the left is called a *histogram*;

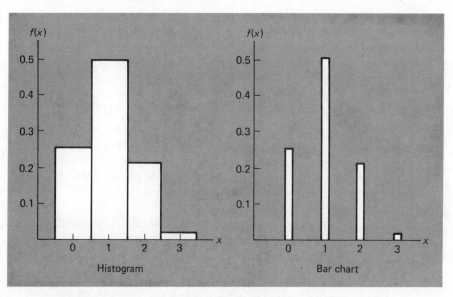

Figure 3.2. Graphs of the probability distribution of the number of preferred ratings.

the heights of the rectangles are proportional to the corresponding probabilities and their bases touch so that there are no gaps between the rectangles representing the successive values of the random variable. The one on the right is called a *bar chart*; the heights of the rectangles are also proportional to the corresponding probabilities, but they are narrow and their width is of no significance.

As we shall see later in this chapter, there are many problems in which we are interested not only in the probability $f(x)$ that the value of a random variable is x, but also in the probability that the value of a random variable is *less than or equal to x*. Writing this probability as $F(x)$, we refer to the function which assigns a value $F(x)$ to each x within the range of the random variable as the *cumulative distribution function*, or simply as the *distribution function*.

Example. Referring again to the lawn-mower-rating example and basing our calculations on the table on page 52, we thus get

Value of x	0	1	2	3
$F(x)$	0.26	0.76	0.98	1.00

for the distribution function of the number of preferred ratings.

3.2   THE BINOMIAL DISTRIBUTION

Many statistical problems deal with situations referred to as "repeated trials." For example, if we want to know the probability that 1 of 5 rivets will rupture in a tensile test, that 9 of 10 vacuum tubes will last at least 1,000 hours, that at least 60 of a shipment of 75 units are effective, we are in each case concerned with a number of "trials" and we are interested in the probability of getting a certain number of "successes."

Letting n represent the number of trials and x the number of successes, problems like these can be solved using the so-called *binomial distribution*, provided that they satisfy the following assumptions:

1. There are only two possible outcomes for each trial, arbitrarily called "success" and "failure" without inferring that a success is necessarily a desirable outcome.*

---

*Thus, getting a defective unit in sampling inspection may be called a "success"; this terminology carries over from the days when probability theory was applied only to games of chance.

2. The probability of a success is the same for each trial.
3. There are n trials, where n is a constant.
4. The n trials are independent.

Trials satisfying these assumptions are also referred to as *Bernoulli trials*; of course, if the assumptions cannot be met, the theory which we shall develop will not apply.

First, let us find the probability of getting x successes in n trials *in a given order*, when the probability of a success is p and, hence, the probability of a failure is $1 - p$. Denoting success and failure, respectively, by the letters S and F, consider the sequence

$$\underbrace{S \quad S \quad \ldots \quad S}_{x} \quad \underbrace{F \quad F \quad \ldots \quad F}_{n - x}$$

consisting of x successes, followed by $n - x$ failures. By the assumption of independence, the probability of this sequence is the product of the probabilities of the individual outcomes; there are x factors p, $n - x$ factors $1 - p$, and, hence, the probability is $p^x(1 - p)^{n-x}$. Note that the identical result would have been obtained for any other sequence containing x successes and $n - x$ failures *in a given order*. There would also have been x factors p and $n - x$ factors $1 - p$, and the desired probability would also have been given by their product.

To obtain the probability of x successes and $n - x$ failures *in any order*, we must now add the probabilities of all sequences containing x successes and $n - x$ failures in some order. Since each of these sequences has the probability $p^x(1 - p)^{n-x}$, we have only to find the number of such sequences and multiply this number by $p^x(1 - p)^{n-x}$. Note that $p^x(1 - p)^{n-x}$ is, thus, multiplied by the number of ways in which the x letters S can be distributed among the n trials. To find this number, suppose that we temporarily assume that the S's are *distinguishable*, say, by giving them subscripts and writing them as $S_1, S_2, \ldots, S_x$. Beginning with S_1, we can assign it to any one of the n trials, that is, we can assign it in n different ways. After that we can assign S_2 in $n - 1$ different ways, S_3 in $n - 2$ ways, $\ldots$, and, finally, S_x in $n - x + 1$ different ways. According to Theorem 2.1 on page 9, all of these S's can, thus, be distributed among the n trials in $n(n - 1)(n - 2) \cdot \ldots \cdot (n - x + 1)$ different ways. If we now omit the subscripts, we find that the various arrangements which we have considered are not all distinct. For instance, $S_1FS_2S_3F$, $S_1FS_3S_2F$, $S_2FS_1S_3F$, $S_2FS_3S_1F$, $S_3FS_1S_2F$, and $S_3FS_2S_1F$ all lead to $SFSSF$. In general, the x letters S with subscripts can be arranged among themselves in $x(x - 1) \cdot \ldots \cdot 2 \cdot 1$ different ways according to Theorem 2.1, and, hence, this many arrangements of the S's with subscripts lead to a single arrangement once the subscripts are dropped. The number of ways in which the x letters S can be distributed among the n trials is, thus, given by

$$\frac{n(n-1)(n-2) \cdot \ldots \cdot (n-x+1)}{x(x-1) \cdot \ldots \cdot 2 \cdot 1}$$

which in the *factorial notation* can be written as $n!/x!(n-x)!$.* It follows that under the assumptions stated on pages 54 and 55 the probability of x successes in n trials is given by

$$b(x;n,p) = \frac{n!}{x!(n-x)!} p^x (1-p)^{n-x} \qquad \text{for } x = 0, 1, \ldots, n$$

The probability distribution defined by this equation is called the *binomial distribution*. In fact, this equation defines a family of probability distributions, with each member of this family being characterized by given values of the *parameters n* and *p*.

Example. To give an example where the binomial distribution can be applied, suppose that a safety engineer claims that only 60 percent of all drivers whose cars are equipped with seat belts use them on short trips and that we want to determine the probability that, under these conditions, 4 of 5 drivers will be using their seat belts. Substituting $n = 5$, $p = 0.60$, and $x = 4$ into the formula for the binomial distribution, we get

$$b(4; 5, 0.60) = \binom{5}{4}(0.60)^4(1-0.60)^{5-4}$$
$$= 0.259$$

and if we wanted to determine also the probability that *at least* 4 of 5 drivers will be using their seat belts, we would get

$$b(4; 5, 0.60) + b(5; 5, 0.60) = 0.259 + 0.078 = 0.337$$

---

*The expression $\dfrac{n!}{x!(n-x)!}$ is often referred to as a *binomial coefficient* since it is the coefficient of a^x in the binomial expansion of $(a+b)^n$; it is often represented by the symbol $\binom{n}{x}$.

The probability that *fewer than* 4 of 5 drivers will be wearing their seat belts is $1 - 0.337 = 0.663$, and the corresponding odds are just about 2 to 1.

If n is large, the calculation of binomial probabilities can become quite tedious; in that case it may be desirable to use numerical approximations or refer to special tables. Values of the binomial distribution and the corresponding cumulative distribution function have been tabulated for $n = 2$ to $n = 49$ by the National Bureau of Standards and for $n = 50$ to $n = 100$ by H. G. Romig (see the Bibliography at the end of the book). Table 1 at the end of this book gives the values of

$$B(x; n, p) = \sum_{k=0}^{x} b(k; n, p) \qquad \text{for } x = 0, 1, 2, \ldots, n$$

for $n = 2$ to $n = 20$ and $p = 0.05, 0.10, 0.15, 0.20, \ldots, 0.50$. We tabulated the cumulative probabilities rather than the values of $b(x; n, p)$, because the values of $B(x; n, p)$ are the ones needed more often in statistical applications. Note, however, that the values of $b(x; n, p)$ can be obtained with the use of Table 1 by making use of the identities in parts (c) and (d) of Theorem 3.1 below. Also note that values of $B(x; n, p)$ are not given in Table 1 for p greater than 0.50, but they can be obtained from this table with the use of the identity given in part (b) of Theorem 3.1.

THEOREM 3.1

(a) $b(x; n, p) = b(n - x; n, 1 - p)$

(b) $B(x; n, p) = 1 - B(n - x - 1; n, 1 - p)$

(c) $b(x; n, p) = B(x; n, p) - B(x - 1; n, p)$

(d) $b(x; n, p) = B(n - x; n, 1 - p) - B(n - x - 1; n, 1 - p)$

The first of these identities can be verified by substituting on both sides the corresponding expressions given by the equation defining the binomial distribution; the second identity follows from (a) and the definition of a cumulative distribution; the third identity follows directly from the definition of a cumulative distribution; and the fourth identity follows by substitution from parts (a) and (c). The reader will be asked to verify these identities in detail in Exercises 5 and 6 on pages 63 and 64.

Example. To illustrate the use of Table 1 and identities (a) and (c) of Theorem 3.1, let us suppose that in a very large shipment of integrated-circuit chips, the probability of failure for any one chip is 0.20. Assuming, furthermore, that the assumptions underlying the binomial distribution are approximately met, the probability that *at most 3 chips fail in a sample of 20* is given by

$$B(3; 20, 0.20) = 0.4114$$

Also, the probability that *5 or more chips fail in a sample of 16* is given by

$$\sum_{x=5}^{16} b(x; 16, 0.20) = 1 - B(4; 16, 0.20) = 1 - 0.7982 = 0.2018$$

the probability that *exactly 12 chips do not fail in a sample of 15* is

$$b(12; 15, 0.80) = b(3; 15, 0.20) = 0.2502$$

and the probability that *exactly 5 chips fail in a sample of 18* is given by

$$b(5; 18, 0.20) = B(5; 18, 0.20) - B(4; 18, 0.20)$$
$$= 0.8671 - 0.7164 = 0.1507$$

To illustrate the use of identities (b) and (d) of Theorem 3.1, let us suppose that somebody claims that 75 percent of all industrial accidents could be prevented by paying strict attention to safety regulations. Then, if the claim is true, the probability that *fewer than 16 of 20 industrial accidents can be prevented by paying strict attention to safety regulations* is given by

$$B(15; 20, 0.75) = 1 - B(4; 20, 0.25)$$
$$= 1 - 0.4148$$
$$= 0.5852$$

and the probability that *exactly 12 of 15 industrial accidents can be prevented by paying strict attention to safety regulations* is given by

$$b(12; 15, 0.75) = B(3; 15, 0.25) - B(2; 15, 0.25)$$
$$= 0.4613 - 0.2361$$
$$= 0.2252$$

Example. To illustrate the use of the binomial distribution in a problem of decision making, suppose that a washing machine manufacturer claims that at most 10 percent of his machines require repairs within the warranty period of twelve months, and that the decision whether to accept or reject his claim has to be made on the basis of a sample of 20 of his machines, of which 5 required repairs within the first year. To justify the claim one way or the other, let us first determine the probability that *5 or more of 20 of these washing machines will require repairs within a year when the probability that any one of them will require repairs within a year is 0.10*. Using Table 1, we get

$$\sum_{x=5}^{20} b(x; 20, 0.10) = 1 - B(4; 20, 0.10) = 0.0432$$

and since this is *very small* (the odds against it are about 22 to 1), it would seem reasonable to reject the washing machine manufacturer's claim. Note that the probability would have been even smaller if we had used a value of p less than 0.10. (See also Exercise 10 on page 64.)

Interesting information about the *shape* of (the graph of) binomial distributions is illustrated in Figures 3.3 and 3.4. First, if $p = 0.50$ the equation for the binomial distribution becomes

$$b(x; n, 0.50) = \frac{n}{x!(n - x)!} \cdot (0.50)^n$$

and it should be noted that $b(x; n, 0.50) = b(n - x; n, 0.50)$. This means that the histogram of a binomial distribution with $p = 0.50$ is *symmetrical*,

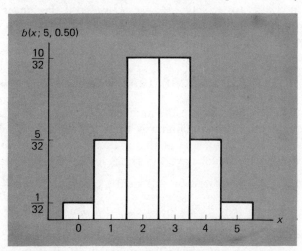

Figure 3.3. Symmetrical binomial distribution with $n = 5$ and $p = 0.50$.

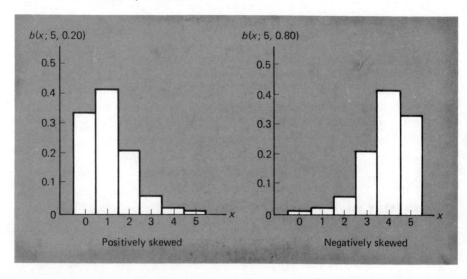

Figure 3.4. Skewed binomial distributions with $n = 5$ and $p = 0.20$ and $p = 0.80$, respectively.

as is illustrated in Figure 3.3. Now, note that if p is *less than* 0.50, it is more likely that x will be small rather than large compared to n, and that the opposite is true if p is *greater than* 0.50. This is illustrated in Figure 3.4, containing the histograms of binomial distributions with $n = 5$ and $p = 0.20$ and $p = 0.80$, respectively. A probability distribution which has a histogram like those of Figure 3.4 is said to be *skewed*; it is said to be *positively skewed* (or to have *positive skewness*) if the longer "tail" is on the right, and it is said to be *negatively skewed* (or to have *negative skewness*) if the longer tail is on the left.

3.3 THE HYPERGEOMETRIC DISTRIBUTION

Suppose we are interested in the number of defectives in a sample of n units drawn from a lot containing N units, of which a are defective. If the sample is drawn in such a way that at each successive drawing whatever units are left in the lot have the same chance of being selected, the probability that the first drawing will yield a defective unit is $\frac{a}{N}$, but for the second drawing it is $\frac{a-1}{N-1}$ or $\frac{a}{N-1}$ *depending on whether or not the first unit drawn was defective.* Thus, the trials are *not independent*, the fourth assumption underlying the binomial distribution is not met, and the binomial distribution does not apply. Note that the binomial distribution would apply if we sampled *with replacement,* namely, if each unit selected for the sample were replaced before the next one is drawn.

To solve the problem of sampling *without replacement* (that is, as we originally formulated the problem), let us refer to the result obtained on page 56, that a subset of x objects can be selected from a set of n objects in

$$\binom{n}{x} = \frac{n!}{x!(n-x)!}$$

ways. We proved this in connection with the problem of the distribution of x successes among n trials, that is, choosing subsets on which the x successes are to occur. We referred to $\binom{n}{x}$ as a binomial coefficient, and it is also called the "number of combinations of n objects taken x at a time."

Returning now to the problem of finding the probability of getting x defectives in a sample of n units taken without replacement, note first that the sample space for this experiment has $\binom{N}{n}$ possible outcomes, namely, the number of ways in which a subset of n objects can be selected from among a set of N objects. Furthermore, the x defectives can be selected from the a defectives in $\binom{a}{x}$ ways, the $n-x$ nondefective units in the sample can be selected from the $N-a$ nondefective units in the lot in $\binom{N-a}{n-x}$ ways, and according to Theorem 2.1, the whole sample can be selected in $\binom{a}{x}\binom{N-a}{n-x}$ ways. Assuming that each of the $\binom{N}{n}$ samples has the same probability of being selected, which is equivalent to the assumption stated in the first paragraph of this section, the required probability for x "successes" in n trials without replacement is

$$h(x; n, a, N) = \frac{\binom{a}{x}\binom{N-a}{n-x}}{\binom{N}{n}} \qquad \text{for } x = 0, 1, \ldots, n$$

by virtue of Theorem 2.3. A probability distribution having this equation is called a *hypergeometric distribution*; the parameters of this family of distributions are the *sample size n,* the *lot size* (or *population size*) N, and the number of "successes" in the lot a, which in our special example was the total number of defectives in the lot (or population).

Example. To illustrate the use of the hypergeometric distribution, suppose that a shipment of 20 tape recorders contains 5 that are defective, and that we want to determine the probability that among 10 of these tape recorders chosen for inspection only 2 will be defective. Substituting $x = 2$, $n = 10$, $a = 5$, and $N = 20$ into the formula for the hypergeometric distribution, we get

$$h(2; 10, 5, 20) = \frac{\binom{5}{2}\binom{15}{8}}{\binom{20}{10}} = \frac{10 \cdot 6{,}435}{184{,}756} = 0.348$$

and it stands to reason that the probability should be fairly high—half the tape recorders contain just slightly less than half the defectives. Probabilities like this are usually calculated using a table of binomial coefficients or a table of the logarithms of factorials, which can be found in most handbooks of mathematical tables. Note that if we had made the *mistake* of using the binomial distribution with $n = 10$ and $p = \frac{5}{20} = 0.25$ to calculate the probability of two defectives, the result would have been 0.282, which is much too small.

In this example, the sample size $n = 10$ was large compared to the lot size $N = 20$, and the effect of not replacing the units was rather severe. However, had the sample size been small compared to the lot size, the composition of the lot would not have been affected very much by drawing the sample, and it would have been reasonable to expect that the binomial distribution with $p = \frac{a}{N}$ should yield a very close approximation.

Continuation of Example. To illustrate this point, suppose that the lot contained 100 units, rather than 20, and that 25 of these units, rather than 5, are defective. The probability of drawing a sample of size 10 with 2 defectives would then be

$$h(2; 10, 25, 100) = \frac{\binom{25}{2}\binom{75}{8}}{\binom{100}{10}} = 0.292$$

so that the approximate value of 0.282 given by the binomial distribution with $n = 10$ and $p = \frac{25}{100} = 0.25$ would have been much closer.

Actually, in many practical situations the sample size is small compared to the lot size (that is, the size of the set from which the sample is drawn), and the binomial distribution very often provides a good approximation to the hypergeometric distribution. In fact, it can be shown formally that $h(x; n, a, N)$ approaches $b(x; n, p)$ with $p = \dfrac{a}{N}$, when $N \longrightarrow \infty$ while $p = \dfrac{a}{N}$ remains fixed. A good rule of thumb is to use the binomial distribution as an approximation to the hypergeometric distribution if $n \leq \dfrac{N}{10}$.

Although we have introduced the hypergeometric distribution in connection with a problem of sampling inspection, it has many other applications. For instance, it can be used to find the probability that 3 out of 12 housewives prefer Brand A detergent to Brand B, if they are selected from among 200 housewives among whom 40 actually prefer Brand A to Brand B. Also, it can be used in connection with a problem of selecting industrial diamonds, some of which have superior qualities and some do not, in connection with a problem of sampling income tax returns, where a among N returns filed contain questionable deductions, and so on.

EXERCISES

1. Suppose that a probability of 0.05 is assigned to each point of the sample space of Figure 2.4 on page 10. Find the probability distribution of the total number of lasers that are suitable for the given task.

2. An experiment consists of four tosses of a coin. Denoting the outcomes *HHTH, THTT, . . .,* and assuming that all 16 outcomes are equally likely, find the probability distribution for the total number of heads.

3. Check whether the following can define probability functions, and explain your answers:
   (a) $f(x) = \frac{1}{3}$ for $x = 0, 1, 2$;
   (b) $f(x) = \dfrac{x}{15}$ for $x = 0, 1, 2, 3, 4, 5$;
   (c) $f(x) = \dfrac{5 - x^2}{6}$ for $x = 0, 1, 2, 3$;
   (d) $f(x) = \frac{1}{4}$ for $x = 3, 4, 5, 6$;
   (e) $f(x) = \dfrac{x + 1}{25}$ for $x = 1, 2, 3, 4, 5$.

4. Given that $f(x) = k(\frac{1}{2})^x$ is a probability distribution for a random variable which can take on the values $x = 0, 1, 2, 3, 4, 5, 6$, find k and also find an expression for the corresponding cumulative probabilities $F(x)$.

5. Prove parts (a) and (b) of Theorem 3.1 on page 57.

6. Prove parts (c) and (d) of Theorem 3.1 on page 57.

7. Prove that

$$\frac{b(x+1; n, p)}{b(x; n, p)} = \frac{p(n-x)}{(1-p)(x+1)}$$

for $x = 0, 1, 2, \ldots, n-1$, and use this *recursion formula* to calculate the values of the binomial distribution with $n = 6$ and $p = 0.30$. Verify the results by means of Table 1.

8. Using Table 1 find
   (a) $B(8; 16, 0.40)$;
   (b) $b(8; 16, 0.40)$;
   (c) $B(9; 12, 0.60)$;
   (d) $b(9; 12, 0.60)$;
   (e) $\sum_{k=6}^{20} b(k; 20, 0.15)$;
   (f) $\sum_{k=6}^{9} b(k; 9, 0.70)$.

9. Using Table 1 find
   (a) $B(7; 19, 0.45)$;
   (b) $b(7; 19, 0.45)$;
   (c) $B(8; 10, 0.95)$;
   (d) $b(8; 10, 0.95)$;
   (e) $\sum_{k=4}^{10} b(k; 10, 0.35)$;
   (f) $\sum_{k=2}^{4} b(k; 9, 0.30)$.

10. Rework the decision problem on page 58, supposing that only 3 of the 20 washing machines required repairs within the first year.

11. Assuming that 2 in 10 automobile accidents are due mainly to driver fatigue, find the probability that among 8 automobile accidents 3 will be due mainly to driver fatigue
    (a) by using the formula for the binomial distribution;
    (b) by using Table 1.

12. A social scientist claims that only 40 percent of all high school seniors capable of doing college work actually go to college. Use Table 1 to find the probabilities that among 14 high school seniors capable of doing college work
    (a) at most 6 will go to college;
    (b) at least 4 will go to college;
    (c) exactly 5 will go to college.

13. If the probability is 0.05 that a hockey game will go into overtime, find the probabilities that among 12 hockey games
    (a) exactly one will go into overtime;
    (b) at most 2 will go into overtime;
    (c) 2 or more will go into overtime.

14. If the probability that a fluorescent light has a useful life of at least 500 hours is 0.85, find the probabilities that among 20 such lights
    (a) exactly 18 will have a useful life of at least 500 hours;
    (b) at least 15 will have a useful life of at least 500 hours;
    (c) at least 2 will *not* have a useful life of at least 500 hours.

15. A quality control engineer wants to check whether (in accordance with specifications) 95 percent of the electronic components shipped by his company are in good working condition. To this end, he randomly selects 15 from each very large lot ready to be shipped and passes it only if they are *all* in good working condition; otherwise, each of the components in the lot is checked.
    (a) What is the probability that he will commit the error of holding a lot for further inspection even though 95 percent of the components are in good working condition?
    (b) What is the probability that he will commit the error of letting a lot pass through without further inspection even though only 90 percent of the components are in good working condition?

16. A food processor claims that at most 10 percent of his jars of instant coffee contain less coffee than claimed on the label. To test this claim, 16 jars of his instant coffee are randomly selected and the contents are weighed; his claim is accepted if fewer than 3 of the jars contain less coffee than claimed on the label. Find the probabilities that the food processor's claim will be accepted when the actual percentage of his jars containing less coffee than claimed on the label is
    (a) 0.05;   (b) 0.10;
    (c) 0.15;   (d) 0.20.

17. A quality control engineer inspects a random sample of three toasters from each lot of 24 toasters that is ready to be shipped. If such a lot contains six toasters with slight defects, find the probabilities that the inspector's sample will
    (a) contain only toasters without defects;
    (b) contain exactly one toaster with a defect;
    (c) contain at least two toasters with defects.

18. If 6 of 18 new buildings in a city violate the building code, what is the probability that a building inspector, who randomly selects 4 of the new buildings, will catch
    (a) none of the new buildings that violate the building code;
    (b) exactly one of the new buildings that violate the building code;
    (c) exactly two of the new buildings that violate the building code;
    (d) at least three of the new buildings that violate the building code?

19. A collection of 15 gold coins contains four counterfeits. If two of these coins are randomly selected to be sold at auction, what are the probabilities that
    (a) neither coin is a counterfeit;
    (b) only one of the two coins is a counterfeit;
    (c) both coins are counterfeits?

20. Among the 16 cities which a professional society is considering for its next three annual conventions, 7 are in the Western part of the United States. If, to avoid arguments, the selection is left to chance, what are the probabilities that

    (a) none of these conventions will be held in the western part of the United States;

    (b) all of these conventions will be held in the western part of the United States?

21. A shipment of 120 burglar alarms contains 5 that are defective. If 3 of these burglar alarms are randomly selected and shipped to a customer,

    (a) find the probability that he will get exactly one bad unit;

    (b) approximate the probability that he will get exactly one bad unit by means of the binomial distribution.

Also, round both answers to three decimals and find the percentage error of the approximation.

22. Among the 300 employees of a company, 240 are union members while the others are not. If 8 of the employees are to be chosen to serve on a committee which administers the pension fund, find the probability that five of them will be union members while the others are not

    (a) by using the formula for the hypergeometric distribution;

    (b) by approximating the probability with the use of the formula for the binomial distribution.

Also, round both answers to four decimals and find the error of the approximation.

3.4 THE MEAN AND THE VARIANCE
OF A PROBABILITY DISTRIBUTION

    Besides the binomial and hypergeometric distributions there are many other probability distributions that have important applications. However, before we go any further, let us discuss some general characteristics of probability distributions. One such characteristic, that of the symmetry or skewness of a distribution, was illustrated by means of Figures 3.3 and 3.4; two others will be presented here with reference to Figure 3.5.

    Example. Figure 3.5 shows the histograms of two binomial distributions; the one which is shaded has the parameters $n = 16$ and $p = \frac{1}{2}$, and the one which is not shaded has the parameters $n = 4$ and $p = \frac{1}{2}$. Essentially, these two distributions differ in two respects. The first distribution (whose histogram is shaded) is "centered" about $x = 8$ while the other is "centered" about $x = 2$, and we thus say that the two distributions differ in their *location*. Another distinction between the two distributions is that the histogram of the first is more "spread out," the probabilities represented by the rectangles are more "dispersed," and we say that the two distributions differ in *variation*.

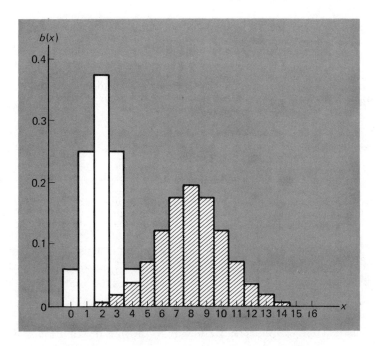

Figure 3.5. Histograms of two binomial distributions.

To make such comparisons more specific, we shall introduce in this section two of the most important measures describing, respectively, the location and the variation of a probability distribution; they are the *mean* and the *variance* of a distribution.

Continuation of Example. Concentrating for the moment on the second of the probability distributions of Figure 3.5, it can easily be verified (see Exercise 2 on page 63) that the probabilities for $x = 0, 1, 2, 3$, or 4 are, respectively, $\frac{1}{16}, \frac{4}{16}, \frac{6}{16}, \frac{4}{16}$, and $\frac{1}{16}$. Hence, we could say that if we repeatedly observed values of this random variable (for instance, if we observed the number of heads in repeated flips of four coins), then *in the long run x* should equal 0 about $\frac{1}{16}$ the time, it should equal 1 about $\frac{4}{16}$ of the time, 2 about $\frac{6}{16}$ of the time, 3 about $\frac{4}{16}$ of the time, and 4 about $\frac{1}{16}$ of the time. Applying the definition of mathematical expectation given on page 44, we obtain

$$0 \cdot \frac{1}{16} + 1 \cdot \frac{4}{16} + 2 \cdot \frac{6}{16} + 3 \cdot \frac{4}{16} + 4 \cdot \frac{1}{16} = 2$$

for the expected number of heads. This result, representing the long-run *average* of the random variable having the binomial distribution with $n = 4$ and $p = \frac{1}{2}$ is what we referred to as the "center" of the distribution; indeed, this average is what we call the *mean* of the probability distribution.

In general, the *mean* of a probability distribution with values $f(x)$ is defined by the equation

$$\mu = \sum_{\text{all } x} x \cdot f(x)$$

and it is denoted by the Greek letter μ (*mu*). It measures the "center" of a probability distribution in the sense of an average or, better, in the sense of a center of gravity. Note that the above formula for μ is, in fact, the *first moment about the origin* of a discrete system of masses $f(x)$ arranged on a straight line at distances x from the origin. We do not have to divide here by $\sum_{\text{all } x} f(x)$, as we do in the usual formula for the x-coordinate of the center of gravity, since this sum is by definition equal to 1.

Continuation of Example. Returning to the first distribution of Figure 3.5, we could find its mean by calculating all the necessary probabilities or by looking them up in Table 1, and substituting them into the formula for μ. However, if we reflect for a moment, we might argue that there is a fifty-fifty chance for a success on each trial; there are 16 trials, and it would seem reasonable to expect *on the average* 8 successes in 16 trials. Similarly, we might argue that if a binomial distribution has the parameters $n = 200$ and $p = 0.20$, we can expect a success about 20 percent of the time and, hence, on the average $200(0.20) = 40$ successes in 200 trials.

Indeed, this argument can be extended to any binomial distribution and we shall now prove that *the mean of the binomial distribution with the parameters n and p* is given by

$$\mu = n \cdot p$$

Substituting the expression which defines $b(x; n, p)$ into the formula for μ, we get

$$\mu = \sum_{x=0}^{n} x \cdot \frac{n!}{x!(n-x)!} p^x (1-p)^{n-x}$$

Then, making use of the fact that $\dfrac{x}{x!} = \dfrac{1}{(x-1)!}$ and $n! = n(n-1)!$, and factoring out n and p, we obtain

$$\mu = np \sum_{x=1}^{n} \frac{(n-1)!}{(x-1)!(n-x)!} p^{x-1}(1-p)^{n-x}$$

where the summation starts with $x = 1$ since the original summand is zero for $x = 0$. If we now let $y = x - 1$ and $m = n - 1$, we obtain

$$\mu = np \sum_{y=0}^{m} \frac{m!}{y!(m-y)!} p^{y}(1-p)^{m-y}$$

and this last sum can easily be recognized as that of all the terms of the binomial distribution with the parameters m and p. Hence, this sum equals 1 and it follows that $\mu = np$.

The result obtained applies, of course, only to the binomial distribution, but given any other special distribution, we can, similarly, express the mean in terms of its parameters. Thus, for *the mean of the hypergeometric distribution with the parameters n, a, and N* we can write

$$\mu = n \cdot \frac{a}{N}$$

and, using a method similar to the one we employed in deriving the formula for the mean of the binomial distribution, the reader will be asked to prove this result in Exercise 6 on page 77.

Example. Referring to the example on page 62, where 5 of 20 tape recorders were defective, we can now say that if 10 of them are chosen for inspection, the distribution of the number of defectives has the mean

$$\mu = 10 \cdot \tfrac{5}{20} = 2.5$$

In other words, if we inspect 10 of the tape recorders, we can *expect* 2.5 defectives, where the word "expect" is to be interpreted as a *mathematical expectation*.

As defined, the mean of *any* probability distribution is the *expected value* of the corresponding random variable—in accordance with the formula for a mathematical expectation, μ is given by the sum of the products obtained by multiplying each value of the random variable by the corresponding probability.

To study the second of the two properties of probability distributions mentioned on page 66, the one concerning their *variation*, let us refer again to the two distributions of Figure 3.5. For the distribution whose histogram is *not shaded* there is a high probability that we will get a value close to the mean; so far as the other distribution is concerned, the one whose histogram is *shaded*, there is a high probability of getting values scattered over considerable distances away from the mean. Using this property, it may seem reasonable to measure the variation of a probability distribution with the quantity

$$\sum_{\text{all } x} (x - \mu) \cdot f(x)$$

namely, the average amount by which the values of the random variable *deviate* from the mean. Unfortunately,

$$\sum_{\text{all } x} (x - \mu) \cdot f(x) = \sum_{\text{all } x} x \cdot f(x) - \sum_{\text{all } x} \mu \cdot f(x)$$
$$= \mu - \mu \cdot \sum_{\text{all } x} f(x) = \mu - \mu = 0$$

so that this expression is always equal to zero. However, since we are really interested only in the *magnitude* of the deviations $x - \mu$ (not in their signs), it suggests itself that we average the *absolute values* of these deviations from the mean. This would, indeed, provide a measure of variation, but on purely theoretical grounds we prefer to work instead with the *squares* of the deviations from the mean. These quantities cannot be negative either, and their average is indicative of the spread or dispersion of a probability distribution. We thus define the *variance* of a probability distribution with values $f(x)$ as

$$\sigma^2 = \sum_{\text{all } x} (x - \mu)^2 \cdot f(x)$$

where σ (*sigma*) is the lower-case Greek letter for s. This measure is not in the same units (or dimension) as the values of the random variable, but we can adjust for this disadvantage by taking its square root, thus defining the *standard deviation*

$$\sigma = \sqrt{\sum_{\text{all } x} (x - \mu)^2 \cdot f(x)}$$

Example. To illustrate how the standard deviation measures the "spread" of a probability distribution, let us compare the values of σ of the two distributions of Figure 3.5. Having already shown on page 67 that for the distribution with $n = 4$ the mean is $\mu = 2$, we find that the variance of this distribution is

$$\sigma^2 = (0 - 2)^2 \cdot \tfrac{1}{16} + (1 - 2)^2 \cdot \tfrac{4}{16} + (2 - 2)^2 \cdot \tfrac{6}{16}$$
$$+ (3 - 2)^2 \cdot \tfrac{4}{16} + (4 - 2)^2 \cdot \tfrac{1}{16}$$
$$= 1$$

and, hence, the standard deviation is $\sigma = 1$. Similarly, it can be shown that for the other distribution $\sigma = 2$, and we find that *the shaded distribution with the greater spread also has the greater standard deviation.*

Given any probability distribution, we can always calculate σ^2 by substituting the corresponding probabilities $f(x)$ into the formula which defines the variance. As in the case of the mean, however, this work can be simplified to a considerable extent when we deal with special kinds of distributions. For instance, it can be shown that *the variance of the binomial distribution with the parameters n and p* is given by the formula

$$\sigma^2 = n \cdot p \cdot (1 - p)$$

and that *the variance of the hypergeometric distribution with the parameters n, a, and N* is given by the formula

$$\sigma^2 = \frac{n \cdot a \cdot (N - a) \cdot (N - n)}{N^2 \cdot (N - 1)}$$

We shall not prove these formulas here, but let us apply them to the two distributions of Figure 3.5.

Continuation of Example. Since the two distributions are binomial distributions with the parameters $n = 4$ and $p = \tfrac{1}{2}$, and $n = 16$ and $p = \tfrac{1}{2}$, substitution into the formula yields

$$\sigma^2 = 4 \cdot \tfrac{1}{2} \cdot \tfrac{1}{2} = 1$$

and

$$\sigma^2 = 16 \cdot \tfrac{1}{2} \cdot \tfrac{1}{2} = 4$$

and this agrees with the results which we gave before.

When we first defined the variance of a probability distribution, it may have occurred to the reader that the formula looked exactly like the one which we use in physics to define second moments, or moments of inertia. Indeed, it is customary in statistics to define the *kth moment about the origin* as

$$\mu'_k = \sum_{\text{all } x} x^k \cdot f(x)$$

and the *kth moment about the mean* as

$$\mu_k = \sum_{\text{all } x} (x - \mu)^k \cdot f(x)$$

Thus, the mean μ is the first moment about the origin, and the variance σ^2 is the second moment about the mean. Higher moments are often used in statistics to give further descriptions of probability distributions. For instance, the third moment about the mean (divided by σ^3 to make this measure independent of the scale of measurement) is used to describe the symmetry or skewness of a distribution; the fourth moment about the mean (divided by σ^4) is, similarly, used to describe its "peakedness," or *kurtosis*. To determine moments about the mean, it is usually easiest to express them in terms of moments about the origin and then to calculate the necessary moment about the origin. For the second moment about the mean we thus have the important formula

$$\sigma^2 = \mu'_2 - \mu^2$$

which the reader will be asked to prove in part (a) of Exercise 11 on page 77; part (b) of that exercise contains a similar formula for expressing μ_3 in terms of moments about the mean.

Example. To illustrate the use of this "short-cut" formula for computing the variance, we return to the distribution of the number of heads in four flips of a coin illustrated in Figure 3.5. For this distribution $\mu = 2$ and, applying the formula for μ'_k with $k = 2$, we get

$$\mu'_2 = 0^2 \cdot \tfrac{1}{16} + 1^2 \cdot \tfrac{4}{16} + 2^2 \cdot \tfrac{6}{16} + 3^2 \cdot \tfrac{4}{16} + 4^2 \cdot \tfrac{1}{16}$$
$$= 5$$

Thus

$$\sigma^2 = 5 - 2^2 = 1$$

which is the result previously obtained.

3.5   CHEBYSHEV'S THEOREM

Earlier in this chapter we used examples to show how the standard deviation measures the variation of a probability distribution, that is, how it controls the concentration of probability in the neighborhood of the mean. *If σ is small, there is a high probability for getting a value close to the mean; if σ is large there is a correspondingly higher probability for getting values further away from the mean.* Formally, this idea is expressed by the following theorem:

THEOREM 3.2 (*Chebyshev's theorem*). *If a probability distribution has the mean μ and the standard deviation σ, the probability of obtaining a value which deviates from the mean by at least k standard deviations is at most $1/k^2$. Symbolically,*

$$P(|x - \mu| \geq k\sigma) \leq \frac{1}{k^2}$$

Example. Thus, the probability that a random variable will take on a value which deviates (differs) from the mean by at least 2 standard deviations is at most $\frac{1}{4}$, the probability that it will take on a value which deviates from the mean by at least 5 standard deviations is at most $\frac{1}{25}$, and the probability that it will take on a value which deviates from the mean by 10 standard deviations or more is less than or equal to $\frac{1}{100}$.

To prove this theorem, consider any probability distribution having the values $f(x)$, the mean μ, and the variance σ^2. Dividing the sum defining the variance into three parts as indicated in Figure 3.6, we have

$$\sigma^2 = \sum_{\text{all } x} (x - \mu)^2 f(x)$$
$$= \sum_{R_1} (x - \mu)^2 f(x) + \sum_{R_2} (x - \mu)^2 f(x) + \sum_{R_3} (x - \mu)^2 f(x)$$

where R_1 is the region for which $x \leq \mu - k\sigma$, R_2 is the region for which $\mu - k\sigma < x < \mu + k\sigma$, and R_3 is the region for which $x \geq \mu + k\sigma$.

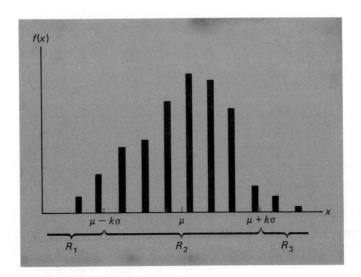

Figure 3.6. Diagram for proof of Chebyshev's theorem.

Since $(x - \mu)^2 f(x)$ cannot be negative, the above sum over R_2 is non-negative, and without it the sum of the summations over R_1 and R_3 is less than or equal to σ^2, that is,

$$\sigma^2 \geq \sum_{R_1} (x - \mu)^2 f(x) + \sum_{R_3} (x - \mu)^2 f(x)$$

But $x - \mu \leq -k\sigma$ in the region R_1 and $x - \mu \geq k\sigma$ in the region R_3, so that in either case $|x - \mu| \geq k\sigma$; hence, in both regions $(x - \mu)^2 \geq k^2\sigma^2$. If we now replace $(x - \mu)^2$ in each sum by $k^2\sigma^2$, a number less than or equal to $(x - \mu)^2$, we obtain the inequality

$$\sigma^2 \geq \sum_{R_1} k^2\sigma^2 f(x) + \sum_{R_3} k^2\sigma^2 f(x)$$

or

$$\frac{1}{k^2} \geq \sum_{R_1} f(x) + \sum_{R_3} f(x)$$

Since $\sum_{R_1} f(x) + \sum_{R_3} f(x)$ represents the probability that x is in the region $R_1 \cup R_3$, namely, that $|x - \mu| \geq k\sigma$, we finally have

$$P(|x - \mu| \geq k\sigma) \leq \frac{1}{k^2}$$

and this completes the proof of Theorem 3.2.

To obtain an alternate form of *Chebyshev's inequality,* as the inequality of Theorem 3.2 is sometimes called, note that the event $|x - \mu| < k\sigma$ is the complement of the event $|x - \mu| \geq k\sigma$; hence,

$$P(|x - \mu| < k\sigma) \geq 1 - \frac{1}{k^2}$$

which states that the probability of getting a value within k standard deviations of the mean is at least $1 - \frac{1}{k^2}$.

Theoretically speaking, the most important feature of Chebyshev's theorem is that it applies to *any* probability distribution for which μ and σ exist. So far as applications are concerned, however, this generality is also its greatest weakness; it only provides an *upper limit* (often a very poor one) to the probability of getting a value which deviates from the mean by more than k standard deviations.

Example. For instance, we can assert *in general* that the probability of getting a value which deviates from the mean by at least two standard deviations is *not more than* 0.25, whereas the corresponding *exact* probability for the binomial distribution with the parameters $n = 16$ and $p = \frac{1}{2}$ is only 0.0768—"not more than 0.25" is correct, but it does not tell us that the actual probability is as small as 0.0768.

An important theoretical result is obtained if we apply Chebyshev's theorem to the binomial distribution. Substituting $\mu = np$ and $\sigma = \sqrt{np(1-p)}$ into the alternate form of the theorem given above, we get

$$P[|x - np| < k\sqrt{np(1-p)}] \geq 1 - \frac{1}{k^2}$$

or

$$P\left(\left|\frac{x}{n} - p\right| < k\sqrt{\frac{p(1-p)}{n}}\right) \geq 1 - \frac{1}{k^2}$$

after we divide both terms of the inequality inside the parentheses by n. If we now substitute ϵ for $k\sqrt{\frac{p(1-p)}{n}}$, we obtain

$$P\left(\left|\frac{x}{n} - p\right| < \epsilon\right) \geq \frac{p(1-p)}{n\epsilon^2}$$

and it follows that

$$\lim_{n \to \infty} P\left(\left|\frac{x}{n} - p\right| < \epsilon\right) = 1$$

for any $\epsilon > 0$. The last equation expresses the *weak law of large numbers*. In words, *as the sample size increases, the probability that the observed proportion of successes differs from p by less than any arbitrary positive constant ϵ approaches 1.* Note that we cannot conclude from this that for large n the actual *number* of successes must necessarily be close to $\mu = np$. In fact, it can be shown that although the probability that the *proportion* of successes is close to p approaches 1 as n becomes large, the probability that the *number* of successes is close to np approaches 0.

Example. For instance, for 40,000 flips of a balanced coin we have $\mu = 40,000 \cdot \frac{1}{2} = 20,000$ and $\sigma = \sqrt{40,000 \cdot \frac{1}{2} \cdot \frac{1}{2}} = 100$, so that Chebyshev's theorem with $k = 10$ tells us that the probability is at least 0.99 that we will get between 19,000 and 21,000 heads, or that the *proportion* of heads will fall between 0.475 and 0.525.

EXERCISES

1. Suppose that the probabilities are, respectively, 0.4, 0.3, 0.2, and 0.1 that there will be 0, 1, 2, or 3 power failures in a certain city during the month of July.
   (a) Calculate the mean of this probability distribution and explain its significance.
   (b) Calculate the variance of this probability distribution.

2. The following table gives the probabilities that a certain computer will malfunction 0, 1, 2, 3, 4, 5, or 6 times on any one day:

Number of malfunctions	x	0	1	2	3	4	5	6
Probability	$f(x)$	0.17	0.29	0.27	0.16	0.07	0.03	0.01

   (a) Find the mean of this probability distribution.
   (b) Find the standard deviation of this probability distribution without using the short-cut formula on page 72.
   (c) Find the standard deviation of this probability distribution using the short-cut formula on page 72.

3. Find the mean and the variance of the *uniform probability distribution* given by

$$f(x) = \frac{1}{n} \qquad \text{for } x = 1, 2, 3, \ldots, n$$

   [*Hint*: The sum of the first n positive integers is $\frac{1}{2}n(n + 1)$, and the sum of their squares is $\frac{1}{6}n(n + 1)(2n + 1)$.]

4. Find μ and σ^2 for the binomial distribution which has the parameters $n = 5$ and $p = 0.70$ by using
   (a) the formulas which define μ and σ^2 on pages 68 and 70 and the probabilities of Table 1;
   (b) the special formulas on pages 68 and 71.

5. Find μ and σ^2 for the hypergeometric distribution which has the parameters $n = 3$, $a = 4$, and $N = 8$
   (a) by first calculating the necessary probabilities and then using the formulas which define μ and σ^2 on pages 68 and 70;
   (b) by using the special formulas on pages 69 and 71.

6. Prove the formula for the mean of the hypergeometric distribution given on page 69, namely, $\mu = n \cdot \dfrac{a}{N}$. [*Hint*: Make use of the identity

$$\sum_{r=0}^{k} \binom{m}{r}\binom{s}{k-r} = \binom{m+s}{k}$$

which can be obtained by equating the coefficients of x^k in $(1 + x)^m(1 + x)^s$ and in $(1 + x)^{m+s}$.]

7. Construct a table showing (a) the upper limits provided by Chebyshev's theorem for the probabilities of obtaining values differing from the mean by at least 1, 2, and 3 standard deviations, and (b) the corresponding probabilities for a binomial distribution with the parameters $n = 16$ and $p = \frac{1}{2}$.

8. What does Chebyshev's theorem tell us about the probability of getting anywhere between 30 and 105 *threes* in 405 rolls of a balanced die?

9. How many times do we have to flip a balanced coin to be able to assert with a probability of at least 0.99 that the difference between the proportion of *tails* and 0.50 will be less than 0.04?

10. If a random variable has a binomial distribution with the parameter $p = \frac{1}{4}$, what can we assert about the proportion of successes with a probability of at least $\frac{15}{16}$ when the other parameter is
    (a) $n = 300$;
    (b) $n = 7,500$;
    (c) $n = 120,000$?

11. Prove that
    (a) $\sigma^2 = \mu_2' - \mu^2$;
    (b) $\mu_3 = \mu_3' - 3\mu_2' \cdot \mu + 2\mu^3$.

3.6   THE POISSON APPROXIMATION
OF THE BINOMIAL DISTRIBUTION

When n is large and p is small, binomial probabilities are often approximated by means of the formula

$$f(x; \lambda) = \frac{\lambda^x e^{-\lambda}}{x!} \qquad \text{for } x = 0, 1, 2, \ldots$$

with λ (*lambda*) equal to the product np. Before we prove that this is reasonable, let us point out that $x = 0, 1, 2, \ldots$ means that there is a *countable infinity* of possibilities, and this requires that we modify the third axiom of probability given on page 18. In its place we substitute

---

AXIOM 3′. *If $A_1, A_2, A_3, \ldots$ is a finite or infinite sequence of mutually exclusive events in S, then*

$$P(A_1 \cup A_2 \cup A_3 \cup \ldots) = P(A_1) + P(A_2) + P(A_3) + \ldots$$

---

while the other postulates remain unchanged; that is, probabilities must still be nonnegative, and for the whole sample space the probability must be 1. To verify the latter for the formula given above, we make use of Axiom 3′ and write

$$P(S) = \sum_{x=0}^{\infty} \frac{e^{-\lambda}\lambda^x}{x!} = e^{-\lambda} \sum_{x=0}^{\infty} \frac{\lambda^x}{x!}$$

Since the infinite series in the expression on the right is the Maclaurin series expansion of e^{λ}, it follows that $P(S) = 1$.

Let us now show that when $n \rightarrow \infty$ and $p \rightarrow 0$, while $np = \lambda$ remains constant, the limiting form of the binomial distribution is as given in the beginning of the preceding paragraph. First let us substitute $\frac{\lambda}{n}$ for p into the formula for the binomial distribution and simplify the resulting expression; thus, we get

$$b(x; n, p) = \frac{n!}{x!(n-x)!} \left(\frac{\lambda}{n}\right)^x \left(1 - \frac{\lambda}{n}\right)^{n-x}$$

$$= \frac{n(n-1)(n-2)\cdot \ldots \cdot(n-x+1)}{x!n^x}(\lambda)^x\left(1 - \frac{\lambda}{n}\right)^{n-x}$$

$$= \frac{\left(1 - \frac{1}{n}\right)\left(1 - \frac{2}{n}\right)\cdot \ldots \cdot\left(1 - \frac{x-1}{n}\right)}{x!}(\lambda)^x\left(1 - \frac{\lambda}{n}\right)^{n-x}$$

If we now let $n \rightarrow \infty$, we find that

$$\left(1 - \frac{1}{n}\right)\left(1 - \frac{2}{n}\right)\cdot \ldots \cdot\left(1 - \frac{x-1}{n}\right) \longrightarrow 1$$

that

$$\left(1 - \frac{\lambda}{n}\right)^{n-x} = \left[\left(1 - \frac{\lambda}{n}\right)^{n/\lambda}\right]^{\lambda}\left(1 - \frac{\lambda}{n}\right)^{-x} \longrightarrow e^{-\lambda}$$

and, hence, that the binomial probability approaches

$$\frac{\lambda^x e^{-\lambda}}{x!} \qquad \text{for } x = 0, 1, 2, \ldots$$

This completes our proof; the distribution at which we have arrived is called the *Poisson distribution*. Since it has many important applications other than approximating binomial probabilities (see Section 3.7), the Poisson distribution has been extensively tabulated. Table 2 at the end of the book gives the values of the probabilities $F(x; \lambda) = \sum_{k=0}^{x} f(k; \lambda)$ for values of λ in varying increments from 0.02 to 25, and its use is very similar to that of Table 1.

The following are some examples illustrating the Poisson approximation of binomial probabilities. An acceptable rule of thumb is to use this approximation if $n \geq 20$ and $p \leq 0.05$; if $n \geq 100$, the approximation is generally excellent so long as $np \leq 10$.

Example. First let us compare the binomial probability of getting 2 successes in 100 trials when the probability of a success is 0.05, with the corresponding Poisson probability of getting 2 successes when $np = 100(0.05) = 5$. Substituting into the appropriate expressions, we get

$$b(2; 100, 0.05) = \binom{100}{2}(0.05)^2(0.95)^{98} = 0.081$$

$$f(2; 5) = \frac{5^2 \cdot e^{-5}}{2!} = 0.084$$

and it is interesting to note that the difference (the error we would make by using the Poisson approximation) is as small as 0.003, or less than 4 percent. Had we used Table 2 (instead of looking up e^{-5} in a mathematical table), we would have obtained $F(2; 5) - F(1; 5) = 0.125 - 0.040 = 0.085$.

Example. Suppose that a fire insurance company has 3,840 policy-holders, and that the probability is $\frac{1}{1,200}$ that any one of these policy-holders will file at least one claim during any given year. To find the probabilities that during a given year 0, 1, 2, 3, 4, ... of the policy-holders will file at least one claim, we can, for practical reasons, forget about using the formula for the binomial distribution. Instead we use the Poisson distribution with $\lambda = 3,840 \cdot \frac{1}{1,200} = 3.2$, and,

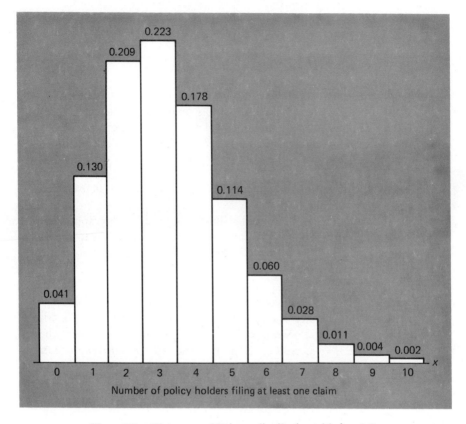

Figure 3.7. Histogram of Poisson distribution with $\lambda = 3.2$.

using Table 2 and the identity $f(x; \lambda) = F(x; \lambda) - F(x - 1; \lambda)$, we get the probabilities shown in the histogram of Figure 3.7.

Using a method similar to that which we used in deriving the formula for the mean of the binomial distribution, it can be shown that *the mean and the variance of the Poisson distribution with the parameter λ are given by*

$$\mu = \lambda \quad \text{and} \quad \sigma^2 = \lambda$$

The first of these results should really have been expected, for in our proof we let λ equal np, which is the mean of the corresponding binomial distribution.

3.7 POISSON PROCESSES

In general, the term "random process" refers to a physical process which is wholly or in part controlled by some sort of chance mechanism. It applies to a sequence of repeated flips of a coin, to measurements of the quality of manufactured products coming off an assembly line, to the vibrations of airplane wings, to the "noise" in a radio signal, and to numerous other phenomena. What characterizes such processes is their "time-dependence," namely, the fact that certain events do or do not take place (depending on chance) at regular intervals of time or throughout a continuous interval of time.

In this section we shall be concerned with processes taking place over continuous intervals of time, such as the occurrence of imperfections on a continuously produced bolt of cloth, the recording of radiation by means of a Geiger counter, the arrival of telephone calls at a switchboard, or the passing by cars of an electronic checking device. We will now show that the mathematical model which we can use to describe many situations like these is that of the Poisson distribution. To find the probability of x *successes during a time interval of length T*, we divide the interval into n equal parts of length Δt, so that $T = n \cdot \Delta t$, and we assume that:

1. The probability of a success during a very small interval of time Δt is given by $\alpha \cdot \Delta t$.
2. The probability of more than one success during such a small time interval Δt is negligible.
3. The probability of a success during such a time interval does not depend on what happened prior to that time.

This means that the assumptions underlying the binomial distribution (see pages 54 and 55) are satisfied, and the probability of x successes in the time interval T is given by the binomial probability $b(x; n, p)$ with $n = \dfrac{T}{\Delta t}$ and $p = \alpha \cdot \Delta t$. Then, following the argument on page 78, we find that when $n \longrightarrow \infty$ the probability of x successes during the time interval T is given by the corresponding Poisson probability with the parameter $\lambda = n \cdot p = \dfrac{T}{\Delta t} \cdot (\alpha \cdot \Delta t) = \alpha T$. Since λ is the mean of this Poisson distribution, note that α is the average (mean) number of successes *per unit time*.

Example. For instance, if a bank can expect to receive 6 bad checks per day, the probability of its getting 4 bad checks on any one day is

$$f(4; 6) = \frac{6^4 \cdot e^{-6}}{4!} = \frac{1{,}296(0.0025)}{24} = 0.135$$

and the probability of its getting 10 bad checks on any two consecutive days is

$$f(10; 12) = \frac{12^{10}(e^{-12})}{10!} = 0.105$$

where we actually obtained this last value by determining $F(10; 12) - F(9; 12)$ with the use of Table 2.

Also, if in the inspection of tin plate produced by a continuous electrolytic process, 0.2 imperfections are spotted on the average per minute, the probability of spotting 1 imperfection in 3 minutes is

$$F(1; 0.6) - F(0; 0.6) = 0.878 - 0.549 = 0.329$$

the probability of spotting *at least* 2 imperfections in 5 minutes is

$$1 - F(1; 1) = 1 - 0.736 = 0.264$$

and the probability of spotting *at most* 1 imperfection in 15 minutes is

$$F(1; 3) = 0.199$$

Example. The Poisson distribution has many important applications in *queueing problems*, where we may be interested, for example, in the number of customers arriving for service at a cafeteria, the number of ships or trucks arriving to be unloaded at a receiving dock, the number of aircraft arriving at an airport, and so forth. Thus, if on the average 0.3 customers arrive per minute at a cafeteria, then the probability that exactly 3 customers will arrive during a 5-minute span is

$$F(3; 1.5) - F(2; 1.5) = 0.934 - 0.809 = 0.125$$

and if on the average 3 trucks arrive per hour to be unloaded at a warehouse, then the probability that at most 20 will arrive during an 8-hour day is

$$F(20; 24) = 0.243$$

Also, if on the average 14.5 planes arrive per day at a private airport, then the probability that the number of arrivals on any one day will be anywhere from 12 to 15, inclusive, is

$$F(15; 14.5) - F(11; 14.5) = 0.619 - 0.220 = 0.399$$

3.8 THE GEOMETRIC DISTRIBUTION

In the example on page 7 we indicated that a countably infinite sample space would be needed if we were interested in the number of cars a person has to inspect until he finds one whose nitrogen-oxide emission does not meet government standards. To treat this problem more generally, suppose that in a sequence of trials we are interested in the number of the trial on which the *first success* occurs, and that the first, second, and fourth assumptions underlying the binomial distribution (pages 54 and 55) are satisfied. Clearly, if the first success is to come on the xth trial, it has to be preceded by $x - 1$ failures, and if the probability of a success is p, the probability of $x - 1$ failures on $x - 1$ trials is $(1 - p)^{x-1}$. Then, if we multiply this expression by the probability p of a success on the xth trial, we find that the probability of getting the first success on the xth trial is given by

$$g(x; p) = p(1 - p)^{x-1} \quad \text{for } x = 1, 2, 3, 4, \ldots$$

This probability distribution is referred to as the *geometric distribution,* and, as the reader will be asked to verify in Exercise 19 on page 88, the mean of this distribution is given by

$$\mu = \frac{1}{p}$$

Example. For instance, if the probability that a burglar will get caught on any given "job" is 0.20, then the probability that he will get caught for the first time on his fourth "job" is

$$g(4; 0.20) = (0.20)(0.80)^{4-1}$$
$$= 0.102$$

Similarly, if the probability is 0.05 that a part produced by a machine will be defective, then the probability that the sixth part produced on a given day is the first one that is defective is

$$g(6; 0.05) = (0.05)(0.95)^{6-1}$$
$$= 0.039$$

Example. In *queueing theory* (see example on page 82), the geometric distribution has important applications in connection with the number of units (customers, trucks, airplanes, etc.) that are being served or waiting to be served at any given time. To illustrate, suppose that the arrival of cars at a gas station with one pump is a Poisson process with a *mean arrival rate* (average number of arrivals per unit time) of α cars per hour, and that the cars can be served at a *mean service rate* of β cars per hour. Since the arrivals constitute a Poisson process, the probability is negligible that there will be more than one arrival during a small time interval Δt, and we shall assume that the same is true also for completions of service.* Furthermore, we shall assume that $\alpha < \beta$, namely, that the mean number of arrivals per unit time is less than the mean number of services that can be completed per unit time.

To arrive at an expression for $P(k)$, the probability that there are k cars in the gas station at any given time being served or waiting to be served, we shall first obtain a set of "recursion relations" expressing $P(k)$ in terms of $P(k-1)$ and $P(k-2)$. To illustrate how these relations are obtained, let us find the probability that there are no cars in the station ($k = 0$) at time $t + \Delta t$. If $k = 0$ at time $t + \Delta t$, by the assumption of Poisson arrivals the only values that k could have had at time t are 0 and 1. If $k = 0$ at time t, then no new cars arrived in the interval from t to $t + \Delta t$ (with probability $1 - \alpha \cdot \Delta t$), but if $k = 1$ at time t, then a service was completed in this interval (with probability $\beta \cdot \Delta t$). Using the rule of elimination (see page 37) we, thus, conclude that

$$P(0) = P(0)(1 - \alpha \cdot \Delta t) + P(1)\beta \cdot \Delta t$$

or

$$P(1) = \frac{\alpha}{\beta} P(0)$$

In general, the situation is typified by the tree diagram of Figure 3.8, and the rule of elimination leads to the following recursion relation:

$$P(k) = P(k-1)\alpha \cdot \Delta t + P(k)[1 - (\alpha + \beta)\Delta t] + P(k+1)\beta \cdot \Delta t$$

for $k \geq 1$. In Exercise 18 on page 88, the reader will be asked to verify that a solution to this set of equations is given by the *geometric*

---

*In fact, it will be assumed that so long as the service station is not empty, completions of service constitute a Poisson process.

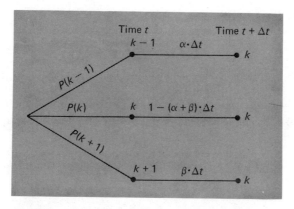

Figure 3.8. Tree diagram for queueing problem.

distribution

$$P(k) = \left(1 - \frac{\alpha}{\beta}\right)\left(\frac{\alpha}{\beta}\right)^k \qquad \text{for } k = 0, 1, 2, \ldots$$

which may be obtained from the expression on page 83 by letting $p = 1 - \frac{\alpha}{\beta}$ and $x = k + 1$.

This distribution reveals many important properties about queues. Note first that since $\alpha < \beta$ by assumption, we have

$$\lim_{k \to \infty} P(k) = 0$$

so that the probability is 1 that the waiting line remains finite. (In fact, it can be shown that if $\alpha \geq \beta$ there is a nonzero probability that the waiting line will eventually grow beyond all bounds.) The mean number of cars in the station (being served and waiting to be served) is given by the mean of the above distribution, which is

$$\frac{1}{p} - 1 = \frac{1}{1 - \frac{\alpha}{\beta}} - 1 = \frac{\alpha}{\beta - \alpha}$$

in accordance with the formula for μ on page 83. (We subtracted the 1 because the values of the random variable started with $k = 0$ instead of $x = 1$.)

Also, the distribution of the *numbers in the queue waiting for service* (thus, excluding the one being served) is given by $P(k + 1)$ for $k \geq 1$ and $P(0) + P(1)$ for $k = 0$. Thus, the mean number of cars in the queue is given by

$$\left(1 - \frac{\alpha}{\beta}\right) \sum_{k=1}^{\infty} k \left(\frac{\alpha}{\beta}\right)^{k+1} = \frac{\alpha^2}{\beta(\beta - \alpha)}$$

To find the mean time a customer spends waiting in the queue, it is necessary first to find the distribution of waiting times, and then to find its mean. We shall omit this work, stating only that the resulting mean waiting time is $\dfrac{\alpha}{\beta(\beta - \alpha)}$.

EXERCISES

1. Prove that for the Poisson distribution

$$\frac{f(x + 1; \lambda)}{f(x; \lambda)} = \frac{\lambda}{x + 1}$$

for $x = 0, 1, 2, \ldots$.

2. Use the recursion formula of Exercise 1 to calculate the values of the Poisson distribution with $\lambda = 3$ for $x = 0, 1, 2, \ldots$, and 9, and draw the histogram of this distribution. Verify your results by referring to Table 2.

3. Using Table 2, find
   (a) $F(4; 7)$; (b) $f(4; 7)$;
   (c) $\sum_{k=6}^{19} f(k; 8)$.

4. Using Table 2, find
   (a) $F(9; 12)$; (b) $f(9; 12)$;
   (c) $\sum_{k=3}^{12} f(k; 7.5)$.

5. Use the Poisson distribution to approximate the binomial probability $b(3; 1,000, 0.03)$.

6. A large shipment of books contains 2 percent with imperfect bindings. Use the Poisson approximation of the binomial distribution to determine the probability that among 600 books (randomly selected from the shipment)
   (a) at most 15 will have imperfect bindings;
   (b) exactly 15 will have imperfect bindings;
   (c) at least 10 will have imperfect bindings.

7. In a given city, 5 percent of all drivers get at least one parking ticket per year. Use the Poisson approximation of the binomial distribution to determine that among 72 drivers (randomly chosen in this city)

(a) exactly 4 will get at least one parking ticket during the next year;
(b) at least 2 will get at least one parking ticket during the next year;
(c) anywhere from 4 to 8, inclusive, will get at least one parking ticket during the next year.

8. The number of weekly breakdowns of a computer is a random variable having a Poisson distribution with $\lambda = 0.3$. What is the probability that the computer will operate without a breakdown for two consecutive weeks?

9. The number of gamma rays emitted per second by a certain radioactive substance is a random variable having the Poisson distribution with $\lambda = 5.8$. If a recording instrument becomes inoperative when there are more than 12 rays per second, what is the probability that this instrument becomes inoperative during any given second?

10. Given that the switchboard of a consultant's office receives on the average 0.9 calls per minute, find the probabilities that
(a) in a given minute there will be at least one incoming call;
(b) between 10 : 00 A.M. and 10 : 02 A.M. there will be exactly 2 incoming calls;
(c) during an interval of 4 minutes there will be at most 3 incoming calls.

11. At a checkout counter customers arrive at an average rate of 1.5 per minute. Find the probabilities that
(a) at most 4 will arrive in any given minute;
(b) at least 3 will arrive during an interval of 2 minutes;
(c) at most 15 will arrive during an interval of 6 minutes.

12. The determination of geometric probabilities is often simplified by making use of the identity

$$g(x; p) = \frac{1}{x} \cdot b(1; x, p)$$

and looking up $b(1; x, p)$ in a table of binomial probabilities. Verify this identity and use it (and Table 1) to evaluate
(a) $g(12; 0.10)$;   (b) $g(10; 0.30)$.

13. An expert shot hits a target 95 percent of the time. What is the probability that he will miss the target for the first time on the 15th shot?

14. In a "torture test" a light switch is turned on and off until it fails. If the probability that the switch will fail any time it is turned on or off is 0.001, what is the probability that the switch will fail *after* it has been turned on or off 1,200 times. Assume that the conditions underlying the geometric distribution are met. [*Hint:* Use the formula for the value of an infinite geometric progression and logarithms.]

15. In the example on page 82 we considered customers arriving at a cafeteria at an average rate of 0.3 per minute. If they are being served at an average rate of 0.5 per minute, find
(a) the average number of customers being served or waiting to be served at any given time;

(b) the average number of customers waiting to be served at any given time;

(c) the average time a customer spends waiting in line.

16. The arrival of trucks at a receiving dock is a Poisson process with a mean arrival rate of 2 per hour. The trucks can be unloaded at a mean rate of 3 per hour, and the time required for unloading has an exponential distribution.

   (a) What is the average number of trucks being unloaded and waiting to be unloaded?

   (b) What is the mean number of trucks in the queue?

   (c) What is the mean time a truck spends waiting in the queue?

   (d) What is the probability that there are no trucks waiting to be unloaded?

17. Referring to Exercise 16, suppose that the cost of keeping a truck in the system is $15 per hour. If it were possible to increase the mean unloading rate to 3.5 trucks per hour at a cost of $100 per day, would this be worthwhile?

18. Verify (by substitution) that the geometric distribution is a solution of the recursion relation given on page 84.

19. Differentiating with respect to p both sides of the equation

$$\sum_{x=1}^{\infty} p(1-p)^{x-1} = 1$$

show that the geometric distribution

$$f(x) = p(1-p)^{x-1} \quad \text{for } x = 1, 2, 3, \ldots$$

has the mean $1/p$.

20. The following is an alternate way of introducing the Poisson distribution. Suppose that $f(x, t)$ is the probability of getting x successes during a time interval of length t when (1) the probability of a success during a very small time interval from t to $t + \Delta t$ is $\alpha \cdot \Delta t$, (2) the probability of more than one success occurring during such a time interval is negligible, and (3) the probability of a success during such a time interval does not depend on what happened prior to time t.

   (a) Show that

$$f(x, t + \Delta t) = f(x, t)[1 - \alpha \Delta t] + f(x - 1, t)\alpha \, \Delta t$$

   (b) Using the result of part (a), show that

$$\frac{d[f(x, t)]}{dt} = \alpha[f(x - 1, t) - f(x, t)]$$

   (c) Verify by substitution that the differential equation obtained in (b) is satisfied by the formula for the Poisson distribution with $\lambda = \alpha t$.

21. Prove that the mean and the variance of the Poisson distribution are both equal to λ

(a) directly, by applying the definitions in Section 3.4;

(b) by using formulas $\mu = np$ and $\sigma^2 = np(1 - p)$ for the binomial distribution and the limit argument on page 78.

22. A company rents out time on a computer for periods of t hours, for which it receives \$600 an hour. The number of times the computer breaks down during t hours is a random variable having the Poisson distribution with $\lambda = (0.8)t$, and if the computer breaks down x times during t hours it costs $50x^2$ dollars to fix it. How should the company select t in order to maximize its expected profit? [*Hint*: Use results obtained in Section 3.6.]

3.9 THE MULTINOMIAL DISTRIBUTION

An immediate generalization of the binomial distribution arises when each trial can have more than two possible outcomes. This happens, for example, when a manufactured product is classified as superior, average, or poor, when a student's performance is judged by giving him an A, B, C, D, or F, or when an experiment is judged successful, unsuccessful, or inconclusive. To treat this kind of problem in general, let us consider the case where there are n independent trials, with each trial permitting k mutually exclusive outcomes whose respective probabilities are $p_1, p_2, \ldots, p_k$ (with $\sum_{i=1}^{k} p_i = 1$). Referring to the outcomes as being of the first kind, the second kind, $\ldots$, and the kth kind, we shall be interested in the probability $f(x_1, x_2, \ldots, x_k)$ of getting x_1 outcomes of the first kind, x_2 outcomes of the second kind, $\ldots$, and x_k outcomes of the kth kind, with $\sum_{i=1}^{k} x_i = n$. Using arguments similar to those which we employed in deriving the equation for the binomial distribution in Section 3.2, it can be shown that the desired probability is given by

$$f(x_1, x_2, \ldots, x_k) = \frac{n!}{x_1! x_2! \cdot \ldots \cdot x_k!} p_1^{x_1} \cdot p_2^{x_2} \cdot \ldots \cdot p_k^{x_k}$$

for $x_i = 0, 1, \ldots, n$ for each i, subject to the restriction $\sum_{i=1}^{k} x_i = n$. The *joint probability distribution* whose values are given by these probabilities is called the *multinomial distribution*; it owes its name to the fact that for the various values of the x_i the probabilities are given by corresponding terms of the multinomial expansion of $(p_1 + p_2 + \cdots + p_k)^n$.

Example. Suppose the probability that the light bulb of a certain kind of slide projector will last fewer than 40 hours of continuous use is 0.30, the probability that it will last anywhere from 40 to 80 hours of continuous use is 0.50, and the probability that it will last more than 80 hours of continuous use is 0.10. To find the probability that among eight such light bulbs two will last fewer than 40 hours, five will last anywhere from 40 to 80 hours, and one will last more than 80 hours, we substitute $x_1 = 2$, $x_2 = 5$, $x_3 = 1$, $p_1 = 0.30$, $p_2 = 0.50$, and $p_3 = 0.20$ into the formula, and we get

$$f(2, 5, 1) = \frac{8!}{2!5!1!}(0.30)^2(0.50)^5(0.20)^1$$

$$= 0.0945$$

EXERCISES

1. Suppose that the probabilities are, respectively, 0.40, 0.40, and 0.20 that in city driving a certain kind of imported car will average less than 22 miles per gallon, anywhere from 22 to 25 miles per gallon, or more than 25 miles per gallon. Find the probability that among 12 such cars tested, four will average less than 22 miles per gallon, six will average anywhere from 22 to 25 miles per gallon, and two will average more than 25 miles per gallon.

2. Suppose that the probabilities are, respectively, 0.60, 0.20, 0.10, and 0.10 that an income tax form will be filled in correctly, that it will contain an error favoring the taxpayer, that it will contain an error favoring the government, or that it will contain both kinds of errors. Find the probability that among 10 of the income tax forms randomly selected for audit five will be correct, three will contain an error favoring the taxpayer, one will contain an error favoring the government, and one will contain both kinds of errors.

3. Using the same sort of reasoning as in the derivation of the formula for the hypergeometric distribution, we can derive a formula which is analogous to the multinomial distribution but applies to *sampling without replacement*. If a set of N objects contains a_1 objects of the first kind, a_2 objects of the second kind, . . . , and a_k objects of the kth kind, so that $a_1 + a_2 + \ldots + a_k = N$, the number of ways in which we can select x_1 objects of the first kind, x_2 objects of the second kind, . . . , and x_k objects of the kth kind is given by the *product* of the number of ways in which we can select x_1 of the a_1 objects of the first kind, x_2 of the a_2 objects of the second kind, . . . , and x_k of the a_k objects of the kth kind. Then, the probability of obtaining that many objects of each kind is simply this product *divided* by the total number of possibilities, namely, the number of ways in which $x_1 + x_2 + \ldots + x_k = n$ objects can be selected from the whole set of N objects.

(a) Write a formula for the probability of thus obtaining x_1 objects of the first kind, x_2 objects of the second kind, . . . , and x_k objects of the kth kind.

(b) If 20 defective glass bricks include 10 that have cracks, seven that are discolored, and three that have cracks and discoloration, what is the probability that among six of the bricks chosen at random for further checks three will have cracks, two will be discolored, and one will have cracks as well as discoloration?

3.10 SIMULATION

In recent years, simulation techniques have been applied to many problems in the various sciences, and if the processes which are being simulated involve an element of chance, these techniques are referred to as *Monte Carlo methods*. Very often, the use of Monte Carlo simulation eliminates the cost of building and operating expensive equipment; it is used, for instance, in the study of collisions of photons with electrons, the scattering of neutrons, and similar complicated phenomena. Monte Carlo methods are also useful in situations where direct experimentation is impossible—say, in studies of the spread of cholera epidemics, which, of course, are not induced experimentally on human populations. In addition, Monte Carlo techniques are sometimes applied to the solution of mathematical problems which actually cannot be solved by direct means, or where a direct solution is too costly or requires too much time.

A classical example of the use of Monte Carlo methods in the solution of a problem of pure mathematics is the determination of π (the ratio of the circumference of a circle to its diameter) by probabilistic means. Early in the eighteenth century George de Buffon, a French naturalist, proved that if a very fine needle of length a is thrown at random on a board ruled with equidistant parallel lines, the probability that the needle will intersect one of the lines is $2a/\pi b$, where b is the distance between the parallel lines. What is remarkable about this fact is that it involves the constant $\pi = 3.1415926 \ldots$, which in elementary geometry is approximated by the circumferences of regular polygons enclosed in a circle of radius $\frac{1}{2}$. Buffon's result implies that if such a needle is actually tossed a great many times, the proportion of the time it crosses one of the lines gives an estimate of $2a/\pi b$ and, hence, an estimate of π since a and b are known. Early experiments of this kind yielded an estimate of 3.1519 (based on 5,000 trials) and an estimate of 3.155 (based on 3,204 trials) in the middle of the nineteenth century.

Although Monte Carlo methods are sometimes based on actual gambling devices (for example, the needle tossing in the estimation of π), it is usually expedient to use so-called tables of random digits or random numbers. Tables of random numbers consist of many pages on which the digits 0, 1, 2, . . . , and 9 are set down in a "random" fashion, much as they would appear if they were generated one at a time by a gambling device giving each digit an equal probability of being selected. Actually, we could construct such tables ourselves—say, by repeatedly drawing numbered slips out of a hat or by using a perfectly constructed spinner—but in practice such tables are usually generated by means of electronic computers. For instance, in one such method the computer begins with a 4-digit number, say, 3571, and squares it, getting 12752041. The middle four digits, in this case 7520, are looked upon as a set of 4 random digits. (If necessary, a 0 is added on the left to make the square have eight digits.) Then 7520 is squared, yielding 56550400, and 5504 is looked upon as the next set of 4 random digits. Continuing in this way, we obtain a table of *pseudo-random digits* which are quite satisfactory for most practical purposes.

Although tables of random numbers are constructed so that the digits can be looked upon as values of a random variable having the discrete uniform distribution $f(x) = \frac{1}{10}$ for $x = 0, 1, 2, \ldots$, or 9, they can be used to simulate values of any discrete random variable, and even continuous random variables.

Example. For instance, with the use of random numbers we can simulate flips of coins in many different ways. One possibility is to let 0, 2, 4, 6, 8 represent heads, 1, 3, 5, 7, 9 represent tails, and to simulate, say, three flips of a coin we can pick sets of three random digits. Thus, if we use the second page of Table VII and the digits in the 9th, 10th, and 11th columns starting from the top, we get

480   280   085   265   303   288   295   388   127   222

and we interpret this as getting, respectively, 3, 3, 2, 2, 1, 3, 1, 2, 1, and 3 heads in 10 flips of 3 coins. Since the probabilities of getting 0, 1, 2, or 3 heads with 3 balanced coins are, respectively, $\frac{1}{8}$, $\frac{3}{8}$, $\frac{3}{8}$, and $\frac{1}{8}$, as can easily be verified, we could also use the scheme shown in the following table:

Number of heads	Probability	Random digits
0	$\frac{1}{8}$	0
1	$\frac{3}{8}$	1, 2, 3
2	$\frac{3}{8}$	4, 5, 6
3	$\frac{1}{8}$	7

Omitting the digits 8 and 9 wherever they may occur, we would, thus, interpret the random digits

4 8 1 1 1 9 3 3 3 7 6 3 6 2 6 5 8 9 3 1 0 6

in the fifth row of the third page of Table 7 as getting, respectively, 2, 1, 1, 1, 1, 1, 1, 3, 2, 1, 2, 1, 2, 2, 1, 1, 0, and 2 heads in 18 tosses of three coins.

Of the two methods used in this example, the first has the disadvantage that we still have to count how many of the digits are even and how many are odd; the second has the disadvantage that some of the digits will have to be omitted, and this may conceivably lead to situations where many random digits will have to be omitted before we find enough that can be used. To avoid such difficulties, it is generally preferable to express the probabilities as *decimals* (rounded, if necessary).

Continuation of Example. Since the probabilities for 0, 1, 2, or 3 heads in three flips of a coin are, respectively, 0.125, 0.375, 0.375, and 0.125, we can use three-digit random numbers and the following scheme:

Number of heads	Probability	Random numbers
0	0.125	000–124
1	0.375	125–499
2	0.375	500–874
3	0.125	875–999

Note that we allocated 125 (or *one-eighth*) of the random numbers from 000 through 999 to 0 heads, 375 (or *three-eighths*) to 1 head, 375 (or *three-eighths*) to 2 heads, and 125 (or *one-eighth*) to 3 heads. If we arbitrarily used the 22nd, 23rd, and 24th columns of the first page of Table 7 starting with the 7th row, we would get

197   565   157   520   946   951   948   586   586   089

and we would interpret this as getting 2, 2, 1, 2, 3, 3, 3, 2, 2, and 0 heads.

The allocation of random numbers to the various values of a random variable can be facilitated by referring to the corresponding *cumulative probabilities*, as is illustrated by the following example:

Example. Suppose that the probabilities are, respectively, 0.082, 0.205, 0.256, 0.214, 0.134, 0.067, 0.028, 0.010, 0.003, and 0.001 that 0, 1, 2, 3, . . . , or 9 cars will arrive at a toll booth of a turnpike in any given minute during the early afternoon. To simulate the arrival of cars at this toll booth, we might use the following scheme:

Number of cars	Probability	Cumulative probability	Random numbers
0	0.082	0.082	000–081
1	0.205	0.287	082–286
2	0.256	0.543	287–542
3	0.214	0.757	543–756
4	0.134	0.891	757–890
5	0.067	0.958	891–957
6	0.028	0.986	958–985
7	0.010	0.996	986–995
8	0.003	0.999	996–998
9	0.001	1.000	999

Note that in each case the *last* random digit is *one less* than the number formed by the last three digits of the corresponding cumulative probability; of course, the cumulative probabilities are the respective probabilities that there will be *at most* 0 cars, *at most* 1 car, *at most* 2 cars, and so on. Now, if we simulate the arrival of cars at the toll booth by using the 5th, 6th, and 7th columns of the fourth page of Table 7 starting with the 11th row, we get

036   417   962   458   778   541   869   379   973   553

325   674   907   710   709   499   493   384   346   301

This means that during 20 one-minute intervals the number of cars arriving at the toll booth is 0, 2, 6, 2, 4, 2, 4, 2, 6, 3, 2, 3, 5, 3, 3, 2, 2, 2, 2, and 2.

EXERCISES

1. Suppose that the probabilities are, respectively, 0.41, 0.37, 0.16, 0.05, and 0.01 that there will be 0, 1, 2, 3, or 4 UFO sightings in a certain region on any one day.
(a) Distribute the two-digit random numbers from 00 through 99 to the five

values of this random variable, so that the corresponding random numbers can be used to simulate the sighting of UFO's in the given region.

(b) Use the result of part (a) to simulate the sighting of UFO's in the given region on 25 days.

2. Referring to the example on page 79 and the probabilities shown in Figure 3.7, distribute the three-digit random numbers from 000 through 999 to the 11 values of the random variable, and simulate the fire insurance company's experience (that is, that number of policyholders who file at least one claim) over a period of 20 years.

3. The following is the probability distribution of the number of telephone calls that are received by the switchboard of a brokerage house per minute:

Number of calls	0	1	2	3	4	5	6	7
Probability	0.2466	0.3452	0.2417	0.1128	0.0395	0.0111	0.0026	0.0005

(a) Distribute the four-digit random numbers from 0000 through 9999 to the eight values of this random variable, so that the corresponding random numbers can be used to simulate the number of calls arriving at the switchboard of this brokerage house.

(b) Use the result of part (a) to simulate the number of calls arriving at the switchboard during 40 consecutive minutes.

4. Depending upon the availability of parts, a company can manufacture 3, 4, 5, or 6 units of a certain item per week with corresponding probabilities of 0.10, 0.40, 0.30, and 0.20. The probabilities that there will be a weekly demand for 0, 1, 2, 3, . . . , or 8 units are, respectively, 0.05, 0.10, 0.30, 0.30, 0.10, 0.05, 0.05, 0.04, and 0.01. If a unit is sold during the week that it is made, it will yield a profit of $100; this profit is reduced by $20 for each week that a unit has to be stored. Use random numbers to simulate the operations of this company for 50 consecutive weeks and estimate their expected weekly profit.

4

Probability Densities

4.1 CONTINUOUS RANDOM VARIABLES

When we first introduced the concept of a random variable in Chapter 3, we presented it as a real-valued function defined over the sample space of an experiment, and we illustrated this idea with the random variable giving the number of preferred ratings which a lawn mower received, assigning the numbers 0, 1, 2, or 3 (whichever was appropriate) to the 18 possible outcomes of the experiment. In the continuous case, where random variables can assume values on a continuous scale, the procedure is very much the same. The outcomes of an experiment are represented by the points on a line segment or a line, and the value of a random variable is a number appropriately assigned to each point by means of some rule or equation. When the value of a random variable is given directly by a measurement or observation, we usually do not bother to differentiate between the value of the random variable, the measurement which we obtain, and the outcome of the experiment, the corresponding point on the real axis. Thus, if an experiment consists of determining what force is required to break a given tensile-test specimen, the result itself, say, 138.4 pounds, is the value of the random variable with which we are concerned. There is no real need in that case to add that the sample space of the experiment consists of all (or part of) the points on the positive real axis.

The problem of defining probabilities in connection with continuous sample spaces and continuous random variables involves some complications. To illustrate the nature of these complications, let us consider the following situation:

Example. Suppose we want to know the probability that if an accident occurred on a freeway whose length is 200 miles, it would happen at some given location or, perhaps, some particular stretch of the road. The outcomes of this experiment can be looked upon as a continuum of points, namely, those on the continuous interval from 0 to 200. Suppose the probability that the accident occurred on any interval of length L is $L/200$, with L measured in miles. Note that this arbitrary assignment of probability is consistent with Axioms 1 and 2 on page 18, since the probabilities are all nonnegative and less than or equal to 1, and $P(S) = \frac{200}{200} = 1$. Of course, we are considering so far only events represented by intervals which form part of the line segment from 0 to 200. Using Axiom 3′ on page 78, we can also obtain probabilities of events which are not intervals but which can be represented by the union of finitely many or countably many intervals. Thus, for two nonoverlapping intervals of length L_1 and L_2 we have a probability of

$$\frac{L_1 + L_2}{200}$$

and for an infinite sequence of nonoverlapping intervals of length L_1, $L_2, L_3, \ldots$, we have a probability of

$$\frac{L_1 + L_2 + L_3 + \cdots}{200}$$

Note that the probability that the accident occurred at any given point is equal to zero because we can look upon a point as an interval of zero length. However, the probability that the accident occurred in a very short interval is positive; for instance, for an interval of length 1 foot the probability is $9.5(10)^{-7}$.

Thus, in extending the concept of probability to the *continuous case*, we again use Axioms 1, 2, and 3′, but we shall have to restrict the meaning of the term "event." So far as practical considerations are concerned, this restriction is of no consequence; we simply do not assign probabilities to some rather abstruse point sets, which cannot be expressed as the unions or intersections of finitely many or countably many intervals.

The way in which we assigned probabilities in the preceding example is, of course, very special; it is similar in nature to the way in which we assign *equal* probabilities to the six faces of a die, heads and tails, the 52 cards in a standard deck, and so forth. To treat the problem of associating probabilities with continuous random variables more generally, suppose we are interested in the probability that a given random variable will take on a value on the interval from a to b, where a and b are constants with $a \leq b$. Suppose, furthermore, that we divide the interval from a to b into n equal sub-intervals of width Δx containing, respectively, the points $x_1, x_2, \ldots, x_n$, and that the probability that the random variable will take on a value on the sub-interval containing x_i is given by $f(x_i) \cdot \Delta x$. Then, the probability that the random variable with which we are concerned will take on a value on the interval from a to b is given by

$$P(a \leq x \leq b) = \sum_{i=1}^{n} f(x_i) \cdot \Delta x$$

Now, if **f** is an integrable function defined for all values of the random variable with which we are concerned ($x \geq 0$ in this case), we shall *define* the probability that the value of the random variable falls between a and b by letting $\Delta x \longrightarrow 0$, namely, as

$$P(a \leq x \leq b) = \int_a^b f(x)dx$$

This definition of probability in the continuous case presupposes the existence of an appropriate function **f** which, integrated from any constant a to any constant b (with $a \leq b$), gives the probability that the corresponding random variable takes on a value on the interval from a to b. Note that a value $f(x)$ of **f** does not give the probability that the corresponding random variable takes on the value x; *in the continuous case probabilities are given by integrals and not by the values of* **f**. The probability that a random variable actually takes on the value x may be obtained by first considering the probability that it takes on a value on the interval from $x - \Delta x$ to $x + \Delta x$. However, if we then let $\Delta x \longrightarrow 0$, it becomes apparent that *the probability is zero that a continuous random variable takes on any given value x.*

The fact that this probability is always zero should not be disturbing. In fact, our definition of probability for the continuous case provides a remarkably good model for dealing with measurements or observations. Owing to the limits of our ability to measure, experimental data never seem to come from a continuous sample space. Thus, while temperatures are fruitfully thought of as points on a continuous scale, any temperature *measurement* actually represents an interval on this scale. If we report a temperature measurement of 74.8 degrees centigrade, we really mean that the temperature lies in the interval from 74.75 to 74.85 degrees centigrade, and not that it is exactly 74.8000. . . . It is important to add that *when we say that there is a zero probability that a random variable will assume any given value x, this does not mean that it is impossible actually to get the value x.* In the continuous case, a zero probability does not imply logical impossibility, but the whole matter is largely academic since, owing to the limitations of our ability to measure and observe, we are always interested in probabilities connected with intervals and not with isolated points.

As an immediate consequence of the fact that in the continuous case probabilities associated with individual points are always zero, we find that if we speak of the probability associated with the interval from *a* to *b*, it does not matter whether either endpoint is included; symbolically,

$$P(a \leq x \leq b) = P(a \leq x < b) = P(a < x \leq b) = P(a < x < b)$$

Drawing an analogy with the concept of a density function in physics, we call the functions **f**, whose existence we stipulated in extending our definition of probability to the continuous case, a *probability density function*, or simply a *probability density*. Whereas density functions are integrated to obtain weights, probability density functions are integrated to obtain probabilities.

Since a probability density, integrated between any two constants *a* and *b*, gives the probability that a random variable assumes a value between these limits, **f** cannot be just any real-valued integrable function. However, if we impose the conditions that:

1. $f(x) \geq 0$ for all x within the domain of **f**.

2. $\int_{-\infty}^{\infty} f(x)\, dx = 1.$

it can be shown that the axioms of probability (with the modification discussed on page 98) are satisfied. Note the similarity between these properties and those given on page 53 for probability distributions.

As in the discrete case, we shall write as $F(x)$ the probability that a random variable with the probability density **f** takes on a value less than or equal to x, and we shall refer to the corresponding function **F** as the *cumulative distribution function*, or simply the *distribution function* of the random variable. Thus, if a random variable with values x has the probability density **f**, the values of its distribution function are given by

$$F(x) = \int_{-\infty}^{x} f(t)\, dt$$

It immediately follows from this definition that

$$P(a \leq x \leq b) = F(b) - F(a)$$

and according to the fundamental theorem of integral calculus that

$$\frac{dF(x)}{dx} = f(x)$$

wherever this derivative exists.

Example. To illustrate the concepts we have introduced, let us consider a random variable whose probability density function is given by

$$f(x) = \begin{cases} 2e^{-2x} & \text{for } x > 0 \\ 0 & \text{for } x \leq 0 \end{cases}$$

Note that, although the random variable cannot assume negative values, we artificially extended the domain of **f** to include all the real numbers. This is a practice we shall follow throughout this book. It is also apparent from the graph of this function in Figure 4.1 that it has a discontinuity at $x = 0$; indeed, a probability density function need not be everywhere continuous, so long as it is integrable between any two limits a and b (with $a \leq b$).

Leaving it to the reader to verify that the given function satisfies the two conditions on page 99 and that it can, therefore, serve as a probability density function, let us now find some probabilities connected with a corresponding random variable. The probability that

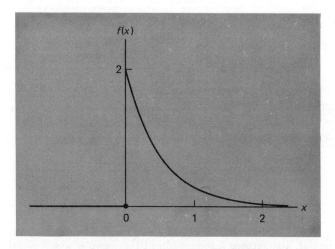

Figure 4.1. Graph of density function.

it will take on a value between 1 and 3 is

$$P(1 \leq x \leq 3) = \int_1^3 2e^{-2x}\, dx = e^{-2} - e^{-6} = 0.135$$

and the probability that it will take on a value greater than 0.5 is

$$P(x > 0.5) = \int_{0.5}^{\infty} 2e^{-2x}\, dx = e^{-1} = 0.368$$

The cumulative distribution function which corresponds to this example is given by

$$F(x) = \begin{cases} 0 & \text{for } x \leq 0 \\ \int_0^x 2e^{-2t}\, dt = 1 - e^{-2x} & \text{for } 0 < x \end{cases}$$

and it can be shown, for example, that

$$P(x \leq 1) = F(1) = 1 - e^{-2} = 0.865$$

Note that the distribution function of this example is *nondecreasing* and that $F(-\infty) = 0$ and $F(\infty) = 1$. Indeed, it can be shown that these properties are shared by *all* cumulative distribution functions.

The measures used to describe probability densities are very similar to the ones which we used to describe probability distributions. Replacing summations with integrals, we define the *kth moment about the origin* as

$$\mu'_k = \int_{-\infty}^{\infty} x^k \cdot f(x) \, dx$$

analogous to the definition we gave on page 72. The first moment about the origin is again referred to as the *mean* and it is denoted as μ; as before, it is the *expected*, or *average value* of a random variable having the probability density **f**. We also define the *kth moment about the mean* as

$$\mu_k = \int_{-\infty}^{\infty} (x - \mu)^k \cdot f(x) \, dx$$

In particular, the second moment about the mean is again referred to as the *variance* and it is written as σ^2; as before, it measures the spread of a probability density (or its graph) in the sense that it gives the *expected* or *average value* of the squared deviations from the mean.

Continuation of Example. Applying these formulas to our example, we find that the mean is

$$\mu = \int_0^{\infty} x \cdot 2e^{-2x} \, dx = \tfrac{1}{2}$$

and that the variance is

$$\sigma^2 = \int_0^{\infty} (x - \tfrac{1}{2})^2 \cdot 2e^{-2x} \, dx = \tfrac{1}{4}$$

In an example like this, the calculation of σ^2 may be simplified by using, as on page 72, the formula $\sigma^2 = \mu'_2 - \mu^2$.

EXERCISES

1. Verify that the function given in the example on page 100 is, in fact, a probability density.

2. If the probability density of a random variable is given by

$$f(x) = \begin{cases} k(1 - x^2) & \text{for } 0 < x < 1 \\ 0 & \text{elsewhere} \end{cases}$$

find

(a) the value of k;

(b) the probability that a random variable having this probability density will take on a value between 0.1 and 0.2;

(c) the probability that a random variable having this probability density will take on a value greater than 0.5;

(d) the distribution function which corresponds to this probability density. [*Hint*: $F(x)$ must be given separately for $x \leq 0$, $0 < x < 1$, and $x \geq 1$.]

Also use the result of part (d) to find

(e) the probability that a random variable having this distribution function will take on a value less than 0.3;

(f) the probability that a random variable having this distribution function will take on a value between 0.4 and 0.6.

3. If the probability density of a random variable is given by

$$f(x) = \begin{cases} x & \text{for } 0 < x < 1 \\ 2 - x & \text{for } 1 \leq x < 2 \\ 0 & \text{elsewhere} \end{cases}$$

find

(a) the probability that a random variable having this probability density will take on a value between 0.2 and 0.8;

(b) the probability that a random variable having this probability density will take on a value between 0.6 and 1.2;

(c) the distribution function which corresponds to this probability density. [*Hint*: $F(x)$ must be given separately for $x \leq 0$, $0 < x < 1$, $1 \leq x < 2$, and $x \geq 2$.]

Also use the result of part (c) to find

(d) the probability that a random variable having this distribution function will take on a value greater than 1.8;

(e) the probability that a random variable having this distribution function will take on a value between 0.4 and 1.6.

4. Given the probability density $f(x) = \dfrac{k}{1 + x^2}$ for $-\infty < x < \infty$, find k.

5. If the distribution function of a random variable is given by

$$F(x) = \begin{cases} 1 - \dfrac{4}{x^2} & \text{for } x > 2 \\ 0 & \text{for } x \leq 2 \end{cases}$$

find

(a) the probability that this random variable will take on a value less than 3;

(b) the probability that this random variable will take on a value between 4 and 5;

(c) the probability density of this random variable. Are there any points at which it is undefined?

Also sketch the graphs of the distribution function and the probability density.

6. In certain experiments, the error made in determining the density of a silicon compound is a random variable having the probability density

$$f(x) = \begin{cases} 25 & \text{for } -0.02 < x < 0.02 \\ 0 & \text{elsewhere} \end{cases}$$

find
(a) the probability that such an error will be between -0.03 and 0.04;
(b) the probability that the *magnitude* of such an error will be less than 0.005.

7. The mileage (in thousands of miles) which car owners get with a certain kind of tire is a random variable having the probability density

$$f(x) = \begin{cases} \dfrac{1}{20}e^{-x/20} & \text{for } x > 0 \\ 0 & \text{for } x \le 0 \end{cases}$$

Find the probabilities that one of these tires will last
(a) at most 10,000 miles;
(b) anywhere from 16,000 to 24,000 miles;
(c) at least 30,000 miles.

8. Prove that the identity $\sigma^2 = \mu_2' - \mu^2$ holds for any probability density for which these moments exist.

9. Find μ and σ^2 for the probability density of Exercise 2.

10. Find μ and σ^2 for the probability density of Exercise 3.

11. Find μ and σ^2 for the distribution of the random variable of Exercise 5.

12. Find μ and σ^2 for the distribution of the errors of measurement of Exercise 6.

13. Show that for the probability density of Exercise 4, μ_2' and, therefore, σ^2 do not exist.

4.2 THE NORMAL DISTRIBUTION

Among the special probability densities we shall study in this chapter, the *normal probability density*, usually referred to simply as the *normal distribution*, is by far the most important.* It was studied first in the eighteenth century when scientists observed an astonishing degree of

---

*The words "density" and "distribution" are often used interchangeably in the literature of applied statistics.

regularity in errors of measurement. They found that the patterns (distributions) they observed were closely approximated by a continuous curve which they referred to as the "normal curve of errors" and attributed to the laws of chance. The equation of the *normal probability density*, whose graph (shaped like the cross section of a bell) is shown in Figure 4.2, is

$$f(x; \mu, \sigma^2) = \frac{1}{\sqrt{2\pi}\,\sigma} e^{-\frac{1}{2}\left(\frac{x-\mu}{\sigma}\right)^2} \qquad -\infty < x < \infty$$

and in Exercises 18 and 19 on page 112, the reader will be asked to verify that its parameters μ and σ are, indeed, its mean and its standard deviation.

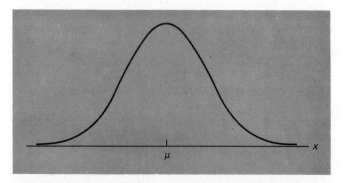

Figure 4.2. Graph of normal probability density.

Since the normal probability density cannot be integrated in closed form between every pair of limits a and b, probabilities relating to normal distributions are usually obtained from special tables, such as Table 3 at the end of this book. This table pertains to the *standard normal distribution*, namely, the normal distribution which has $\mu = 0$ and $\sigma = 1$, and its entries are the values of

$$F(z) = \frac{1}{\sqrt{2\pi}} \int_{-\infty}^{z} e^{-\frac{1}{2}t^2}\, dt$$

for $z = 0.00, 0.01, 0.02, \ldots, 3.49$. To find any probability $P(a \leq z \leq b)$, where z is the value of a random variable having the *standard* normal distribution, we use the equation $P(a \leq z \leq b) = F(b) - F(a)$, and if either *a* or *b* is *negative*, we also have to use the identity $F(-z) = 1 - F(z)$, which the reader will be asked to verify in Exercise 17 on page 112.

Example. For instance, the probability that a random variable having the standard normal distribution will take on a value between 0.87 and 1.28 is

$$P(0.87 \leq z \leq 1.28) = F(1.28) - F(0.87)$$
$$= 0.8997 - 0.8078$$
$$= 0.0919$$

and the probability that it will take on a value between -0.34 and 0.62 is

$$P(-0.34 \leq z \leq 0.62) = F(0.62) - F(-0.34)$$
$$= 0.7324 - (1 - 0.6331)$$
$$= 0.3655$$

Also, the probability that it will take on a value greater than 0.85 is

$$P(0.85 < z) = 1 - F(0.85)$$
$$= 1 - 0.8023$$
$$= 0.1977$$

and the probability that it will take on a value greater than -0.65 is

$$P(-0.65 < z) = 1 - F(-0.65)$$
$$= 1 - [1 - F(0.65)]$$
$$= F(0.65)$$
$$= 0.7422$$

To use Table 3 in connection with normal distributions other than the standard normal distribution, we have only to make use of the following theorem:

THEOREM 4.1 *If a random variable has the values x and a distribution with the mean μ and the variance σ^2, the corresponding standardized random variable has the values*

$$z = \frac{x - \mu}{\sigma}$$

and its distribution has zero mean and unit variance.

To prove this result, we have only to set up the integral for the probability $P(x \leq a)$, make the change of variable $z = \dfrac{x - \mu}{\sigma}$, and we will find that $P(x \leq a)$ equals $F\left(\dfrac{a - \mu}{\sigma}\right)$, which is the corresponding value of the cumulative distribution function of the standard normal distribution.

Thus, if we want to find the probability that a random variable having a normal distribution with the mean μ and the variance σ^2 will take on a value between a and b, we have only to calculate the probability that a random variable having the *standard* normal distribution will take on a value between $\dfrac{a - \mu}{\sigma}$ and $\dfrac{b - \mu}{\sigma}$; symbolically,

$$P(a \leq x \leq b) = F\left(\frac{b - \mu}{\sigma}\right) - F\left(\frac{a - \mu}{\sigma}\right)$$

where $F\left(\dfrac{b - \mu}{\sigma}\right)$ and $F\left(\dfrac{a - \mu}{\sigma}\right)$ can be obtained from Table 3.

Example. It has been found through experience that a large variety of physical measurements have approximately normal distributions. Thus, suppose that the amount of cosmic radiation to which a person is exposed while flying by jet across the United States is a random variable having a normal distribution with mean of 4.35 mrem* and a standard deviation of 0.59 mrem. Then, the probability that a person will be exposed to anywhere from 4.00 to 5.00 mrem on such a flight is

$$F\left(\frac{5.00 - 4.35}{0.59}\right) - F\left(\frac{4.00 - 4.35}{0.59}\right) = F(1.10) - F(-0.59)$$
$$= 0.8643 - (1 - 0.7224)$$
$$= 0.5867$$

and the probability that a person will be exposed to at least 5.50 mrem on such a flight is

$$1 - F\left(\frac{5.50 - 4.35}{0.59}\right) = 1 - F(1.95)$$
$$= 1 - 0.9744$$
$$= 0.0256$$

---

*This unit of radiation stands for "milli roentgen equivalent man."

Example. To consider a somewhat different example, suppose that the actual amount of instant coffee which a filling machine puts into 4-ounce jars varies somewhat from jar to jar, and that it can be looked upon as a random variable having a normal distribution with a standard deviation of 0.04 ounce. If only 2 percent of the jars are to contain less than 4.00 ounces, let us determine *the average amount of coffee which the filling machine will have to put into these jars.* This example differs from the preceding one insofar as we are given a percentage (or probability), we are given $\sigma = 0.04$, and we are asked to find μ. Since the value of z for which the entry in Table 3 comes closest to $1.0000 - 0.0200 = 0.9800$ is 2.05, we have

$$-2.05 = \frac{4.00 - \mu}{0.04}$$

and, solving for μ, we get $\mu = 4.082$ ounces. Since this may be unsatisfactory so far as the food processors are concerned, the reader will be asked to show in Exercise 14 on page 112 that if the variability of the filling machine is reduced so that $\sigma = 0.025$ ounce, this will lower the required average amount of coffee to $\mu = 4.05$ ounces, yet keep just about 98 percent of the jars above four ounces.

4.3 THE NORMAL APPROXIMATION
OF THE BINOMIAL DISTRIBUTION

In some books the normal distribution is introduced as a probability density which can be used to approximate the binomial distribution when n is large and P, the probability of a success, is close to 0.50 and, hence, not small enough to use the Poisson approximation. Thus, let us state, without proof, the following theorem:

THEOREM 4.2. *If x is a value of a random variable having the binomial distribution with the parameters n and p, and if*

$$z = \frac{x - np}{\sqrt{np(1 - p)}}$$

then the limiting form of the distribution function of this standardized random variable as $n \to \infty$ is given by

$$F(z) = \int_{-\infty}^{z} \frac{1}{\sqrt{2\pi}} e^{-\frac{1}{2}t^2} dt \qquad -\infty < z < \infty$$

Note that although x takes on only the values $0, 1, 2, \ldots, n$, in the limit as $n \to \infty$ the distribution of the corresponding *standardized* random variable is continuous, and the corresponding density is the standard normal density. It will be explained in the examples which follow how we account for the fact that a probability distribution is approximated by means of a continuous probability density.

Example. Suppose that 20 percent of the diodes made in a certain plant are defective, so that they have to be inspected carefully before being shipped. Let us find the probability of obtaining at most 15 rejects (defectives) in a lot of 100 chosen for inspection. Assuming that the assumptions underlying the binomial distribution are met, the exact answer is given by

$$\sum_{x=0}^{15} b(x; 100, 0.20) = B(15; 100, 0.20)$$

Since Table 1 does not include values of n as large as 100, let us approximate this probability by using the normal distribution with the mean $\mu = np = 100(0.20) = 20$ and the variance $\sigma^2 = np(1 - p) = 100(0.20)(0.80) = 16$. Observe from Figure 4.3 that we are actually approximating the sum of the areas of the first 16 rectangles (cor-

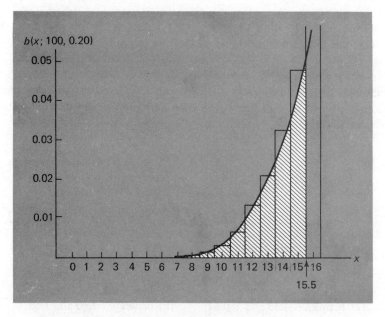

Figure 4.3. Normal approximation of binomial distribution.

responding to $x = 0, 1, 2, \ldots, 15$) of the histogram of the binomial distribution by means of the shaded area under a continuous curve. Thus, we must find the area under the approximating normal curve not between 0 and 15, but between -0.5 and 15.5 or, what is practically the same, between $-\infty$ and 15.5. Making this *continuity correction*, we obtain

$$B(15; 100, 0.20) \simeq F\left(\frac{15.5 - 20}{4}\right)$$

$$= F(-1.13) = 0.1292$$

which is close to the exact value $B(15; 100, 0.20) = 0.1285$ (obtained from the H. G. Romig table referred to in the Bibliography at the end of the book).

Had we been interested in the probability of getting exactly 15 defectives, we would make the *continuity correction* of representing 15 by means of the interval from 14.5 to 15.5, thus getting

$$b(15; 100, 0.20) \simeq F\left(\frac{15.5 - 20}{4}\right) - F\left(\frac{14.5 - 20}{4}\right)$$

$$= F(-1.13) - F(-1.38)$$

$$= 0.0454$$

This is fairly close to the exact value $b(15; 100, 0.20) = 0.0481$ (obtained again from the H. G. Romig table).

Note that whenever the normal distribution is used to approximate the binomial distribution, each x is replaced by the corresponding interval from $x - \frac{1}{2}$ to $x + \frac{1}{2}$; this is the *continuity correction* referred to in the above examples.

EXERCISES

1. If a random variable has the standard normal distribution, find the probability that it will take on a value
   (a) less than 1.50;
   (b) less than -1.20;
   (c) greater than 2.16;
   (d) greater than -1.75;
   (e) between 0 and 3.00;
   (f) between 1.22 and 2.35;
   (g) between -1.33 and -0.33;
   (h) between -1.60 and 1.80.

2. Find z if the probability that a random variable having the standard normal distribution will take on a value
   (a) less than z is 0.9911;
   (b) greater than z is 0.1093;
   (c) greater than z is 0.6443;
   (d) less than z is 0.0217;
   (e) between $-z$ and z is 0.9298.

3. If a random variable has a normal distribution, what is the probability that it will take on a value
   (a) within one standard deviation of the mean;
   (b) within two standard deviations of the mean;
   (c) within three standard deviations of the mean?

4. Given a random variable having the normal distribution with $\mu = 65$ and $\sigma = 20$, find the probabilities that it will take on a value
   (a) less than 53.2;   (b) greater than 70.0;
   (c) between 83.2 and 95.7;   (d) between 61.2 and 68.8;
   (e) between 35.6 and 45.6;
   (f) less than 58.6 or greater than 70.9.

5. Given a random variable having the normal distribution with the mean 16.2 and the variance 1.5625, find the probabilities that it will take on a value
   (a) greater than 16.8;   (b) less than 14.9;
   (c) between 13.6 and 18.8;   (d) between 16.5 and 16.7.

6. If a random variable has the binomial distribution with $n = 20$ and $p = 0.60$, use the normal curve approximation to determine
   (a) the probability that it will take on the value 14;
   (b) the probability that it will take on a value less than 12.
   Compare both results with the corresponding probabilities obtained with the use of Table 3.

7. A manufacturer knows that on the average 2 percent of the electric toasters which he makes will require repairs within 90 days after they are sold. Use the normal approximation of the binomial distribution to determine the probability that among 1,200 of these toasters at least 30 will require repairs within the first 90 days after they are sold.

8. The probability that an electronic component will fail in less than 1,000 hours of continuous use is 0.25. Use the normal approximation to find the probability that among 200 such components fewer than 45 will fail in less than 1,000 hours of continuous use.

9. A safety engineer feels that 30 percent of all industrial accidents in his plant are caused by failure of employees to follow instructions. If this figure is correct, find, approximately, the probability that among 84 industrial accidents in this plant anywhere from 20 to 30 (inclusive) will be due to failure of employees to follow instructions.

10. If $\frac{2}{3}$ of all clouds seeded with silver iodide show spectacular growth, what is

the probability that among 45 clouds thus seeded at most 25 will show spectacular growth?

11. The burning time of an experimental rocket is a random variable having the normal distribution with $\mu = 4.76$ seconds and $\sigma = 0.04$ second. What is the probability that this kind of rocket will burn
    (a) less than 4.66 seconds;
    (b) more than 4.80 seconds;
    (c) anywhere from 4.70 to 4.82 seconds?

12. In a photographic process, the developing time of prints may be looked upon as a random variable having the normal distribution with a mean of 16.28 seconds and a standard deviation of 0.12 second. Find the probability that it will take
    (a) anywhere from 16.00 to 16.50 seconds to develop one of the prints;
    (b) at least 16.20 seconds to develop one of the prints;
    (c) at most 16.35 seconds to develop one of the prints.

13. Specifications for a certain job call for washers with an inside diameter of 0.300 ± 0.005 inch. If the inside diameters of the washers supplied by a given manufacturer may be looked upon as a random variable having the normal distribution with $\mu = 0.302$ inch and $\sigma = 0.003$, what percentage of these washers will meet specifications?

14. Verify the claim made in the example on page 108, namely, that if the variability of the filling machine is reduced so that $\sigma = 0.025$ ounce, this will lower the required average amount of coffee to 4.05 ounces, yet keep 98 percent of the jars above four ounces.

15. A stamping machine produces can tops whose diameters are normally distributed with a standard deviation of 0.01 inch. At what "nominal" (mean) diameter should the machine be set so that no more than 5 percent of the can tops produced have diameters exceeding 3 inches?

16. Extruded plastic rods are automatically cut into nominal lengths of 6 inches. Actual lengths are normally distributed about a mean of 6 inches and their standard deviation is 0.06 inch.
    (a) What proportion of the rods exceeds tolerance limits of 5.9 inches to 6.1 inches?
    (b) To what value does the standard deviation need to be reduced if 99 percent of the rods must be within tolerance?

17. Verify the identity $F(-z) = 1 - F(z)$ given on page 106.

18. Verify that the parameter μ in the expression for the normal density on page 105 is, in fact, its mean.

19. Verify that the parameter σ^2 in the expression for the normal density on page 105 is, in fact, its variance.

20. Show that the normal density has a relative maximum at $x = \mu$ and inflection points at $x = \mu \pm \sigma$.

4.4   OTHER PROBABILITY DENSITIES

In the application of statistics to problems in engineering and physical science, we shall encounter many probability densities other than the normal distribution. Among these are the *t*, *F*, and chi-square distributions, the fundamental *sampling distributions* which we shall introduce in Chapter 6, and the exponential and Weibull distributions, which we shall apply to problems of reliability and life testing in Chapter 15. In the remainder of this chapter we shall discuss five continuous distributions, the *uniform distribution*, the *log-normal distribution*, the *gamma distribution*, the *beta distribution*, and the *Weibull distribution*, for the twofold purpose of giving further examples of probability densities and of laying the foundation for future applications.

4.5   THE UNIFORM DISTRIBUTION

The *uniform distribution* with the parameters α and β is defined by the equation

$$f(x) = \begin{cases} \dfrac{1}{\beta - \alpha} & \text{for } \alpha < x < \beta \\ 0 & \text{elsewhere} \end{cases}$$

and its graph is shown in Figure 4.4. Note that all values of x from α to β are "equally likely" in the sense that the probability that x lies in a

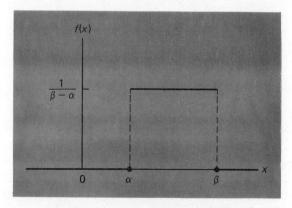

Figure 4.4.   Uniform density function.

narrow interval of width Δx entirely contained in the interval from α to β is equal to $\Delta x/(\beta - \alpha)$, regardless of the exact location of the interval.

Example. To illustrate how a physical situation might give rise to a uniform distribution, suppose that a wheel of a locomotive has the radius r and that x is the location of a point on its circumference measured along the circumference from some reference point 0. When the brakes are applied, some point will make sliding contact with the rail, and heavy wear will occur at that point. For repeated application of the brakes, it would seem reasonable to assume that x is a value of a random variable having the uniform distribution with $\alpha = 0$ and $\beta = 2\pi r$. If this assumption were incorrect, that is, if some set of points on the wheel made contact more often than others, the wheel would eventually exhibit "flat spots" or wear out of round.

To determine the mean and the variance of the uniform distribution, we first evaluate the two integrals

$$\mu'_1 = \int_\alpha^\beta x \cdot \frac{1}{\beta - \alpha} dx = \frac{\alpha + \beta}{2}$$

and

$$\mu'_2 = \int_\alpha^\beta x^2 \cdot \frac{1}{\beta - \alpha} dx = \frac{\alpha^2 + \alpha\beta + \beta^2}{3}$$

Thus,

$$\mu = \frac{\alpha + \beta}{2}$$

and, making use of the formula $\sigma^2 = \mu'_2 - \mu^2$, we find that

$$\sigma^2 = \tfrac{1}{12}(\beta - \alpha)^2$$

4.6 THE LOG-NORMAL DISTRIBUTION

The *log-normal distribution* occurs in practice whenever we encounter a random variable which is such that its logarithm has a normal distribution. This probability density function is given by

$$f(x) = \begin{cases} \dfrac{1}{\sqrt{2\pi}\,\beta}\,x^{-1}e^{-(\ln x-\alpha)^2/2\beta^2} & \text{for } x > 0,\ \beta > 0 \\ 0 & \text{elsewhere} \end{cases}$$

where "ln x" stands for the natural logarithm of x. A graph of the log-normal distribution with $\alpha = 0$ and $\beta = 1$ is shown in Figure 4.5, and it can be seen from the figure that this distribution is positively skewed.

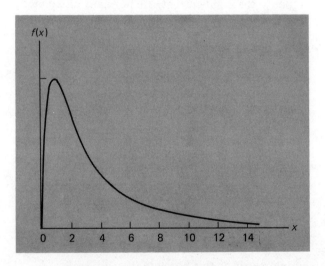

Figure 4.5. Log-normal density function.

To find the probability that a random variable having the log-normal distribution assumes a value between a and b ($0 < a < b$), we write

$$P(a \le x \le b) = \int_a^b \frac{1}{\sqrt{2\pi}\,\beta}\,x^{-1}e^{-(\ln x-\alpha)^2/2\beta^2}\,dx$$

Changing variables by letting $y = \ln x$ and, hence, $dy = x^{-1}\,dx$, we obtain

$$P(a \le x \le b) = \int_{\ln a}^{\ln b} \frac{1}{\sqrt{2\pi}\,\beta}\,e^{-(y-\alpha)^2/2\beta^2}\,dy$$

and it can be seen that this probability equals the probability that a random variable having the normal distribution with $\mu = \alpha$ and $\sigma = \beta$ assumes a value between ln a and ln b. Thus,

$$P(a \leq x \leq b) = F\left(\frac{\ln b - \alpha}{\beta}\right) - F\left(\frac{\ln a - \alpha}{\beta}\right)$$

where $F(z)$ is the probability that a random variable having the standard normal distribution assumes a value less than or equal to z.

Example. To illustrate the use of the log-normal distribution, suppose that a series of experiments have shown that the current gains of certain transistors (which are proportional to the logarithm of I_o/I_i, the ratio of the output to the input current) are normally distributed. If current gain is measured in units for which it equals ln (I_o/I_i), and if it is normally distributed with $\mu = 2$ and $\sigma^2 = 0.01$, then the probability that I_o/I_i, which thus has the log-normal distribution with $\alpha = 2$ and $\beta = 0.1$, will take on a value between 6.1 and 8.2 is given by

$$P\left(6.1 \leq \frac{I_o}{I_i} \leq 8.2\right) = F\left(\frac{\ln 8.2 - 2}{0.1}\right) - F\left(\frac{\ln 6.1 - 2}{0.1}\right)$$
$$= F(1.0) - F(-2.0)$$
$$= 0.8185$$

To find the mean of the *log-normal distribution*, we write

$$\mu = \frac{1}{\sqrt{2\pi}\beta} \int_0^\infty x \cdot x^{-1} e^{-(\ln x - \alpha)^2/2\beta^2} \, dx$$

and upon letting $y = \ln x$, this becomes

$$\mu = \frac{1}{\sqrt{2\pi}\beta} \int_{-\infty}^\infty e^y e^{-(y-\alpha)^2/2\beta^2} \, dy$$

This integral can be evaluated by completing the square on the exponent $y - (y - \alpha)^2/2\beta^2$, thus obtaining an integrand which has the form of a normal density. The final result, which the reader will be asked to verify in Exercise 3 on page 123, is

$$\mu = e^{\alpha + \beta^2/2}$$

Similar, but more lengthy, calculations yield

$$\sigma^2 = e^{2\alpha + \beta^2}(e^{\beta^2} - 1)$$

for the variance of the log-normal distribution.

Continuation of Example. For the distribution of I_o/I_i (the ratio of the output to the input current), we thus find that

$$\mu = e^{2 + (0.1)^2/2} = 7.4$$

and

$$\sigma^2 = e^{4 + (0.1)^2} e^{(0.1)^2} - 1 = 0.56$$

4.7 THE GAMMA DISTRIBUTION

Several important probability densities for which we shall discuss applications later on are special cases of the *gamma distribution*, whose equation is given by

$$f(x) = \begin{cases} \dfrac{1}{\beta^\alpha \Gamma(\alpha)} x^{\alpha - 1} e^{-x/\beta} & \text{for } x > 0, \alpha > 0, \beta > 0 \\ 0 & \text{elsewhere} \end{cases}$$

where $\Gamma(\alpha)$ is a value of the *gamma function*, defined by

$$\Gamma(\alpha) = \int_0^\infty x^{\alpha - 1} e^{-x} \, dx$$

Integration by parts shows that

$$\Gamma(\alpha) = (\alpha - 1)\Gamma(\alpha - 1)$$

for any $\alpha > 0$ and, hence, that $\Gamma(\alpha) = (\alpha - 1)!$ when α is a positive integer. Graphs of several gamma distributions are shown in Figure 4.6 and they exhibit the fact that these distributions are positively skewed. In fact, the skewness *decreases* as α *increases* for any fixed value of β.

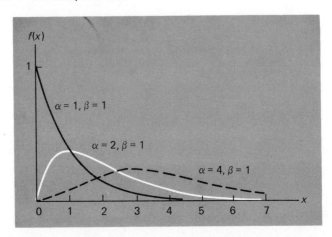

Figure 4.6. Gamma density functions.

The mean and the variance of the *gamma distribution* may be obtained by making use of the gamma function and its special properties mentioned above. For the mean we have

$$\mu = \frac{1}{\beta^\alpha \Gamma(\alpha)} \int_0^\infty x \cdot x^{\alpha-1} e^{-x/\beta} \, dx$$

and after letting $y = x/\beta$ and, hence, $dy = dx/\beta$, we get

$$\mu = \frac{\beta}{\Gamma(\alpha)} \int_0^\infty y^\alpha e^{-y} \, dy = \frac{\beta \Gamma(\alpha + 1)}{\Gamma(\alpha)}$$

Then, making use of the identity $\Gamma(\alpha + 1) = \alpha \cdot \Gamma(\alpha)$, we arrive at the result that

$$\mu = \alpha\beta$$

Using similar methods and the identity $\sigma^2 = \mu_2' - \mu^2$, it can also be shown that the variance of the gamma distribution is given by

$$\sigma^2 = \alpha\beta^2$$

In the special case where $\alpha = 1$, we get the *exponential density*, whose equation is thus

$$f(x) = \begin{cases} \dfrac{1}{\beta} e^{-x/\beta} & \text{for } x > 0, \beta > 0 \\ 0 & \text{elsewhere} \end{cases}$$

and whose mean and variance are $\mu = \beta$ and $\sigma^2 = \beta^2$. Note that the distribution of the example on page 100 was an exponential distribution with $\beta = \frac{1}{2}$.

The exponential distribution has many important applications; for instance, it can be shown that in connection with Poisson processes (see Section 3.7) the *waiting time* between successive arrivals (successes) have exponential distributions. More specifically, it can be shown that if in a Poisson process the *mean arrival rate* (average number of arrivals per unit time) is α, the time until the *first arrival*, or the waiting time between successive arrivals, has an exponential distribution with $\beta = \dfrac{1}{\alpha}$ (see Exercise 11 on page 124).

Example. With reference to the example on page 82, where on the average 3 trucks arrived per hour to be unloaded at a warehouse, we can now say that on the average there is a waiting time of $\frac{1}{3}$ of an hour, or 20 minutes, between the arrival of successive trucks. Also, the probability that the time between the arrival of successive trucks is less than 5 minutes is

$$\int_0^{1/12} 3e^{-3x} \, dx = 1 - e^{-1/4} = 0.221$$

and the probability that it will exceed 45 minutes is

$$\int_{3/4}^{\infty} 3e^{-3x} \, dx = e^{-9/4} = 0.105$$

4.8 THE BETA DISTRIBUTION

In Chapter 9 we shall need a probability density for a random variable which takes on values on the interval from 0 to 1, and most appropriate for this purpose is the *beta distribution*, whose equation is given by

$$f(x) = \begin{cases} \dfrac{\Gamma(\alpha + \beta)}{\Gamma(\alpha) \cdot \Gamma(\beta)} x^{\alpha-1}(1 - x)^{\beta-1} & \text{for } 0 < x < 1,\ \alpha > 0,\ \beta > 0 \\ 0 & \text{elsewhere} \end{cases}$$

The mean and the variance of this distribution are given by

$$\mu = \frac{\alpha}{\alpha + \beta} \quad \text{and} \quad \sigma^2 = \frac{\alpha\beta}{(\alpha + \beta)^2(\alpha + \beta + 1)}$$

Note that for $\alpha = 1$ and $\beta = 1$ we obtain a special case of the *uniform density* of Section 4.5, defined on the interval from 0 to 1. The following example, pertaining to a percentage or proportion, illustrates a typical application of the beta distribution:

Example. Suppose that in a certain county the *proportion* of highway sections that require repairs in any given year is a random variable having the beta distribution with $\alpha = 3$ and $\beta = 2$. The mean of this distribution, whose graph is shown in Figure 4.7, is $\mu = \dfrac{3}{3 + 2} = 0.60$, which means that on the average 60 percent of the highway sections require repairs in any given year. To find a probability connected with this distribution, let us determine, say, the probability that *at most 50 percent* of the highway sections will require repairs in a given year. Substituting $\alpha = 3$ and $\beta = 2$ into the formula for the beta distribution and making use of the fact that $\Gamma(5) = 4! = 24$, $\Gamma(3) = 2! = 2$, and $\Gamma(2) = 1! = 1$, we get

$$f(x) = \begin{cases} 12x^2(1 - x) & \text{for } 0 < x < 1 \\ 0 & \text{elsewhere} \end{cases}$$

Thus, the desired probability is given by

$$\int_0^{1/2} 12x^2(1 - x)\, dx = \tfrac{5}{16}$$

In general, probabilities relating to beta distributions are determined with the use of special tables.

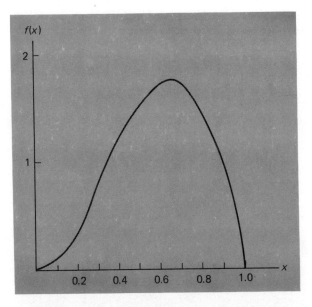

Figure 4.7.   Beta distribution with $\alpha = 3$ and $\beta = 2$.

4.9   THE WEIBULL DISTRIBUTION

Closely related to the exponential distribution is the *Weibull distribution*, whose probability density is given by

$$f(x) = \begin{cases} \alpha\beta x^{\beta-1}e^{-\alpha x^{\beta}} & \text{for } x > 0, \alpha > 0, \beta > 0 \\ 0 & \text{elsewhere} \end{cases}$$

To demonstrate this relationship, let us write the probability that a random variable having the Weibull distribution will take on a value less than a as

$$P(x < a) = \int_0^a \alpha\beta x^{\beta-1}e^{-\alpha x^{\beta}}\, dx$$

Then making the change of variable $y = x^{\beta}$ so that $dy = \beta x^{\beta-1}\, dx$, we get

$$P(x < a) = P(y < a^{\beta}) = \int_0^{a^{\beta}} \alpha e^{-\alpha y}\, dy = 1 - e^{-\alpha a^{\beta}}$$

and it can be seen that y is a value of a random variable having an expo-

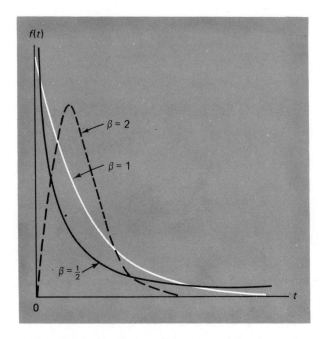

Figure 4.8. Weibull distributions with $\alpha = 1$ and $\beta = \frac{1}{2}$, 1, and 2.

nential density. The graphs of several Weibull distributions having $\alpha = 1$ and, respectively, $\beta = \frac{1}{2}$, $\beta = 1$, and $\beta = 2$ are shown in Figure 4.8.

The mean of the Weibull distribution having the parameters α and β may be obtained by evaluating the integral

$$\mu = \int_0^\infty x \cdot \alpha\beta x^{\beta-1} e^{-\alpha x^\beta}\, dx$$

Making the change of variable $u = \alpha x^\beta$, we get

$$\mu = \alpha^{-1/\beta} \int_0^\infty u^{1/\beta} e^{-u}\, du$$

and recognizing the integral as $\Gamma\left(1 + \dfrac{1}{\beta}\right)$, namely, as a value of the gamma function which we defined on page 117, we find that the mean of the Weibull distribution is given by

$$\mu = \alpha^{-1/\beta}\Gamma\left(1 + \frac{1}{\beta}\right)$$

Using a similar method to determine first μ_2', the reader will be asked to show in Exercise 18 on page 125 that the variance of this distribution is given by

$$\sigma^2 = \alpha^{-2/\beta}\left\{\Gamma\left(1 + \frac{2}{\beta}\right) - \left[\Gamma\left(1 + \frac{1}{\beta}\right)\right]^2\right\}$$

Example. Suppose that the lifetime of a certain kind of battery (in hours) may be looked upon as a random variable having the Weibull distribution with $\alpha = 0.1$ and $\beta = 0.5$. We thus find that such batteries last on the average

$$\mu = (0.1)^{-2}\Gamma(3) = 200 \text{ hours}$$

and if we wanted to know the probability that such a battery will last more than 300 hours, we would get

$$P(x > 300) = \int_{300}^{\infty} (0.05)x^{-0.5}e^{-0.1x^{0.5}}\,dx$$
$$= e^{-0.1(300)^{0.5}}$$
$$= 0.177$$

EXERCISES

1. Find the cumulative distribution function which corresponds to the uniform density.

2. From experience Mr. Harris has found that the low bid on a construction job can be regarded as a random variable having the uniform density

$$f(x) = \begin{cases} \dfrac{3}{4C} & \text{for } \dfrac{2C}{3} < x < 2C \\ 0 & \text{elsewhere} \end{cases}$$

where C is his own estimate of the cost of the job. What percentage should Mr. Harris add to his cost estimate when submitting bids to maximize his expected profit?

3. Verify the expression given on page 116 for the mean of the log-normal distribution.

4. If a random variable has the log-normal distribution with $\alpha = -1$ and $\beta = 2$, find
(a) the mean and the standard deviation of this distribution;
(b) the probability that this random variable will take on a value between 3.2 and 8.4;
(c) the probability that this random variable will take on a value greater than 5.0.

5. Verify the expression for the variance of the gamma distribution given on page 118.

6. Show that when $\alpha > 1$, the graph of the gamma density has a relative maximum at $x = \beta(\alpha - 1)$. What happens when $0 < \alpha < 1$ and when $\alpha = 1$?

7. If a random variable has the gamma distribution with $\alpha = 2$ and $\beta = 2$, find
(a) the mean and the standard deviation of this distribution;
(b) the probability that the random variable will take on a value less than 4.

8. In a certain city, the daily consumption of electric power (in millions of kilowatt-hours) can be treated as a random variable having a gamma distribution with $\alpha = 3$ and $\beta = 2$. If the power plant of this city has a daily capacity of 12 million kilowatt-hours, what is the probability that this power supply will be inadequate on any given day?

9. If n salesmen are employed in a door-to-door selling campaign, the gross sales volume in thousands of dollars may be regarded as a random variable having the gamma distribution with $\alpha = 100\sqrt{n}$ and $\beta = \frac{1}{2}$. If the sales costs are $5,000 per salesman, how many salesmen should be employed to maximize the profit?

10. If the number of days a watch will run without having to be reset is a random variable having the exponential distribution with $\beta = 50$, find the probability that
(a) such a watch will have to be reset in less than 20 days;
(b) such a watch will not have to be reset in at least 60 days.

11. Given a Poisson process with on the average α arrivals per unit time, find
(a) the probability that there will be no arrivals during a time interval of length t;
(b) the probability that the waiting time between arrivals will be at least of length t.
Having thus obtained the distribution function for the waiting time between successive arrivals, find
(c) the corresponding probability density by differentiating with respect to t.

12. With reference to Exercise 8 on page 87, find the percent of the time that the interval between breakdowns of the computer will
(a) be less than one week;
(b) exceed 5 weeks.

13. With reference to Exercise 10 on page 87, find the probability that the time between successive calls arriving at the switchboard of the consulting firm will
    (a) be less than half a minute;
    (b) exceed 3 minutes.

14. Verify for $\alpha = 3$ and $\beta = 3$ that the integral of the beta density, from 0 to 1, is equal to 1.

15. Show that when $\alpha > 1$ and $\beta > 1$, the beta density has a relative maximum at

$$x = \frac{\alpha - 1}{\alpha + \beta - 2}$$

16. Suppose that the proportion of defectives shipped by a vendor, which varies somewhat from shipment to shipment, may be looked upon as a random variable having the beta distribution with $\alpha = 1$ and $\beta = 4$.
    (a) Find the mean of this beta distribution, namely, the average proportion of defectives in a shipment from this vendor.
    (b) Find the probability that a shipment from this vendor will contain 25 percent or more defectives.

17. If the annual proportion of erroneous income tax returns filed with the I.R.S. can be looked upon as a random variable having a beta distribution with $\alpha = 2$ and $\beta = 9$, what is the probability that in any given year there will be fewer than 10 percent erroneous returns?

18. Verify the formula for the variance of the Weibull distribution given on page 123.

19. Suppose that the time to failure (in minutes) of certain electronic components subjected to continuous vibrations may be looked upon as a random variable having the Weibull distribution with $\alpha = \frac{1}{5}$ and $\beta = \frac{1}{3}$.
    (a) How long can such a component be expected to last?
    (b) What is the probability that such a component will fail in less than 5 hours.

20. Referring to the example on page 123, find the probability that such a battery will not last 100 hours.

4.10 JOINT PROBABILITY DENSITIES

There are many situations in which we describe the outcome by giving the values of *several* random variables. For instance, we may measure the weight and the hardness of a rock, the volume, pressure, and temperature of a gas, or the thickness, color, compressive strength, and potassium content of a piece of glass. If $x_1, x_2, \ldots, x_k$ are the values of k random variables, we shall refer to a function $\mathbf{f}$ with values $f(x_1, x_2, \ldots, x_k)$ as the *joint probability density* of these random variables if the probability that $a_1 \leq x_1 \leq b_1$, $a_2 \leq x_2 \leq b_2$, ..., and $a_k \leq x_k \leq b_k$ is given by the multiple integral

$$\int_{a_1}^{b_1} \int_{a_2}^{b_2} \cdots \int_{a_k}^{b_k} f(x_1, x_2, \ldots, x_k) \, dx_1 \, dx_2 \ldots dx_k$$

Thus, not every function **f** with values $f(x_1, x_2, \ldots, x_k)$ can serve as a joint probability density, but if we impose the conditions that:

1. $f(x_1, x_2, \ldots, x_k) \geq 0$ *for all values of* $x_1, x_2, \ldots, x_k$ *for which the joint probability density is defined*

2. $\int_{-\infty}^{\infty} \int_{-\infty}^{\infty} \cdots \int_{-\infty}^{\infty} f(x_1, x_2, \ldots, x_k) \, dx_1 \, dx_2 \ldots dx_k = 1$

it can be shown that the axioms of probability (with the modifications discussed in Section 4.1) are satisfied.

To extend the concept of a cumulative distribution function to the k-variable case, we write as $F(x_1, x_2, \ldots, x_k)$ the probability that the first random variable will take on a value less than or equal to x_1, the second random variable will take on a value less than or equal to x_2, ..., and the kth random variable will take on a value less than or equal to x_k, and we refer to the corresponding function **F** as the *joint distribution function* of the k random variables.

Example. To illustrate the various concepts we have introduced, let us consider the joint probability density given by

$$f(x_1, x_2) = \begin{cases} 6e^{-2x_1 - 3x_2} & \text{for } x_1 > 0, \, x_2 > 0 \\ 0 & \text{elsewhere} \end{cases}$$

Integrating between the appropriate limits, we thus find that the probability of the first random variable taking on a value between 1 and 2 and the second taking on a value between 2 and 3 is given by

$$\int_1^2 \int_2^3 6e^{-2x_1 - 3x_2} \, dx = (e^{-2} - e^{-4})(e^{-6} - e^{-9})$$

$$= 0.0003$$

and that the probability of the first random variable taking on a value less than 2 and the second taking on a value greater than 2 is given by

$$\int_0^2 \int_2^\infty 6e^{-2x_1-3x_2}\, dx = (1 - e^{-4})e^{-6}$$
$$= 0.0025$$

Also, by definition the joint distribution function of the two random variables is given by

$$F(x_1, x_2) = \begin{cases} \int_0^{x_1} \int_0^{x_2} 6e^{-2u-3v}\, du\, dv & \text{for } x_1 \geq 0, x_2 \geq 0 \\ 0 & \text{elsewhere} \end{cases}$$

or

$$F(x_1, x_2) = \begin{cases} (1 - e^{-2x_1})(1 - e^{-3x_2}) & \text{for } x_1 \geq 0, x_2 \geq 0 \\ 0 & \text{elsewhere} \end{cases}$$

and the probability that both random variables will take on a value less than 1, for example, is given by

$$F(1, 1) = (1 - e^{-2})(1 - e^{-3})$$
$$= 0.8216$$

Given the joint probability density of k random variables, the probability density of the ith random variables can be obtained by integrating out the other variables; symbolically, we thus write

$$f_i(x_i) = \int_{-\infty}^\infty \cdots \int_{-\infty}^\infty \int_{-\infty}^\infty \cdots \int_{-\infty}^\infty f(x_1, x_2, \ldots, x_k)\, dx_1 \ldots dx_{i-1}\, dx_{i+1} \ldots dx_k$$

and, in this context, the function $\mathbf{f}_i$ is called the *marginal density* of the ith random variable. Integrating out only some of the k random variables, we can similarly define *joint marginal densities* of any two, three, or more of the k random variables.

Continuation of Example. For instance, in our example the marginal density of the first random variable is given by

$$f_1(x_1) = \begin{cases} \int_0^\infty 6e^{-2x_1-3x_2}\, dx_2 & \text{for } x_1 > 0 \\ 0 & \text{elsewhere} \end{cases}$$

or

$$f_1(x_1) = \begin{cases} 2e^{-2x_1} & \text{for } x_1 > 0 \\ 0 & \text{elsewhere} \end{cases}$$

To explain what we mean by the *independence* of continuous random variables, we could proceed as in Section 2.5 and define *conditional densities* first; however, it will be easier to say that

---

k random variables are independent if and only if

$$F(x_1, x_2, \ldots, x_k) = F_1(x_1) \cdot F_2(x_2) \cdot \ldots \cdot F_k(x_k)$$

for all values of these random variables for which the respective functions are defined.

In this notation, $F(x_1, x_2, \ldots, x_k)$ is, as before, a value of the joint distribution function of the k random variables, while $F_i(x_i)$ for $i = 1, 2, \ldots, k$ are the corresponding values of the individual distribution functions of the respective random variables.

Continuation of Example. In our example we have

$$F(x_1, x_2) = \begin{cases} (1 - e^{-2x_1})(1 - e^{-3x_2}) & \text{for } x_1 > 0 \text{ and } x_2 > 0 \\ 0 & \text{elsewhere} \end{cases}$$

$$F_1(x_1) = \begin{cases} 1 - e^{-2x_1} & \text{for } x_1 > 0 \\ 0 & \text{elswehere} \end{cases}$$

$$F_2(x_2) = \begin{cases} 1 - e^{-3x_2} & \text{for } x_2 > 0 \\ 0 & \text{elsewhere} \end{cases}$$

and it can be seen that the two random variables of our example *are* independent.

Analogous to the special multiplication law for independent events and its extension on page 36, it follows immediately from our definition that *if k random variables are independent, any value of their joint probability density equals the product of the corresponding values of the marginal densities of the k random variables*; symbolically,

---

$$f(x_1, x_2, \ldots, x_k) = f_1(x_1) \cdot f_2(x_2) \cdot \ldots \cdot f_k(x_k)$$

Given two random variables with values x_1 and x_2, we shall define the *conditional probability density* of the first given that the second takes on the value x_2 as

$$g_1(x_1 \mid x_2) = \frac{f(x_1, x_2)}{f_2(x_2)} \qquad f_2(x_2) \neq 0$$

where $f(x_1, x_2)$ and $f_2(x_2)$ are, as before, values of the joint density of the two random variables and the marginal density of the second. Note that this definition parallels that of conditional probabilities on page 34.

Example. If two random variables have the joint probability density

$$f(x_1, x_2) = \begin{cases} \frac{2}{3}(x_1 + 2x_2) & \text{for } 0 < x_1 < 1, 0 < x_2 < 1 \\ 0 & \text{elsewhere} \end{cases}$$

the marginal density of the second is given by

$$f_2(x_2) = \int_0^1 \tfrac{2}{3}(x_1 + 2x_2)\, dx_1 = \tfrac{1}{3}(1 + 4x_2) \qquad \text{for } 0 < x_2 < 1$$

and $f_2(x_2) = 0$ elsewhere. Hence, the conditional density of the first given that the second takes on the value x_2 is given by

$$g_1(x_1 \mid x_2) = \frac{\tfrac{2}{3}(x_1 + 2x_2)}{\tfrac{1}{3}(1 + 4x_2)} = \frac{2x_1 + 4x_2}{1 + 4x_2} \qquad \text{for } 0 < x_1 < 1$$

and $g_1(x_1 \mid x_2) = 0$ elsewhere. We should add that this conditional density holds only for $0 < x_2 < 1$.

The following is another concept which is of importance when dealing with k random variables. If a random variable takes on the value $g(x_1, x_2, \ldots, x_k)$ whenever k random variables take on the values $x_1, x_2, \ldots, x_k$, then the *expected*, or *average* value of the first random variable (namely, the mean of its distribution) is given by

$$\int_{-\infty}^{\infty} \int_{-\infty}^{\infty} \ldots \int_{-\infty}^{\infty} g(x_1, \ldots, x_k) f(x_1, \ldots, x_k)\, dx_1\, dx_2 \ldots dx_k$$

where $f(x_1, \ldots, x_k)$ is, as before, a value of the joint density of the k random variables.

Example. For instance, referring to the joint density of the example on page 126, we find that the expected value of the *product* of the two random variables is

$$\int_0^\infty \int_0^\infty x_1 x_2 \cdot 6e^{-2x_1 - 3x_2} \, dx_1 \, dx_2 = \tfrac{1}{6}$$

EXERCISES

1. If two random variables have the joint density

$$f(x_1, x_2) = \begin{cases} x_1 x_2 & \text{for } 0 < x_1 < 1, 0 < x_2 < 2 \\ 0 & \text{elsewhere} \end{cases}$$

find
   (a) the probability that both random variables will taken on values less than 1;
   (b) the probability that the *sum* of the values taken on by the two random variables will be less than 1;
   (c) the joint distribution function;
   (d) both marginal densities.
   Also check whether the two random variables are independent.

2. If two random variables have the joint density

$$f(x, y) = \begin{cases} \tfrac{6}{5}(x + y^2) & \text{for } 0 < x < 1, 0 < y < 1 \\ 0 & \text{elsewhere} \end{cases}$$

find
   (a) the probability that $0.2 < x < 0.5$ and $0.4 < y < 0.6$;
   (b) the joint distribution function;
   (c) both marginal densities;
   (d) the probability that $y < 0.5$, using the result of (c);
   (e) an expression for $f_1(x \mid y)$ for $0 < y < 1$;
   (f) an expression for $f_1(x \mid \tfrac{1}{2})$ and the mean of the conditional density of the first random variable when the second takes on the value $y = \tfrac{1}{2}$.

3. With reference to the example on page 129, find
   (a) an expression for the conditional density of the first random variable when the second takes on the value $x_2 = 0.25$;
   (b) an expression for the conditional density of the second random variable when the first takes on the value x_1.

4. If three random variables have the joint density

$$f(x, y, z) = \begin{cases} k(x + y)e^{-z} & \text{for } 0 < x < 1, 0 < y < 2, z > 0 \\ 0 & \text{elsewhere} \end{cases}$$

(a) find k;

(b) find the probability that $x < y$ and $z > 1$;

(c) check whether the three random variables are independent;

(d) check whether any two of the three random variables are *pairwise* independent.

5. A pair of random variables has the "circular normal distribution" if their joint density is given by

$$f(x_1, x_2) = \frac{1}{2\pi\sigma^2} e^{-\frac{1}{2\sigma^2}[(x_1-\mu_1)^2 + (x_2-\mu_2)^2]}$$

for $-\infty < x_1 < \infty$ and $-\infty < x_2 < \infty$.

(a) If $\mu_1 = 2$ and $\mu_2 = -2$, and $\sigma = 10$, use Table 3 to find the probability that $-8 < x_1 < 14$ and $-9 < x_2 < 3$.

(b) If $\mu_1 = \mu_2 = 0$ and $\sigma = 3$, find the probability that (x_1, x_2) is contained in the region between the two circles $x_1^2 + x_2^2 = 9$ and $x_1^2 + x_2^2 = 36$.

6. A bomb aimed at a point target has a lethal radius of 500 feet. Using the target as the origin of a rectangular system of coordinates, assume that the coordinates (x, y) of the point of impact are values of a pair of random variables having the circular normal distribution (see Exercise 5) with $\mu_1 = \mu_2 = 0$ and $\sigma = 300$ feet. What is the probability that the target will be destroyed?

7. Referring to the random variables of Exercise 1, find the expected value of the random variable whose values are given by $g(x_1, x_2) = x_1 + x_2$.

8. Referring to the random variables of Exercise 2, find the expected value of the random variable whose values are given by $g(x, y) = x^2 y$.

9. If measurements of the length and the width of a rectangle have the joint density

$$f(x, y) = \begin{cases} \dfrac{1}{ab} & \text{for } L - \dfrac{a}{2} < x < L + \dfrac{a}{2}, \ W - \dfrac{b}{2} < y < W + \dfrac{b}{2} \\ 0 & \text{elsewhere} \end{cases}$$

find the mean and the variance of the corresponding distribution of the area of the rectangle.

10. Establish a relationship between $g_1(x_1 | x_2)$, $g_2(x_2 | x_1)$, $f_1(x_1)$, and $f_2(x_2)$.

4.11 SIMULATION

To simulate the observation of continuous random variables with the use of random numbers, we refer to their *cumulative distribution functions*, given either graphically or as tables. (The theory on which this method is based involves the so-called *probability integral transformation*, which is treated in all of the textbooks on mathematical statistics referred to in the Bibliography at the end of the book.)

To illustrate the graphical method, suppose we want to observe the values of a random variable having a normal distribution with a given mean and standard deviation. Taking two- or three-digit random numbers (depending on the accuracy that can be attained) and treating them as decimal fractions by putting a decimal point to the left of the first digit, we refer to a graph of the cumulative distribution function of the *standard* normal distribution (see, for instance, Figure 4.9), mark the decimal fractions corresponding to the random digits on the $F(z)$ scale, and then read the corresponding values of a random variable having the standard normal distribution off the horizontal scale. Finally, we obtained the simulated values of the original random variable by means of the equation $z = \dfrac{x - \mu}{\sigma}$ (see page 106) or $x = \mu + \sigma z$.

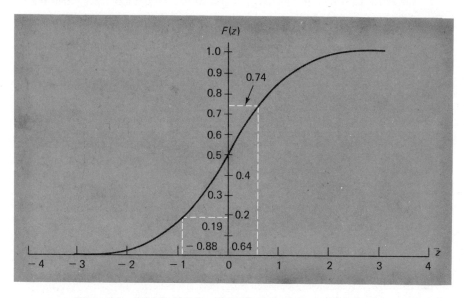

Figure 4.9. Distribution function of standard normal distribution.

Example. Suppose we want to simulate observations of a random variable having the normal distribution with $\mu = 18.3$ and $\sigma = 20$. Using the 3rd and 4th columns of the first page of Table 7, starting with the 36th row, we obtain the two-digit random numbers 74, 19, 78, 82, 06, 11, ... Treating the first of these as the decimal fraction

0.74 and marking it on the vertical scale of Figure 4.9, we find that the corresponding value of z is 0.64 and, hence, that the simulated value of the original random variable is

$$x = 18.3 + 20(0.64) = 31.1$$

Similarly, treating the random number 19 as a decimal fraction and marking 0.19 on the vertical scale of Figure 4.9, we find that the corresponding value of z is -0.88 and, hence, that the simulated value of the original random variable is

$$x = 18.3 + 20(-0.88) = -0.7$$

In Exercise 1 on page 134 the reader will be asked to verify that the next four random digits, namely, 78, 82, 06, and 11 yield $x = 33.7$, $x = 36.7$, $x = -12.7$, and $x = -6.3$.

The graphical method we have described has the advantage that it can be used to simulate *any* continuous distribution, but it is generally very tedious as it requires that we carefully draw the graph of the cumulative distribution function. This can be avoided if a suitable table of cumulative probabilities is available. For instance, to simulate the observation of values of a random variable having the standard normal distribution (and, hence, any normal distribution), we refer to Table 3 and proceed as follows: given a two- or three-digit random number, we change it, as before, into a *random decimal* between 0 and 1. Then, if the random decimal is greater than or equal to 0.5000, we simply look for the value of z for which the entry in Table 3 is closest to the random decimal; if the random decimal is less than 0.5000, we look for the value of z for which the entry in Table 3 is closest to 1 *minus* the random decimal and record $-z$ as the simulated observation.

Continuation of Example. Let us use this method to rework the example on page 132, where we simulated observations of a random variable having the normal distribution with $\mu = 18.3$ and $\sigma = 20$, using the two-digit random numbers 74, 19, 78, 82, 06, 11, ... Checking 0.7400 in Table 3, we find that the closest entry corresponds to $z = 0.64$, and, proceeding as before, we get the simulated observation

$$x = \mu + \sigma z = 18.3 + 20(0.64) = 31.1$$

Then, checking $1 - 0.1900 = 0.8100$ in Table 3, we find that the closest entry corresponds to 0.88, and, proceeding as on page 133 with $z = -0.88$, we get the simulated observation

$$x = 18.3 + 20(-0.88) = -0.7$$

EXERCISES

1. Continue the example in the text by using
   (a) the diagram of Figure 4.9,
   (b) Table 3,
   to find the simulated values of the random variable which correspond to the random numbers 78, 82, 06, and 11.

2. Suppose that the number of hours it takes a person to learn how to operate a certain machine is a random variable having a normal distribution with $\mu = 5.8$ and $\sigma = 1.2$. Simulate the length of time it takes eight persons to learn how to operate the machine using
   (a) the diagram of Figure 4.9;
   (b) Table 3.

3. Suppose that the increase in the pulse rate of a person performing a certain task is a random variable whose distribution can be approximated closely with a normal distribution having the mean 28.6 and the standard deviation 5.15. Simulate the increase in the pulse rate of 20 persons performing this task, using Table 3 and rounding the answers to the nearest positive integer.

4. Suppose that the durability of a paint (in years) is a random variable having an exponential density (see page 119) with the mean $\beta = 2$.
   (a) Draw a graph of the distribution function of this random variable.
   (b) Use the graph obtained in (a) to simulate an experiment in which the paint is applied to 12 houses and it is observed in each case how long the paint lasts (that is, how long it takes until there are certain agreed-upon signs of deterioration).

5. Suppose that the occurrence of breakdowns of a radar tracking system is a Poisson process with $\lambda = 0.02$ per hour, and the time required to repair any breakdown is fixed at 10 hours. Simulate 1,000 hours of operation of this system and estimate the proportion of "down time." [*Hint*: Simulate the waiting times between successive breakdowns using an appropriate exponential distribution.]

6. Since many simulation experiments entail sampling from normal populations, statisticians have constructed special tables of so-called *random normal deviates* (see Bibliography at the end of the book), which are looked upon as values of a random variable having the standard normal distribution.

(a) Using the method directly based on Table 3, construct a table of 100 such random normal deviates.

(b) Use the table obtained in (a) to simulate 20 observations of a random variable having a normal distribution with $\mu = 14.25$ and $\sigma = 2.16$.

7. Suppose that the annual number of collision-insurance claims in a given automobile insurance policy class has the log-normal distribution with $\alpha = 1.5$ and $\beta = 0.6$. Simulate 10 years of experience with this policy class.

5

Treatment
of Data

5.1 FREQUENCY DISTRIBUTIONS

Statistical data, obtained from surveys, experiments, or any series of measurements, are often so numerous that they are virtually useless unless they are condensed, or reduced, into a more suitable form. Thus, the first step of a statistical analysis often consists of the construction of a *frequency table*, that is, a table which divides the data into a relatively small number of classes (categories), listing the number of observations belonging to each class. A frequency table, or a *frequency distribution* as it is also called, sacrifices some of the information contained in a set of data; instead of knowing the exact value of each item, we only know that it belongs in a certain class. On the other hand, this kind of grouping often brings out important features of the data, and the gain in "legibility" of the data usually more than compensates for the loss of information. In what follows, we shall consider mainly *numerical* frequency distributions, that is, frequency distributions where the data are grouped according to

136

their numerical size. If the data are grouped according to some quality, or attribute, we shall refer to such a distribution as a *categorical distribution*.

The first step in constructing a frequency table consists of *deciding how many classes to use and choosing the limits for each class*. Generally speaking, the number of classes we use depends on the number of observations, but it is seldom profitable to use fewer than five or more than fifteen classes. Among other things, we base this decision on the *range* of the data, that is, the difference between the largest observation and the smallest. Then, we tally the observations (much like the tally of votes in an election or, for that matter, any kind of count) and, thus, determine the *class frequencies*, namely, the number of observations in each class.

Example. To illustrate the construction of a frequency table, let us consider the following 80 determinations of the daily emission of sulfur oxides (in tons) of an industrial plant:

15.8	26.4	17.3	11.2	23.9	24.8	18.7	13.9	9.0	13.2
22.7	9.8	6.2	14.7	17.5	26.1	12.8	28.6	17.6	23.7
26.8	22.7	18.0	20.5	11.0	20.9	15.5	19.4	16.7	10.7
19.1	15.2	22.9	26.6	20.4	21.4	19.2	21.6	16.9	19.0
18.5	23.0	24.6	20.1	16.2	18.0	7.7	13.5	23.5	14.5
14.4	29.6	19.4	17.0	20.8	24.3	22.5	24.6	18.4	18.1
8.3	21.9	12.3	22.3	13.3	11.8	19.3	20.0	25.7	31.8
25.9	10.5	15.9	27.5	18.1	17.9	9.4	24.1	20.1	28.5

Since the largest observation is 31.8, the smallest is 6.2, and the range is 25.6, we might choose the *six* classes having the limits 5.0–9.9, 10.0–14.9, . . . , 30.0–34.9, we might choose the *seven* classes 5.0–8.9, 9.0–12.9, . . . , 29.0–32.9, or we might choose the *nine* classes 5.0–7.9, 8.0–10.9, . . . , 29.0–31.9. Note that in each case *the classes do not overlap, they will accommodate all of the data*, and *they are all of the same size*.

Deciding upon the second of these classifications, we now tally the 80 observations and obtain the results shown in the following table:

Class limits	Tally	Frequency
5.0– 8.9	/ / /	3
9.0–12.9	ℳℳ ℳℳ	10
13.0–16.9	ℳℳ ℳℳ / / / /	14
17.0–20.9	ℳℳ ℳℳ ℳℳ ℳℳ ℳℳ	25
21.0–24.9	ℳℳ ℳℳ ℳℳ / /	17
25.0–28.9	ℳℳ / / / /	9
29.0–32.9	/ /	2
	Total	80

Note that the class limits we used here contain *as many decimals as the original data*. If the original data had been given to two decimals we would have used the class limits 5.00–8.99, 9.00–12.99, . . . , 29.00–32.99, and if they had been rounded to the nearest ton we would have used the class limits 5–8, 9–12, . . . , 29–32.

In this example, the actual daily emissions of sulfur oxides can be thought of as values of a continuous random variable, but if we used classes such as 5.0–9.0, 9.0–13.0, . . . , 29.0–33.0, we would run into difficulties: 9.0 could go into the first class or into the second, and 17.0 could go into the third class or into the fourth. To avoid this difficulty, we can let the first class go from 4.95 to 8.95, the second from 8.95 to 12.95, . . . , and the last from 28.95 to 32.95. We refer to these new limits as the *class boundaries*, and it should be noted that there will be no ambiguities even though the boundaries overlap. These class boundaries are, so to speak, "impossible" values, since the data were given to only one decimal. In practice, we use the class boundaries rather than the original class limits chiefly when it is essential to stress the fact that the data are values assumed by a continuous random variable (or that they are otherwise continuous kinds of measurements).

As we have pointed out earlier, once data are grouped, each observation in a given class has lost its identity in the sense that its exact value is no longer known. This leads to some difficulties when we want to give further descriptions of the data, but we get around this by representing each observation in a class by its midpoint, called the *class mark*. In general, the class marks of a frequency distribution are obtained by averaging successive class limits or successive class boundaries. If the classes are all of equal length, as in our example, we refer to the common interval between any successive class marks as the *class interval*. Note that the class interval can also be obtained from the difference of successive class boundaries, but *not* from the difference of successive class limits. If the classes are *not* all of equal length, we cannot speak of *the* class interval (of the distribution), but we can refer to the difference between the boundaries of a given class as its *length*.

Continuation of Example. For the distribution of the sulfur dioxide emission data, the class marks are 6.95, 10.95, 14.95, 18.95, 22.95, 26.95, and 30.95, and the class interval is 4.

There are several alternate forms of distributions into which data are sometimes grouped. Foremost among these are the "less than" or "more than" *cumulative frequency distributions*. A cumulative "less than" distribution shows the total number of observations that are less than given values, and it is easily constructed from the corresponding (ordinary) frequency table. We do not know how many observations are less than the individual class marks, but we do know how many are less than each class boundary or each lower class limit.

Continuation of Example. Using the class boundaries of the distribution of the sulfur oxides emission data, we get

Tons of sulfur oxides	Cumulative frequency
less than 4.95	0
less than 8.95	3
less than 12.95	13
less than 16.95	27
less than 20.95	52
less than 24.95	69
less than 28.95	78
less than 32.95	80

where the cumulative frequencies were obtained by adding the frequencies, one by one, starting from the top of the frequency table. Note that instead of "less than 4.95" we could also have written "less than 5.0" or "4.9 or less."

Cumulative "more than" distributions are constructed, similarly, by adding the frequencies, one by one, starting at the bottom of the frequency table. In practice, "less than" cumulative distributions are used more widely, and it is not uncommon to refer to cumulative "less than" distributions simply as cumulative distributions.

If it is desired to compare two or more frequency distributions whose total frequencies are not equal, it may be necessary (or at least advantageous) to convert them into so-called *percentage distributions*. We simply divide each class frequency by the total number of observations in the distribution and multiply by 100; in this way we indicate what percentage of the data falls into each class of a distribution. The same principle can also be applied to cumulative distributions, by replacing each cumulative frequency by the corresponding cumulative percentage.

Example. As a further illustration let us construct a frequency distribution as well as a percentage distribution for the following data representing the number of workers absent from a factory on 50 working days:

13	5	13	37	10	16	2	11	6	12
8	21	12	11	7	7	9	16	49	18
3	11	19	6	15	10	14	10	7	24
11	3	6	10	4	6	32	9	12	7
29	12	9	19	8	20	15	5	17	10

Since these data consist of integers, there will be no ambiguities if we use the class limits 0–4, 5–9, 10–14, ..., and we thus obtain the following frequency and percentage distributions:

Number of absences	Frequency	Number of absences	Percentage
0– 4	4	0– 4	8
5– 9	15	5– 9	30
10–14	16	10–14	32
15–19	8	15–19	16
20–24	3	20–24	6
25 or more	4	25 or more	8

Note that in these tables we replaced the five classes 25–29, 30–34, 35–39, 40–44, and 45–49, which would ordinarily have been needed, by a single class having no upper limit. This device of using an *open class* is thus used (especially in connection with skewed distributions) to eliminate the trouble of having to list a large number of classes that are either empty or have very small frequencies.

Note also that when distributions of data are presented in their final form, the tally is practically always omitted, and we show only the class limits, the class boundaries, or the class marks, and, of course, the corresponding class frequencies, percentages, cumulative frequencies, or cumulative percentages.

5.2   GRAPHS OF FREQUENCY DISTRIBUTIONS

Many important properties of frequency distributions, such as their symmetry or skewness, the number of their modes (maxima), and so on, are best exhibited by means of graphs. In this section we shall introduce some of the most widely used forms of graphical presentation of frequency distributions, percentage distributions, and cumulative distributions.

Referring to the frequency table of the sulfur oxides emission data on page 137, we might present the information it contains graphically by first plotting the points (x_i, f_i), where x_i is the class mark of the ith class and f_i is the corresponding frequency, and then drawing straight-line segments to connect successive points. The resulting *frequency polygon* is shown in Figure 5.1; it should be noted that we added classes with zero frequencies at both ends of the distribution in order to "tie down" the graph to the horizontal axis. It is apparent from Figure 5.1 that the distribution of the sulfur oxides emission data is nearly symmetrical and bell-shaped, somewhat resembling the graph of the normal distribution shown in Figure 4.2.

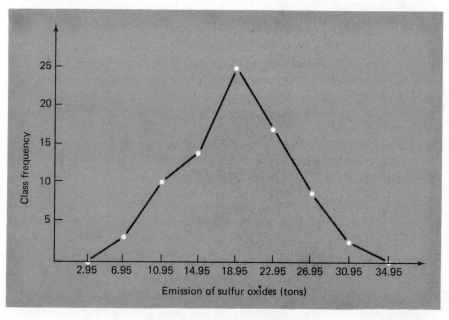

Figure 5.1. Frequency polygon.

An alternate way of presenting a frequency distribution in graphical form is the *histogram*, first mentioned on page 53 in connection with probability distributions. The histogram of a frequency distribution is constructed like that of a probability distribution; the heights of the rectangles represent the class frequencies and the bases of the rectangles extend between successive class boundaries. A histogram of the sulfur oxides emission data is shown in Figure 5.2.

It is sometimes preferable to look upon the areas of the rectangles, rather than their heights, as representing the class frequencies of a histogram. This applies in particular to situations where we wish to approximate histograms with smooth curves or where there are classes of unequal length (see Exercise 17 on page 149).

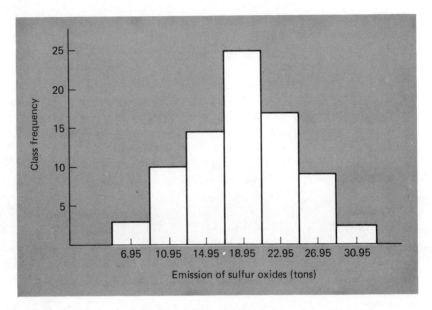

Figure 5.2.   Histogram.

Inspection of the graph of a frequency distribution often brings out features that are not immediately apparent from the data themselves. Aside from the fact that such a graph presents a good overall picture of the data, it can also emphasize irregularities and unusual features. For instance, outlying observations which somehow do not fit the overall picture, that is, the overall pattern of the data, may be due to errors of measurement, equipment failure, and similar causes. Also, the fact that a histogram or a frequency polygon exhibits two or more *modes* (maxima) can provide pertinent information. The appearance of two modes may imply, for example, a shift in the process that is being measured, or it may imply that the data come from several sources. With some experience one learns to spot such irregularities or anomalies, and an experienced engineer would find it just as surprising if the histogram of a distribution of integrated-circuit failure times were symmetrical as if a distribution of American men's hat sizes were bimodal.

Cumulative distributions are usually represented graphically by means of *ogives*. An ogive is similar to a frequency polygon, except that we plot the cumulative frequencies at the class boundaries instead of the ordinary frequencies at the class marks. The resulting points are again connected by means of straight-line segments, as shown in Figure 5.3, which represents the cumulative "less than" distribution of the sulfur oxides emission data.

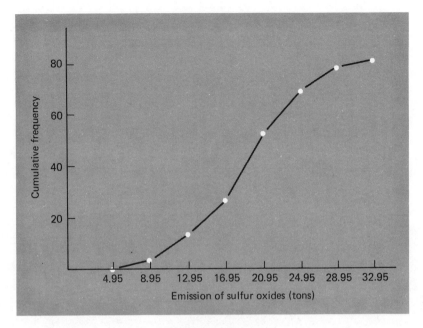

Figure 5.3. Ogive.

Ogives are nearly always S-shaped, with the relative size of the two tails of the S determined by the symmetry or lack of symmetry of the distribution. If a distribution follows closely the pattern of a normal curve, it is possible to "straighten out" the S by using a special vertical scale, called a *probability scale*. This scale is so designed that any cumulative normal distribution will graph as a straight line. Special graph paper, called *arithmetic probability graph paper*, is commercially available; it has one arithmetic (ordinary) scale and one probability scale (see Figure 5.4).

Example. To illustrate, the cumulative "less than" percentage distribution of the sulfur oxides emission data has been plotted on probability graph paper in Figure 5.4. Note that the cumulative percentages are plotted at the corresponding class boundaries. As can be seen from this diagram, the points all lie close to the dashed line, and we can say that the data are approximately normally distributed.

To check for "normality," probability graph paper can also be used directly with ungrouped data. In that case we first arrange the observations according to size, and if there are n observations we plot $100 \dfrac{(i - \frac{1}{2})}{n}$ or

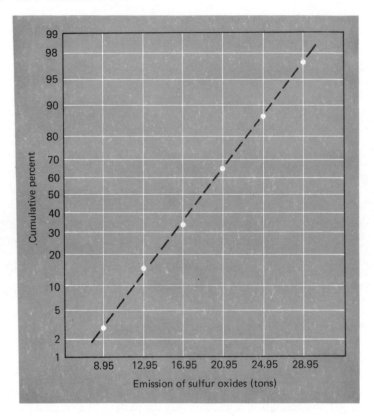

Figure 5.4.   Normal probability graph — grouped data.

$\dfrac{(100i - 50)}{n}$ percent on the vertical scale corresponding to the ith largest observation (see Exercise 16 on page 148). If n is large, it is not necessary to plot every point; every fifth or tenth point will generally suffice to check whether the distribution of the data follows closely the pattern of a normal distribution.

   Continuation of Example.   Referring again to the sulfur dioxides emission data on page 137 (arranged according to size), a probability graph plotted for every tenth observation, and also the smallest observation, is shown in Figure 5.5. Again, it is apparent that the points fall close to a straight line and, hence, that the data are approximately normally distributed.

   In view of the symmetry of the normal distribution, the mean of the normal distribution which approximates a given distribution can be obtain-

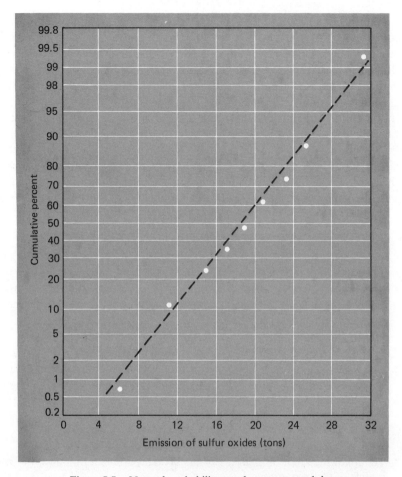

Figure 5.5. Normal probability graph — ungrouped data.

ed by reading off the horizontal scale the value on the line (fitted to the points) which corresponds to 50 percent on the vertical scale. Similarly, the standard deviation of the distribution can be approximated by taking the difference between the values on the horizontal scale which correspond to 84 percent and 50 percent on the vertical scale (or *half* the difference of the values which correspond to 84 percent and 16 percent on the vertical scale).

Continuation of Example. Thus, it can be seen from Figures 5.4 or 5.5 that the mean of the normal distribution which approximates the distribution of the sulfur oxides emission data is about 18.5, and that its standard deviation is about $24 - 18.5 = 5.5$.

It must be understood that the use of probability graph paper, which should really be called *normal* probability graph paper, is only an approximate (and highly subjective) device for checking whether a distribution follows the pattern of a normal curve. Only large and obvious departures from linearity in such a graph are real evidence that the data do not follow the pattern of a normal curve.

Sometimes other special scales are used to graph cumulative distributions. For example, if it is suspected that a set of data follows the pattern of a log-normal distribution, probability graph paper with a logarithmic scale can be used. Such paper is also commercially available. Various other scales have been devised for checking whether data follow the patterns of corresponding theoretical distributions.

EXERCISES

1. Measurements of the ignition temperature of a gas vary from 1,161 to 1,319 degrees Fahrenheit. Construct a table with eight equal classes into which these data might be grouped. Give
   (a) the class limits;   (b) the class marks;
   (c) the class boundaries;   (d) the class interval.

2. The weights of certain mineral specimens, given to the nearest tenth of an ounce, are grouped into a table having the classes 10.5–11.4, 11.5–12.4, 12.5–13.4, and 13.5–14.4 ounces. Find
   (a) the class marks;
   (b) the class boundaries;
   (c) the class interval.

3. The following are measurements of the breaking strength (in ounces) of a sample of 60 linen threads:

32.5	15.2	35.4	21.3	28.4	26.9	34.6	29.3	24.5	31.0
21.2	28.3	27.1	25.0	32.7	29.5	30.2	23.9	23.0	26.4
27.3	33.7	29.4	21.9	29.3	17.3	29.0	36.8	29.2	23.5
20.6	29.5	21.8	37.5	33.5	29.6	26.8	28.7	34.8	18.6
25.4	34.1	27.5	29.6	22.2	22.7	31.3	33.2	37.0	28.3
36.9	24.6	28.9	24.8	28.1	25.4	34.5	23.6	38.4	24.0

   Group these measurements into a distribution having the classes 15.0–19.9, 20.0–24.9, . . . , 35.0–39.9 and construct a histogram.

4. Convert the distribution of Exercise 3 into a cumulative "less than" distribution and graph its ogive.

5. Convert the cumulative distribution of Exercise 4 into a cumulative percentage distribution and, plotting its graph on arithmetic probability paper,

check whether it is reasonable to approximate the distribution of the original data with a normal distribution.

6. The following are the ignition times of certain upholstery materials exposed to a flame, given to the nearest hundredth of a second:

2.58	2.51	4.04	6.43	1.58	4.32	2.20	4.19
4.79	6.20	1.52	1.38	3.87	4.54	5.12	5.15
5.50	5.92	4.56	2.46	6.90	1.47	2.11	2.32
6.75	5.84	8.80	7.40	4.72	3.62	2.46	8.75
2.65	7.86	4.71	6.25	9.45	12.80	1.42	1.92
7.60	8.79	5.92	9.65	5.09	4.11	6.37	5.40
11.25	3.90	5.33	8.64	7.41	7.95	10.60	3.81
3.78	3.75	3.10	6.43	1.70	6.40	3.24	1.79
4.90	3.49	6.77	5.62	9.70	5.11	4.50	2.50
5.21	1.76	9.20	1.20	6.85	2.80	7.35	11.75

Group these figures into a table with a suitable number of equal classes and construct a frequency polygon.

7. Convert the distribution of Exercise 6 into a cumulative "less than" percentage distribution and plot its ogive.

8. Plot the cumulative percentage distribution of Exercise 7 on arithmetic as well as logarithmic probability graph paper. What do these graphs suggest about the form of the distribution? (If logarithmic probability paper is not available, plot the logarithms of the class boundaries on arithmetic probability paper.)

9. The following are 15 measurements of the boiling point of a silicon compound (in degrees Centigrade): 166, 141, 136, 153, 170, 162, 155, 146, 183, 157, 148, 132, 160, 175, and 150. Plot the data directly on arithmetic probability paper and indicate whether they follow the pattern of a normal distribution.

10. In a two-week study of the productivity of workers, the following data were obtained on the total number of acceptable pieces which 100 workers produced:

65	36	49	84	79	56	28	43	67	36
43	78	37	40	68	72	55	62	22	82
88	50	60	56	57	46	39	57	73	65
59	48	76	74	70	51	40	75	56	45
35	62	52	63	32	80	64	53	74	34
76	60	48	55	51	54	45	44	35	51
21	35	61	45	33	61	77	60	85	68
45	53	34	67	42	69	52	68	52	47
62	65	55	61	73	50	53	59	41	54
41	74	82	58	26	35	47	50	38	70

Group these data into a suitable frequency table and plot its histogram.

11. Convert the distribution obtained in Exercise 10 into a cumulative "less than" percentage distribution and plot its ogive.

12. Plot the cumulative percentage distribution of Exercise 11 on arithmetic probability paper and interpret the result.

13. The following are the number of automobile accidents which occurred at 60 major intersections in a certain city during the Fourth of July weekend:

```
0  2  5  0  1  4  1  0  2  1
5  0  1  3  0  0  2  1  3  1
1  4  0  2  4  1  2  4  0  4
3  5  0  1  3  6  4  2  0  2
0  2  3  0  4  2  5  1  1  2
2  1  6  5  0  3  3  0  0  4
```

(a) Group these data into a frequency distribution showing how often each of the numbers occurred.

(b) Construct a cumulative "or more" distribution and draw its ogive.

14. Categorical frequency distributions are often represented graphically by means of *pie charts* in which a circle is divided into sectors proportional in size to the frequencies with which the data are distributed among the categories. Construct a pie chart to represent the following distribution:

	Number of B.A. degrees in engineering conferred in the U.S. in 1970–71
Aeronautical engineering	2,443
Chemical engineering	3,579
Civil engineering	6,526
Electrical engineering	12,198
Mechanical engineering	8,858
Industrial engineering	3,171
General engineering	2,864

15. The *pictogram* of Figure 5.6 is intended to illustrate the fact that in a certain area, average family income has doubled from $5,000 in 1955 to $10,000 in 1970. Does this pictogram convey a "fair" impression of the actual increase in family income? If not, state how it should be modified.

16. Given a set of observations $x_1, x_2, \ldots$, and x_n, we define their *empirical cumulative distribution* as the function whose values $F(x)$ equal the proportion of the observations less than or equal to x.

(a) Graph the empirical cumulative distribution for the 15 measurements of Exercise 9.

(b) Verify that the method of curve fitting used on page 144 (the check for "normality" for ungrouped data) amounts to plotting points corresponding to the *midpoints* of the steps of the empirical cumulative distribution.

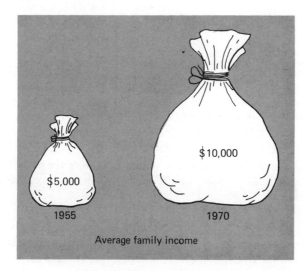

Figure 5.6. Pictogram for Exercise 15.

17. Convert the distribution of the sulfur oxides emission data on page 137 into a distribution having the classes 5.0–8.9, 9.0–20.9, 21.0–28.9, and 29.0–32.9. Draw two histograms of this distribution, one in which the class frequencies are given by the *heights* of the rectangles and one in which the class frequencies are given by the *areas* of the rectangles. Explain why the first of these histograms gives a very misleading picture.

5.3 DESCRIPTIVE MEASURES

In Section 3.4 we introduced the mean and the variance of a probability distribution as parameters which measure, respectively, its center and its spread. In this section we shall define corresponding measures to describe a set of data or its distribution.

Given a set of n measurements or observations, $x_1, x_2, \ldots, x_n$, there are several ways in which we can describe their center (middle, or central location). Foremost among these are the *arithmetic mean* and the *median*, although other kinds of "averages" are sometimes used for special purposes. The arithmetic mean or, more succinctly, the *mean* is defined by the formula

$$\bar{x} = \frac{\sum\limits_{i=1}^{n} x_i}{n}$$

Note that we have written the mean of the x's as $\bar{x}$ and not as μ, the symbol we used for the mean of a probability distribution (or probability density). In this connection, we shall follow the general practice of using Latin letters to denote descriptions of actual data and Greek letters to denote descriptions of theoretical distributions. It is also of interest to note that the above formula would yield the mean of the distribution of a random variable which takes on the values x_i with equal probabilities of $\frac{1}{n}$.

Sometimes it is preferable to use the *median* as a descriptive measure of the center, or location, of a set of data. This is true, particularly, if it is desired to reduce all calculations to a minimum or if it is desired to eliminate the effect of extreme (very large or very small) values. The median of n observations $x_1, x_2, \ldots, x_n$ can be defined loosely as the "middlemost" value once the data are arranged according to size. More precisely, if the observations are arranged according to size and n is an odd number, the median is the value of the observation numbered $\frac{n+1}{2}$; if n is an even number, the median is defined as the mean (average) of the observations numbered $\frac{n}{2}$ and $\frac{n+2}{2}$.

Example. For instance, the median of the five observations 15, 14, 2, 27, 13 is 14 (the value of the third largest observation), and the median of the six observations 17, 9, 15, 19, 4, 16 is 15.5 (the mean of the third and fourth largest observations).

Although the mean and the median each provides a single number to represent an entire set of data, the mean is usually preferred in problems of estimation and other problems of statistical inference. An intuitive reason for preferring the mean is that the median does not utilize all the information contained in the observations. A related reason is that the median is generally subject to greater chance fluctuations, that is, it is apt to vary more from sample to sample. This important concept of "sampling variability" will be explored in detail in Chapter 6.

Example. To give an example where the median actually gives a *better* description than the mean, suppose an employer claims that the mean 'salary paid to engineers in his firm is $24,000. This gives the impression that this firm is a good place to work. However on further examination it turns out that it is a small company which employs four young engineers at $12,000 each, plus the owner whose income is $72,000. Thus, the income distribution is highly skewed; its mean of $24,000 represents nothing useful, while the median income of $12,000 is at least representative of what a young engineer can expect to earn with this firm.

Like the variance of a probability distribution, the *variance* of n observations $x_1, x_2, \ldots, x_n$ measures essentially the average of their squared deviations from their mean, $\bar{x}$, and it is defined by the formula

$$s^2 = \frac{\sum\limits_{i=1}^{n} (x_i - \bar{x})^2}{n - 1}$$

There are several reasons for using the divisor $n - 1$ instead of n. First, only $n - 1$ of the deviations from the mean, $x_i - \bar{x}$, are *independent*, since their sum is always equal to zero (see Exercise 1 on page 156). In other words, $n - 1$ of the deviations from the mean automatically determine the nth. Another reason is that if we look upon the x's as values assumed by some random variable, division by $n - 1$ in the formula for s^2 makes this variance a *better* estimate of σ^2, the variance of the distribution of this random variable. We shall say more about this in Chapters 7 and 8.

Consistent with the terminology of Chapters 3 and 4, we define the *standard deviation* of n observations $x_1, x_2, \ldots, x_n$ as the square root of their variance, namely, as

$$s = \sqrt{\frac{\sum\limits_{i=1}^{n} (x_i - \bar{x})^2}{n - 1}}$$

Note that if we had divided by n instead of by $n - 1$, the resulting formula could have been used for the standard deviation of the distribution of a random variable which assumes the values x_i with equal probabilities of $1/n$.

The standard deviation and the variance are measures of *absolute variation*, that is, they measure the actual amount of variation present in a set of data, and they are dependent on the scale of measurement. To compare the variation in several sets of data, it is generally desirable to use measures of *relative variation;* for this purpose we use the so-called *coefficient of variation*

$$CV = \frac{s}{\bar{x}} \cdot 100$$

Note that this measure, which gives the standard deviation as a percentage of the mean, is independent of the scale of measurement.

Example. If a set of data has the mean $\bar{x} = 12.0$ and the standard deviation $s = 1.5$, the coefficient of variation is $CV = 12.5$ percent; in other words, the standard deviation is 12.5 percent of the mean.

In this section, we have limited the discussion to the mean, the median, the variance, and the standard deviation. There are, of course, many other ways of describing sets of data. For instance, if we want to determine a value below which we find the lowest 25 percent of data, we calculate the *first quartile* Q_1; if we want to determine a value above which we find the highest 5 percent of the data, we calculate the *ninety-fifth percentile* P_{95}; and so on. For grouped data, the calculation of such *fractiles* is similar to that of the median. There are also alternate ways of describing the variability of a set of data. For instance, there is the *sample range* (the largest value minus the smallest, which we shall have the occasion to use is Section 8.1), the *mean deviation*, and the *interquartile range*. As the need arises, new ways and new methods of describing statistical data are constantly being developed.

5.4   THE CALCULATION OF $\bar{x}$ AND s

In this section we shall discuss methods for calculating $\bar{x}$ and s for *raw* (ungrouped) as well as grouped data. The methods we shall employ are particularly well-suited for small calculators, and they are both rapid and accurate.

The calculation of $\bar{x}$ for *ungrouped data* does not pose any problems; we have only to add the values of the observations and divide by n. On the other hand, the calculation of s^2 is usually too cumbersome if we directly use the formula on page 151. Instead, we shall use the algebraically equivalent form

$$s^2 = \frac{n \cdot \sum\limits_{i=1}^{n} x_i^2 - \left(\sum\limits_{i=1}^{n} x_i\right)^2}{n(n-1)}$$

which has the dual advantage of requiring less labor and giving better accuracy. (In Exercise 5 on page 157 the reader will be asked to show that this formula is equivalent to the one on page 151.) The decrease in labor results from the fact that we do not actually have to calculate the deviations from the mean; the increase in accuracy is due to the reduction of the number of divisions and subtractions, and their postponement until the last two steps in the calculation.

Example. To illustrate the calculation of $\bar{x}$ and s let us find the mean and the standard deviation of the following miles per gallon obtained in 20 test runs performed with an imported sub-compact car:

$$\begin{array}{ccccc}
19.7 & 21.5 & 22.5 & 22.2 & 22.6 \\
21.9 & 20.5 & 19.3 & 19.9 & 21.7 \\
22.8 & 23.2 & 21.4 & 20.8 & 19.4 \\
22.0 & 23.0 & 21.1 & 20.9 & 21.3
\end{array}$$

Using a calculator, we find that the sum of these figures is 427.7 and that the sum of their squares is 9,173.19. Consequently,

$$\bar{x} = \frac{427.7}{20} = 21.38$$

and

$$s^2 = \frac{20(9,173.19) - (427.7)^2}{20.19} = 1.412$$

and it follows that $s = 1.19$. In computing the necessary sums we usually retain all decimal places, but as in this example, at the end we usually round to one more decimal than we had in the original data.

The calculations we have illustrated in this example can be made even simpler by first *coding* the observations; that is, by performing a suitable linear transformation.

Continuation of Example. If we multiply each of the mileages by 10 to eliminate decimal points and then (arbitrarily) subtract 200 to make the numbers easier to work with, we now have the coded values

$$
\begin{array}{rrrrr}
-3 & 15 & 25 & 22 & 26 \\
19 & 5 & -7 & -1 & 17 \\
28 & 32 & 14 & 8 & -6 \\
20 & 30 & 11 & 9 & 13
\end{array}
$$

which we shall refer to as u's. Calculating the mean and the standard deviation of the u's by the method that we used before, we obtain 277 for the sum of the u's, 6,519 for the sum of their squares, and, hence,

$$\bar{u} = 13.8 \quad \text{and} \quad s_u = 11.9$$

In order to *undo* the coding, let us observe that the x's and the u's are related by the equation

$$u_i = 10x_i - 200 \quad \text{or} \quad x_i = 0.1u_i + 20$$

Continuing from here, simple algebra will show that $\bar{x} = 0.1\bar{u} + 20$ and $s_x = 0.1s_u$, where s_x and s_u are, respectively, the standard deviations of the x's and the u's. Thus, we get

$$\bar{x} = 0.1(13.8) + 20 = 21.38 \quad \text{and} \quad s_x = 0.1(11.9) = 1.19$$

which agrees with the values previously obtained.

In Exercise 6 on page 157, the reader will be asked to verify that, in general, when data are coded so that

$$x_i = c \cdot u_i + a$$

the corresponding formulas for the mean and the standard deviation are

$$\boxed{\bar{x} = c \cdot \bar{u} + a \quad \text{and} \quad s_x = c \cdot s_u}$$

To calculate $\bar{x}$ and s from *grouped data*, we shall need to make some assumption about the distribution of the values within each class. If we represent all values within a class by the corresponding class mark, the sum of the x's and the sum of their squares can now be written

$$\sum_{i=1}^{k} x_i f_i \quad \text{and} \quad \sum_{i=1}^{k} x_i^2 f_i$$

where x_i is the class mark of the ith class, f_i is the corresponding class frequency, and k is the number of classes in the distribution. Substituting these sums into the computing formulas for $\bar{x}$ and s^2, we obtain

$$\bar{x} = \frac{\sum\limits_{i=1}^{k} x_i f_i}{n}$$

$$s^2 = \frac{n \cdot \sum\limits_{i=1}^{k} x_i^2 f_i - \left(\sum\limits_{i=1}^{k} x_i f_i\right)^2}{n(n-1)}$$

Example. To illustrate, let us return to the frequency table of the sulfur oxides emission data on page 137. Recording the class marks x_i in the first column, the class frequencies f_i in the second column, and the products $x_i f_i$ and $x_i^2 f_i$ in the third and fourth columns, we obtain

x_i	f_i	$x_i f_i$	$x_i^2 f_i$
6.95	3	20.85	144.9075
10.95	10	109.50	1,199.0250
14.95	14	209.30	3,129.0350
18.95	25	473.75	8,977.5625
22.95	17	390.15	8,953.9425
26.95	9	242.55	6,536.7225
30.95	2	61.90	1,915.8050
	80	1,508.00	30,857.0000

Substituting $n = 80$ and the necessary sums into the formulas for $\bar{x}$ and s, we obtain

$$\bar{x} = \frac{1,508}{80} = 18.85$$

and

$$s^2 = \frac{80(30,857) - (1,508)^2}{80 \cdot 79} = 30.77$$

These calculations were very tedious, and we went through them mainly to impress the reader with the simplifications that can be attained by *coding*, in this case coding the class marks. The coding we use in connection with distributions is to represent the class marks with successive integers, preferably with 0 near the center of the distribution or near the class which has the highest frequency.

Continuation of Example. Coding the class marks of our distribution as $-3, -2, -1, 0, 1, 2, 3$ and referring to these coded class marks as u's, we find that the above computing table reduced to

u_i	f_i	$u_i f_i$	$u_i^2 f_i$
-3	3	-9	27
-2	10	-20	40
-1	14	-14	14
0	25	0	0
1	17	17	17
2	9	18	36
3	2	6	18
	80	-2	152

and the mean and the variance of the u's are

$$\bar{u} = \frac{-2}{80} = -0.025$$

and

$$s_u^2 = \frac{80(152) - (-2)^2}{80 \cdot 79} = 1.923$$

Since the coding we used is such that $x_i = 4u_i + 18.95$, the two formulas on page 154 yield $\bar{x} = 4(-0.025) + 18.95 = 18.85$ and $s_x^2 = 4^2(1.923) = 30.77$, which agree with the results previously obtained.

In general, if c is the class interval of a distribution and x_0 is the class mark to which we assign 0 in the new scale, the coding is given by $x_i = c \cdot u_i + x_0$ and, hence, the formulas for the mean and the standard deviation are

$$\bar{x} = c \cdot \bar{u} + x_0 \quad and \quad s_x = c \cdot s_u$$

EXERCISES

1. Show that $\sum_{i=1}^{n} (x_i - \bar{x}) = 0$.

2. The following are the number of twists that were required to break 12 forged alloy bars: 33, 24, 39, 48, 26, 35, 38, 54, 23, 34, 29, and 37. Find

(a) the mean;

(b) the median;

(c) s using the formula on page 151;

(d) s using the computing formula on page 153.

3. The following are the I.Q.'s of 15 persons selected for jury duty by a court: 108, 97, 89, 111, 127, 103, 92, 88, 110, 94, 118, 96, 103, 107, and 112.

(a) Find the mean and comment on a lawyer's claim that "the average juror has an I.Q. over 100."

(b) Find s and determine the percentage of these I.Q.'s that are within one standard deviation on either side of the mean.

4. The following are 12 temperature readings taken at various locations in a large kiln (in degrees Fahrenheit): 475, 500, 460, 425, 460, 410, 470, 475, 460, 510, 450, 415.

(a) Find the mean.

(b) Find the variance.

(c) Find the coefficient of variation.

5. Show that the computing formula for s^2 on page 153 is *equivalent* to the one used to define it on page 151.

6. If data are coded so that $x_i = cu_i + a$ (see page 154), show that

$$\bar{x} = c \cdot \bar{u} + a \quad \text{and} \quad s_x = c \cdot s_u$$

7. Rework Exercise 4 by using the coding $u = \dfrac{x}{5} - 90$ and compare the results.

8. Calculate $\bar{x}$ and s for the data of Exercise 9 on page 147

(a) without coding;

(b) after suitably coding the 15 measurements.

9. Referring to the example on page 139, find the mean and the standard deviation of the data on absenteeism

(a) by using the raw (ungrouped) data;

(b) by using the frequency distribution obtained on page 140.

10. Find the mean and the variance of the accident data of Exercise 13 on page 148

(a) by using the ungrouped data;

(b) by using the frequency distribution obtained in that exercise.

11. Use the distribution obtained in Exercise 3 on page 146 to find the mean and the standard deviation of the breaking strengths.

12. Use the distribution obtained in Exercise 6 on page 147 to find the mean and the variance of the ignition times. Also determine the coefficient of variation.

13. To find the *median* of a distribution obtained for n observations, we first determine the class into which the median must fall. Then, if there are j

values in this class and k values below it, the median is located $\dfrac{(n/2) - k}{j}$ of the way into this class, and to obtain the median we multiply this fraction by the class interval and add the result to the lower class boundary of the class into which the median must fall. This method is based on the assumption that the observations in each class are "spread uniformly" throughout the class interval, and this is why we count $\dfrac{n}{2}$ of the observations instead of $\dfrac{n+1}{2}$ as on page 150. To illustrate, let us refer to the distribution of the sulfur oxides emission data on page 139. Since $n = 80$, it can be seen that the median must fall into the class 17.0–20.95, and since $j = 25$ and $k = 27$, it follows that the median is $16.95 + \dfrac{40 - 27}{25} \cdot 4 = 19.03$.

(a) Find the median of the distribution of the absenteeism data obtained on page 140.

(b) Find the median of the distribution of breaking strengths obtained in Exercise 3 on page 146.

(c) Find the median of the distribution of ignition times obtained in Exercise 6 on page 147.

14. If k sets of data consist, respectively, of $n_1, n_2, \ldots, n_k$ observations and have the means $\bar{x}_1, \bar{x}_2, \ldots, \bar{x}_k$, then the overall mean of all the data is given by the formula

$$\bar{x} = \frac{\displaystyle\sum_{i=1}^{k} n_i \bar{x}_i}{\displaystyle\sum_{i=1}^{k} n_i}$$

(a) The average annual salaries paid to top-level management in three companies are \$54,000, \$62,000, and \$59,000. If the respective numbers of top-level executives in these companies are 4, 15, and 11, find the average salary paid to these 30 executives.

(b) If the mean weight of the 4 offensive starting backs of a professional football team is 214 pounds and the mean weight of the 7 offensive starting linemen is 258 pounds, what is the mean weight of this starting eleven?

(c) Prove this formula for the overall mean of several sets of data.

15. The formula of Exercise 14 is a special case of the following formula for computing *weighted means:*

$$\bar{x}_w = \frac{\displaystyle\sum_{i=1}^{k} w_i x_i}{\displaystyle\sum_{i=1}^{k} w_i}$$

where w_i is a weight indicating the relative importance of the ith observation.

(a) If an instructor counts the final examination in a course four times as much as each one-hour examination, what is the weighted average grade of a student who received grades of 69, 75, 56, and 72 in four one-hour examinations and a final examination grade of 78?

(b) If someone invests $4,000 at 5.5 percent in a savings account, $10,000 at 7.25 percent in a certificate of deposit, and $18,000 at 9 percent in second mortgages, find the average percentage yield of these investments.

6

Sampling
Distributions

6.1 POPULATIONS AND SAMPLES

Usage of the term "population" in statistics is a carry-over from the days
when statistics was applied mainly to sociological and economic phenomena. Nowadays, it is applied to any set or collection of objects, actual
or conceptual, and mainly to sets of numbers, measurements, or observations. For example, if we are interested in determining the average number
of television sets per household in the United States, the totality of these
figures, one for each household, constitutes the population for this study.
Similarly, the population from which inspectors draw a sample to determine some quality characteristic of a manufactured product may be the
corresponding measurements for all units in a given lot; depending on the
objectives of the inspection, it may also consist of the corresponding
measurements for *all* units that may conceivably be manufactured.

In some cases, such as the above example concerning the number of television sets per household, the population is *finite;* in other cases, such as the determination of some characteristic of all units, past, present, and future, that might conceivably be manufactured by a given process, it is convenient to think of the population as *infinite*. Similarly, we look upon the results obtained in a series of flips of a coin as a sample from the hypothetically *infinite* population consisting of all conceivably possible flips of the coin.

Populations are often described by the distributions of their values, and it is common practice to refer to a population in terms of this distribution. (For finite populations, we are referring here to the actual distribution of its values; for infinite populations, we are referring to the corresponding probability distribution or probability density.) For example, we may refer to a number of flips of a coin as a sample from a "binomial population" or to certain measurements as a sample from a "normal popuation." Hereafter, when referring to a "population $f(x)$" we shall mean a population such that its elements have a frequency distribution, a probability distribution, or a density with values given by $f(x)$.

If a population is infinite it is impossible to observe all its values, and even if it is finite it may be impractical or uneconomical to observe it in its entirety. Thus, it is usually necessary to use a sample, a part of a population, and infer from it results pertaining to the entire population. Clearly, such results can be useful only if the sample is in some way "representative" of the population. It would be unreasonable, for instance, to expect useful generalizations about the population of 1974 family incomes in the United States on the basis of data pertaining to home owners only. Similarly, we can hardly expect reasonable generalizations about the performance of a tire if it is tested only on smooth roads. To assure that a sample is representative of the population from which it is obtained, and to provide a framework for the application of probability theory to problems of sampling, we shall confine the use of the term "sample" to so-called *random samples*. For sampling from finite populations, random samples are defined as follows:

A set of observations $x_1, x_2, \ldots, x_n$ constitutes a random sample of size n from a finite population of size N, if it is chosen so that each subset of n of the N elements of the population has the same probability of being selected.

Note that this definition of randomness pertains essentially to the manner in which the sample values are selected. This holds also for the following definition of a random sample from an infinite population:

A set of observations $x_1, x_2, \ldots, x_n$ *constitutes a random sample of size n from the infinite population* $f(x)$ *if:*

1. Each x_i *is a value of a random variable whose distribution is given by* $f(x)$.

2. These n random variables are independent.

There are several ways of assuring the selection of a sample that is at least approximately random. When dealing with a finite population, we can serially number the elements of the population and then select a sample with the aid of a table of random digits (see discussion on page 92).

Example. For instance, if a population nas $N = 500$ elements and we wish to select a random sample of size $n = 10$, we can use three arbitrarily selected columns of Table 7 to obtain 10 different three-digit numbers less than or equal to 500, which will then serve as the serial numbers of the elements to be included in the sample.

When the population size is large, the use of random numbers can become very laborious and at times practically impossible. For instance, if a sample of five cartons of canned peaches is to be chosen for inspection from among the many thousands stored in a warehouse, one can hardly expect to number all the cartons, make a selection with the use of random numbers, and then pull out the ones that were chosen. In a situation like this, one really has very little choice but to make the selection relatively haphazard, hoping that this will not seriously violate the assumption of randomness which is basic to most statistical theory.

When dealing with infinite populations, the situation is somewhat different since we cannot physically number the elements of the population; but efforts should be made to approach conditions of randomness by the use of artificial devices. For example, in selecting a sample from a production line we may be able to approximate conditions of randomness by choosing one unit each half hour; when tossing a coin we can try to flip it in such a way that neither side is intentionally favored, and so forth. The proper use of artificial or mechanical devices for selecting random samples is always preferable to human judgment, as it is extremely difficult to avoid unconscious biases when making almost any kind of selection.

Even with the careful choice of artificial devices, it is all too easy to commit gross errors in the selection of a random sample. To illustrate some of these pitfalls, suppose we have the task of selecting logs being fed into a sawmill by a constant-speed conveyer belt, for the purpose of obtaining a random sample of their lengths. One sampling device, which at first sight would seem to assure randomness, consists of measuring the logs which pass a given point at the end of a certain number of ten-minute intervals. However, further thought reveals that this method of selection favors the longer logs, since they require more time to pass the given point. Thus, the sample is not random since the longer logs have a better chance of being included. Another common mistake in selecting a sample is that of sampling from the wrong population or from a poorly specified population. As we have pointed out earlier, we would hardly get a sample from which we could generalize about family incomes in the United States if we limited our sample to home owners. Similarly, if we wanted to determine the effect of vibrations on a structural member, we should be careful to delineate the frequency band of vibrations that is of relevance, and to vibrate test specimens only at frequencies selected randomly from this band.

The purpose of most statistical investigations is to generalize from information contained in random samples about the population from which the samples were obtained. In particular, we are usually concerned with the problem of making inferences about the *parameters* of populations, such as the mean μ and the standard deviation σ. In making such inferences, we usually use *statistics* such as $\bar{x}$ and s, namely, quantities calculated on the basis of sample observations. Since the selection of a random sample is controlled largely by chance, so are the values we obtain for these statistics. The remainder of this chapter will be devoted to a discussion of *sampling distributions*—distributions which describe the chance fluctuations of statistics calculated on the basis of random samples.

6.2   THE SAMPLING DISTRIBUTION OF THE MEAN (σ KNOWN)

Suppose a random sample of n observations has been taken from some population and that $\bar{x}$ has been computed, say, to estimate the mean of the population. It should be clear that if we took a second random sample of size n from this population, it would be quite unreasonable to expect the identical value for $\bar{x}$, and if we took several more samples, probably no two of the $\bar{x}$'s would be alike. The differences among such $\bar{x}$'s are generally attributed to chance, and this raises important questions concerning their distribution, specifically concerning the extent of their chance fluctuations. To approach this question experimentally, let us consider the experiment described in the following example:

Example. Suppose that 50 random samples of size $n = 10$ are to be taken from a population having the *discrete uniform distribution*

$$f(x) = \begin{cases} \frac{1}{10} & \text{for } x = 0, 1, 2, \ldots, 9 \\ 0 & \text{elsewhere} \end{cases}$$

A convenient way of obtaining these samples is to use a table of random digits, like Table 7, letting each sample be a set of 10 consecutive digits in arbitrarily chosen rows or columns. The means of 50 samples thus obtained are

4.4	3.2	5.0	3.5	4.1	4.4	3.6	6.5	5.3	4.4
3.1	5.3	3.8	4.3	3.3	5.0	4.9	4.8	3.1	5.3
3.0	3.0	4.6	5.8	4.6	4.0	3.7	5.2	3.7	3.8
5.3	5.5	4.8	6.4	4.9	6.5	3.5	4.5	4.9	5.3
3.6	2.7	4.0	5.0	2.6	4.2	4.4	5.6	4.7	4.3

and the following is a frequency table showing their distribution:

$\bar{x}$	Frequency
2.0–2.9	2
3.0–3.9	14
4.0–4.9	19
5.0–5.9	12
6.0–6.9	3
	50

It is apparent from the histogram representing the above distribution in Figure 6.1, that the distribution of the $\bar{x}$'s is fairly bell-shaped, even though the population itself has a uniform distribution. This raises the question whether this kind of result is typical of what we might expect in the long run; in other words, would we get a similar distribution if we took 100 samples, 1,000 samples, or perhaps even more? To answer this kind of question, we shall have to investigate the *theoretical sampling distribution of $\bar{x}$* which, for the given example, provides us with the probabilities of getting $\bar{x}$'s between 2.0 and 2.9, between 3.0 and 3.9, . . . , between 6.0 and 6.9, and perhaps values less than 2.0 or greater than 6.9. Although we

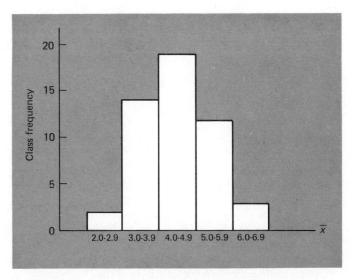

Figure 6.1. Experimental sampling distribution of the mean.

could actually evaluate these probabilities for this example, it is usually sufficient to refer to some general theorems concerning sampling distributions. The first of these, stated as follows, gives expressions for the mean and the variance of sampling distributions of $\bar{x}$:

THEOREM 6.1. *If a random sample of size n is taken from a population having the mean μ and the variance σ^2, then $\bar{x}$ is a value of a random variable whose distribution has the mean μ. For samples from* infinite *populations the variance of this distribution is σ^2/n; for samples from* finite *populations of size N the variance is*
$$\frac{\sigma^2}{n} \cdot \frac{N-n}{N-1}.$$

Writing $\mu_{\bar{x}}$ for the mean of the sampling distribution of $\bar{x}$, we shall first prove for the continuous case that $\mu_{\bar{x}} = \mu$. (The proof for the discrete case follows the identical steps, with integral signs replaced by $\sum$'s.) Using the definition on page 129, we have

$$\mu_{\bar{x}} = \int_{-\infty}^{\infty} \int_{-\infty}^{\infty} \cdots \int_{-\infty}^{\infty} \sum_{i=1}^{n} \frac{x_i}{n} f(x_1, x_2, \ldots, x_n) \, dx_1 \, dx_2 \, \ldots \, dx_n$$

$$= \frac{1}{n} \sum_{i=1}^{n} \int_{-\infty}^{\infty} \int_{-\infty}^{\infty} \cdots \int_{-\infty}^{\infty} x_i f(x_1, x_2, \ldots, x_n) \, dx_1 \, dx_2 \, \ldots \, dx_n$$

where $f(x_1, x_2, \ldots, x_n)$ is a value of the joint density of the random variables whose values constitute the random sample. Using the assumption of randomness, we can write

$$f(x_1, x_2, \ldots, x_n) = f(x_1)f(x_2) \cdot \ldots \cdot f(x_n)$$

and we now have

$$\mu_{\bar{x}} = \frac{1}{n} \sum_{i=1}^{n} \int_{-\infty}^{\infty} f(x_1) \, dx_1 \cdot \ldots \cdot \int_{-\infty}^{\infty} x_i f(x_i) \, dx_i \cdot \ldots \cdot \int_{-\infty}^{\infty} f(x_n) \, dx_n$$

Since each integral except the one with the integrand $x_i f(x_i)$ equals 1, and the one with the integrand $x_i f(x_i)$ equals μ, we finally obtain

$$\mu_{\bar{x}} = \frac{1}{n} \sum_{i=1}^{n} \mu = \mu$$

and this completes the proof of the first part of the theorem.

To prove that for random samples from infinite populations $\sigma_{\bar{x}}^2$, the variance of the sampling distribution of $\bar{x}$, equals σ^2/n, we shall make the simplifying assumption that $\mu = 0$. As the reader will be asked to show in Exercise 10 on page 172, this does *not* involve any loss of generality; as a matter of fact, we stated a similar result on page 154, which showed that addition of a constant to each value does not affect the standard deviation (or the variance) of a set of data. Using the definition on page 129, we thus have

$$\sigma_{\bar{x}}^2 = \int_{-\infty}^{\infty} \int_{-\infty}^{\infty} \ldots \int_{-\infty}^{\infty} \bar{x}^2 f(x_1, x_2, \ldots, x_n) \, dx_1 \, dx_2 \ldots dx_n$$

and making use of the fact that

$$\bar{x}^2 = \frac{1}{n^2} \left(\sum_{i=1}^{n} x_i \right)^2 = \frac{1}{n^2} \left(\sum_{i=1}^{n} x_i^2 + \sum_{i \neq j} \sum x_i x_j \right)$$

we obtain

$$\sigma_{\bar{x}}^2 = \frac{1}{n^2} \sum_{i=1}^{n} \int_{-\infty}^{\infty} \int_{-\infty}^{\infty} \ldots \int_{-\infty}^{\infty} x_i^2 f(x_1, x_2, \ldots, x_n) \, dx_1 \, dx_2 \ldots dx_n$$

$$+ \frac{1}{n^2} \sum_{i \neq j} \sum \int_{-\infty}^{\infty} \int_{-\infty}^{\infty} \ldots \int_{-\infty}^{\infty} x_i x_j f(x_1, x_2, \ldots, x_n) \, dx_1 \, dx_2 \ldots dx_n$$

where $\sum_{i \neq j} \sum$ extends over all i and j from 1 to n, not including the terms where $i = j$. Again using the fact that $f(x_1, x_2, \ldots, x_n) = f(x_1) f(x_2) \cdot \ldots$

$\cdot f(x_n)$, we can write each of the above multiple integrals as a product of simple integrals, with each simple integral that has an integrand of the form $f(x)$ equalling 1. We thus obtain

$$\sigma_{\bar{x}}^2 = \frac{1}{n^2} \sum_{i=1}^{n} \int_{-\infty}^{\infty} x_i^2 f(x_i)\, dx_i + \frac{1}{n^2} \sum_{i \neq j} \sum \int_{-\infty}^{\infty} x_i f(x_i)\, dx_i \cdot \int_{-\infty}^{\infty} x_j f(x_j)\, dx_j$$

and since each integral in the first sum equals σ^2 while each integral in the second sum equals 0, we finally have

$$\sigma_{\bar{x}}^2 = \frac{1}{n^2} \sum_{i=1}^{n} \sigma^2 = \frac{\sigma^2}{n}$$

This completes the proof of the second part of the theorem. We shall not prove the corresponding result for random samples from finite populations, but it should be noted that in the resulting formula for $\sigma_{\bar{x}}^2$ the factor $\dfrac{N-n}{N-1}$, often called the *correction factor for finite populations*, is close to 1 (and can be omitted for most practical purposes) unless the sample constitutes a substantial portion of the population.

Example. For instance, if a random sample of size $n = 10$ is drawn from a population of size $N = 1,000$, then $\dfrac{N-n}{N-1}$ is approximately 0.991.

Although it is not very surprising that the mean of the theoretical sampling distribution of $\bar{x}$ equals the mean of the population, the fact that its variance equals σ^2/n, for random samples from infinite populations, is interesting and important. To point out its implications, let us apply Chebyshev's theorem to the sampling distribution of $\bar{x}$, substituting $\bar{x}$ for x and $\sigma/\sqrt{n}$ for σ in the inequality given on page 75. We thus obtain

$$P\left(|\bar{x} - \mu| < \frac{k\sigma}{\sqrt{n}}\right) \geq 1 - \frac{1}{k^2}$$

and letting $k\sigma/\sqrt{n} = \epsilon$, we get

$$P(|\bar{x} - \mu| < \epsilon) \geq 1 - \frac{\sigma^2}{n\epsilon^2}$$

Thus, for any given $\epsilon > 0$, the probability that $\bar{x}$ differs from μ by less than ϵ can be made arbitrarily close to 1 by choosing n sufficiently large. In less rigorous language, the larger the sample size, the closer we can expect $\bar{x}$ to be to the mean of the population. In this sense we can say that $\bar{x}$ becomes more and more *reliable* as an estimate of μ as the sample size is increased. The reliability of $\bar{x}$ as an estimate of μ is often measured by the expression $\sigma/\sqrt{n}$, also called the *standard error of the mean*. Note that this measure of the reliability of $\bar{x}$ decreases in proportion to the *square root* of n; for example, it is necessary to *quadruple* the size of the sample in order to *halve* the standard deviation of the sampling distribution of the mean. This also indicates what might be called a "law of diminishing returns" so far as increasing the sample size is concerned. Usually it does not pay to take excessively large samples since the extra labor and expense is not accompanied by a proportional gain in reliability. For instance, if we increase the size of a sample from 25 to 2,500, the errors to which we are exposed are reduced only by a factor of 10.

Example. Let us now return to the experimental sampling distribution on page 164, and let us check how closely its mean and variance correspond to the values we should expect in accordance with Theorem 6.1. Since the population from which the 50 samples of size 10 were obtained has the mean

$$\mu = \sum_{x=0}^{9} x \cdot \tfrac{1}{10} = 4.5$$

and the variance

$$\sigma^2 = \sum_{x=0}^{9} (x - 4.5)^2 \tfrac{1}{10} = 8.25$$

Theorem 6.1 leads us to expect a mean of $\mu_{\bar{x}} = 4.5$ and a variance of $\sigma_{\bar{x}}^2 = 8.25/10 = 0.825$. Calculating the mean and the variance from the frequency table on page 164, we obtain $\bar{x}_{\bar{x}} = 4.45$ and $s_{\bar{x}}^2 = 0.939$, which are reasonably close to the theoretical values.

Theorem 6.1 provides only partial information about the theoretical sampling distribution of the mean. In general, it is impossible to determine such a distribution exactly without knowledge of the actual form of the population, but it is possible to find the limiting distribution as $n \longrightarrow \infty$ of a statistic closely related to $\bar{x}$ assuming only that the population has a finite variance σ^2. The statistic we are referring to here is the *standardized mean*

$$z = \frac{\bar{x} - \mu}{\sigma/\sqrt{n}}$$

namely, the difference between $\bar{x}$ and μ divided by the standard deviation of the sampling distribution of $\bar{x}$. With reference to this statistic we have the following form of the *central limit theorem:*

THEOREM 6.2. *If $\bar{x}$ is the mean of a random sample of size n taken from a population having the mean μ and the finite variance σ^2, then*

$$z = \frac{\bar{x} - \mu}{\sigma/\sqrt{n}}$$

is the value of a random variable whose distribution function approaches that of the standard normal distribution as $n \to \infty$.

We shall not be able to prove this theorem in this text, but at least partial (experimental) verification may be obtained by plotting the cumulative percentage distribution corresponding to the distribution of Figure 6.1 on arithmetic probability paper. As can be seen from Figure 6.2, the

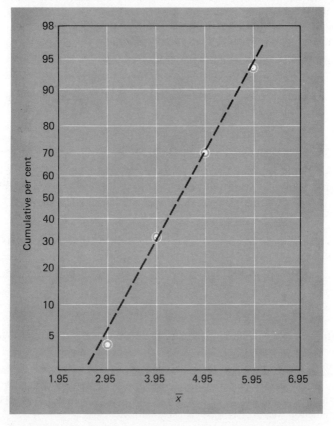

Figure 6.2. Experimental verification of central limit theorem.

points fall close to a straight line, and it seems that even for $n = 10$ the sampling distribution of $\bar{x}$ for this example follows the overall pattern of a normal distribution. In practice, the normal distribution provides an excellent approximation to the sampling distribution of $\bar{x}$ for n as small as 25 or 30, with hardly any restrictions on the shape of the population. As we saw in our example, the sampling distribution of $\bar{x}$ had the general shape of a normal distribution even for samples of size 10 from a discrete uniform distribution, and this is true in general provided the shape of the population distribution is not too skewed. In fact, it can be shown that the sampling distribution of $\bar{x}$ is exactly normal (regardless of the size of the sample), when the sample is obtained from a normal population.

Example. To illustrate the actual use of Theorem 6.2, suppose it is known that a one-gallon can of a certain kind of paint will cover on the average 513.3 square feet with a standard deviation of 31.5 square feet. Suppose, furthermore, we want to know the probability that the mean area covered by a sample of 40 of these one-gallon cans will be anywhere from 510.0 to 520.0 square feet. By Theorem 6.2 we shall have to find the normal curve area between

$$z = \frac{510.0 - 513.3}{31.5/\sqrt{40}} = -0.66 \quad \text{and} \quad z = \frac{520.0 - 513.3}{31.5/\sqrt{40}} = 1.34$$

and checking these values in Table 3 we obtain a probability of 0.6553. Note that if $\bar{x}$ turned out to be much less than 513.3, say, less than 500.0, this might cause serious doubt whether the sample actually came from a population having $\mu = 513.3$ and $\sigma = 31.5$; the probability of obtaining such a small value (a z-value less than 2.67) is only 0.0038.

EXERCISES

1. An inspector examines every 20th piece coming off an assembly line. List some of the conditions under which this method might not yield a random sample.

2. In 1932 the *Literary Digest* predicted the presidential election by random sampling from telephone directories and from its list of subscribers. The prediction was grossly incorrect; explain why.

3. A market research organization wants to try a new product in 8 of the 50 states. Use random numbers (Table 7) to make this selection.

4. Take 30 slips of paper and label five each -4 and 4, four each -3 and 3, three each -2 and 2, and two each -1, 0, and 1.
   (a) If each slip of paper has the same probability of being drawn, find the probability of getting -4, -3, -2, -1, 0, 1, 2, 3, 4, and find the mean and the variance of this distribution.
   (b) Draw 50 samples of size 10 from this population, each sample being drawn without replacement, and calculate their means.
   (c) Calculate the mean and the variance of the 50 means obtained in part (b).
   (d) Compare the results obtained in part (c) with the corresponding values expected according to Theorem 6.1. [Note that μ and σ^2 were obtained in part (a).]

5. Repeat Exercise 4, but select each sample *with replacement;* that is, replace each slip of paper and reshuffle before the next one is drawn.

6. Given the infinite population whose distribution is given by

x	$f(x)$
1	0.25
2	0.25
3	0.25
4	0.25

list the 16 possible samples of size 2 and use this list to construct the distribution of $\bar{x}$ for random samples of size 2 from the given population. Verify that the mean and the variance of this sampling distribution are identical with the corresponding values expected according to Theorem 6.1.

7. When we sample from an infinite population, what happens to the standard error of the mean if
   (a) the sample size is increased from 80 to 720;
   (b) the sample size is increased from 100 to 225;
   (c) the sample size is decreased from 750 to 30?

8. What is the value of the *finite population correction factor* in the formula for $\sigma_{\bar{x}}$ when
   (a) $n = 5$ and $N = 200$;
   (b) $n = 10$ and $N = 400$;
   (c) $n = 100$ and $N = 5,000$?

9. Show that there is a *fifty-fifty chance* that the mean of a random sample of size n from an infinite population with the standard deviation σ will deviate

from μ by less than $0.6745 \cdot \dfrac{\sigma}{\sqrt{n}}$. It has been the custom to refer to this quantity as the *probable error of the mean.*

10. If x is a value of a continuous random variable and $y = ax + b$, show that
    (a) $\mu_x = a\mu_x + b$;
    (b) $\sigma_y^2 = a^2\sigma_x^2$.

11. Prove that $\mu_{\bar{x}} = \mu$ for random samples from discrete (finite or countably infinite) populations.

12. A random sample of size 36 is taken from an infinite population having the mean $\mu = 63$ and the variance $\sigma^2 = 81$. What can we assert about the probability of getting a sample mean greater than 66.75 using
    (a) Chebyshev's theorem;
    (b) the central limit theorem given on page 169?

13. A random sample of size 100 is taken from an infinite population having the mean $\mu = 76$ and the variance $\sigma^2 = 256$. What is the probability of getting an $\bar{x}$ between 75 and 78?

14. If measurements of the specific gravity of a metal can be looked upon as a sample from a normal population having a standard deviation of 0.04, what is the probability that the mean of a random sample of size 25 will be "off" by at most 0.02?

15. A wire-bonding process is said to be in control if the mean pull strength is 10 pounds. It is known that the pull-strength measurements are normally distributed with a standard deviation of 1.5 pounds. Periodic random samples of size 4 are taken from this process and the process is said to be "out of control" if a sample mean is less than 7.75 pounds. Comment.

16. If the distribution of the weights of all men traveling by air between Dallas and El Paso has a mean of 163 pounds and a standard deviation of 18 pounds, what is the probability that the combined gross weight of 36 men traveling on a plane between these two cities is more than 6,000 pounds?

6.3 THE SAMPLING DISTRIBUTION OF THE MEAN (σ UNKNOWN)

Application of the theory of the preceding section requires knowledge of the population standard deviation σ. If n is large, this does not pose any problems as it is reasonable to use the theory also when σ is unknown, substituting for it the sample standard deviation s. So far as the statistic $\dfrac{\bar{x} - \mu}{s/\sqrt{n}}$ is concerned, little is known about its exact distribution for small values of n unless we make the assumption that the sample comes from a *normal population*. Under this assumption, it is possible to prove the following theorem:

THEOREM 6.3. *If $\bar{x}$ is the mean of a random sample of size n taken from a normal population having the mean μ and the variance σ^2, then*

$$t = \frac{\bar{x} - \mu}{s/\sqrt{n}}$$

is the value of a random variable having the Student-t distribution* with the parameter $\nu = n - 1$.

This theorem is *more general* than Theorem 6.2 in the sense that it does not require knowledge of σ; on the other hand, it is *less general* than Theorem 6.2 in the sense that it requires the assumption of a normal population.

As can be seen from Figure 6.3, the overall shape of a t distribution is similar to that of a normal distribution—both are bell-shaped and symmetrical about the mean. Like the standard normal distribution, the t distribution has the mean 0, but its variance depends on the parameter ν (nu), called the *number of degrees of freedom*. The variance of the t distribution exceeds 1, but it approaches 1 as $n \longrightarrow \infty$. In fact, it can be shown that the t distribution with ν degrees of freedom approaches the standard normal distribution as $\nu \longrightarrow \infty$.

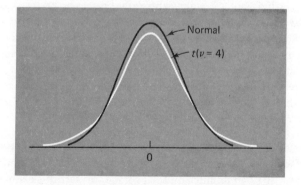

Figure 6.3. Student-t and standard normal distributions.

---

*The Student-t distribution was first investigated in 1908 by W. S. Gosset, who published his findings under the pen name of "Student" because his employers did not permit the publication of research done by their staff.

Table 4 at the end of the book contains selected values of t_α for various values of ν, where t_α is such that the area under the t distribution to its right is equal to α. In this table the left-hand column contains values of ν, the column headings are areas α in the right-hand tail of the t distribution, and the entries are values of t_α (see also Figure 6.4). It is not necessary to tabulate values of t_α for $\alpha > 0.50$, as it follows from the symmetry of the t distribution that $t_{1-\alpha} = -t_\alpha$; thus, the value of t that corresponds to a left-hand tail area of α is $-t_\alpha$.

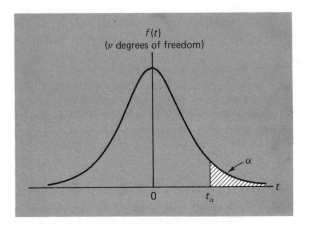

Figure 6.4. Tabulated values of t.

Note that in the bottom row of Table 4 the entries correspond to the values of z which cut off right-hand tails of area α under the standard normal curve. Using the notation z_α for such a value of z, it can be seen, for example, that $z_{.025} = 1.96 = t_{.025}$ for $\nu = \infty$. This result should really have been expected since the t distribution approaches the standard normal distribution as $\nu \to \infty$. In fact, observing that the values of t_α for 29 or more degrees of freedom are close to the corresponding values of z_α, we conclude that *the standard normal distribution provides a good approximation to the t distribution for samples of size 30 or more.*

Example. To illustrate the use of the t distribution, suppose that a manufacturer of fuses claims that with a 20 percent overload, his fuses will blow in 12.40 minutes on the average. To test this claim, a sample of 20 of the fuses was subjected to a 20 percent overload, and the times it took them to blow had a mean of 10.63 minutes and a standard deviation of 2.48 minutes. On the assumption that these times may be looked upon as a random sample from a normal population with

$\mu = 12.40$, the statistic

$$t = \frac{10.36 - 12.40}{2.48/\sqrt{20}} = -3.19$$

is a value of a random variable having the t distribution with $\nu = 20 - 1 = 19$ degrees of freedom. Now, from Table 4 we find that for $\nu = 19$ the probability that t will exceed 2.861 is 0.005, and hence we conclude that the probability that t will be less than -2.861 is also 0.005. Since the value we obtained for our example is $t = -3.19$, we are faced with the following two alternatives: either it is true that $\mu = 12.40$ minutes and we have observed a relatively rare event, or $\mu \neq 12.40$ minutes (in fact, it is less than 12.40). If we actually had to make this choice, we might well be inclined to accept the second alternative, namely, to reject the manufacturer's claim.

The assumption that the sample must come from a normal population is not so severe a restriction as it may first seem. Studies have shown that the distribution of the statistic $t = \dfrac{\bar{x} - \mu}{s/\sqrt{n}}$ is fairly close to a t distribution even for samples from certain nonnormal populations. In practice, it is necessary to make sure primarily that the population from which we are sampling is approximately bell-shaped and not too skewed. A practical way of checking this assumption is to plot the observations on arithmetic probability paper as described on page 143. (If such a plot shows a distinct curve rather than a straight line, it may be possible to "straighten it out" by transforming the data—say, by taking their logarithms or their square roots.)

6.4 THE SAMPLING DISTRIBUTION OF THE VARIANCE

So far we have discussed only sampling distributions of the mean, but if we had taken the medians or the standard deviations of the 50 samples mentioned in the example on page 164, we would similarly have obtained experimental sampling distributions of these statistics. In this section we shall be concerned with the theoretical sampling distribution of s^2 for random samples from normal populations. Since s^2 cannot be negative, we should expect that this sampling distribution is *not* a normal curve; in fact, it is related to a *gamma distribution* with the parameters $\alpha = \nu/2$ and

$\beta = 2$ (see page 117), called the *chi-square distribution*. Specifically, we have

---

THEOREM 6.4. *If s^2 is the variance of a random sample of size n taken from a normal population having the variance σ^2, then*

$$\chi^2 = \frac{(n-1)s^2}{\sigma^2}$$

is a value of a random variable having the chi-square distribution with the parameter $v = n - 1$.

Table 5 at the end of the book contains selected values of χ_α^2 for various values of v, again called the *number of degrees of freedom*, where χ_α^2 is such that the area under the chi-square distribution to its right is equal to α. In this table the left-hand column contains values of v, the column headings are areas α in the right-hand tail of the chi-square distribution, and the entries are values of χ_α^2 (see also Figure 6.5.) Unlike the t distribution, it is necessary to tabulate values of χ_α^2 for $\alpha > 0.50$, because the chi-square distribution is not symmetrical.

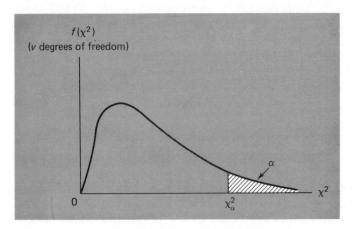

Figure 6.5. Tabulated values of chi-square.

Example. To illustrate the use of Theorem 6.4, suppose an optical firm purchases glass to be ground into lenses, and that past experience has shown that the variance of the refractive index of this kind of glass is $1.26 \cdot 10^{-4}$. To grind the glass into lenses having a given focal length, it is important that the various pieces of glass have nearly the same index of refraction; hence, let us suppose that such a shipment of glass is to be rejected if the sample variance of 20 pieces selected at random exceeds $2.00 \cdot 10^{-4}$. Assuming that the sample values may be looked upon as coming from a normal population with $\sigma^2 = 1.26 \cdot 10^{-4}$, the probability that a shipment will be *erroneously* rejected may be computed as follows. First we obtain

$$\chi^2 = \frac{19(2.00 \cdot 10^{-4})}{1.26 \cdot 10^{-4}} = 30.2$$

and then we find from Table 5 that for 19 degrees of freedom $\chi^2_{.05} = 30.1$. Thus, the probability that a good shipment will erroneously be rejected by this criterion is less than 0.05.

A problem closely related to that of finding the distribution of the sample variance is that of finding the distribution of the *ratio* of the variances of two independent random samples. This problem is important because it arises in tests in which we want to determine whether two samples come from populations having equal variances. If they do, the two sample variances should be nearly the same; that is, their ratio should be close to 1. To determine whether the ratio of two sample variances is too small or too large, we use the theory given in the following theorem:

THEOREM 6.5. *If s_1^2 and s_2^2 are the variances of independent random samples of size n_1 and n_2, respectively, taken from two normal populations having the same variance, then*

$$F = \frac{s_1^2}{s_2^2}$$

is a value of a random variable having the F distribution with the parameters $v_1 = n_1 - 1$ and $v_2 = n_2 - 1$.

The F distribution is related to the *beta distribution* (page 120), and it has the two parameters v_1, the *degrees of freedom for the sample variance in the numerator*, and v_2, the *degrees of freedom for the sample variance in the denominator*; when referring to a particular F distribution, we always give first the degrees of freedom for the numerator. As it would require too large a table to give values of F_α corresponding to many different right-hand tail areas α, and since $\alpha = 0.05$ and $\alpha = 0.01$ are most commonly

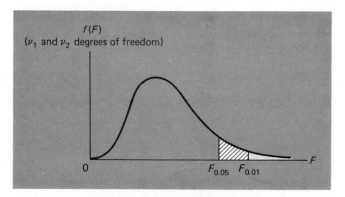

Figure 6.6. Tabulated values of F.

used, Table 6 contains only values of $F_{.05}$ and $F_{.01}$ for various combinations of values of v_1 and v_2 (see also Figure 6.6).

It is possible to use Table 6 also to find values of F corresponding to left-hand tails of area 0.05 and 0.01. Writing $F_\alpha(v_1, v_2)$ for F_α with v_1 and v_2 degrees of freedom, we simply use the identity

$$F_{1-\alpha}(v_1, v_2) = \frac{1}{F_\alpha(v_2, v_1)}$$

Example. For instance, to find F_α for 10 and 20 degrees of freedom and $\alpha = 0.95$, we have only to write

$$F_{.95}(10, 20) = \frac{1}{F_{.05}(20, 10)} = \frac{1}{2.77} = 0.36$$

Note that Theorem 6.4 as well as Theorem 6.5 requires the assumption that we are sampling from normal populations. As in the case of the t distribution, this assumption can be loosened somewhat in actual practice without materially altering the respective sampling distributions. Again, the use of arithmetic probability paper is suggested to investigate whether it is reasonable to treat the samples as coming from normal populations.

EXERCISES

1. A random sample of size 25 from a normal population has the mean $\bar{x} = 47.5$ and the standard deviation $s = 8.4$. Basing the decision on the t statistic, is it reasonable to say that this information supports the claim that the mean of the population is 42.1?

2. The following are the times between six calls for an ambulance (in a certain city) and the patient's arrival in the hospital: 27, 15, 20, 32, 18, and 26 minutes. Use the t statistic to judge the reasonableness of the ambulance service's claim that it takes *on the average* 20 minutes between the call for an ambulance and the patient's arrival at the hospital.

3. A process for making certain bearings is under control if the diameters of the bearings have a mean of 0.5000 cm. What can we say about this process if a sample of ten of these bearings has a mean diameter of 0.5060 cm and a standard deviation of 0.0040 cm?

4. The claim that the variance of a normal population is $\sigma^2 = 21.3$ is rejected if the variance of a random sample of size 15 exceeds 39.74. What is the probability that the claim will be rejected even though $\sigma^2 = 21.3$?

5. A random sample of 10 observations is taken from a normal population having the variance $\sigma^2 = 42.5$. Find the approximate probability of obtaining a sample standard deviation between 3.14 and 8.94.

6. If two independent random samples of size $n_1 = 9$ and $n_2 = 16$ are taken from a normal population, what is the probability that the variance of the first sample will be at least 4 times as large as the variance of the second sample?

7. With reference to Exercise 6, find the probability that *either* sample variance will be 4 times as large as the other.

8. The chi-square distribution with 4 degrees of freedom is given by

$$f(x) = \begin{cases} \frac{1}{4} \cdot x \cdot e^{-x/2} & x > 0 \\ 0 & x \leq 0 \end{cases}$$

Find the probability that the variance of a random sample of size 5 from a normal population with $\sigma = 12$ exceeds 180.

9. The t distribution with 1 degree of freedom is given by

$$f(t) = \frac{1}{\pi}(1 + t^2)^{-1} \qquad -\infty < t < \infty$$

Verify the corresponding value of $t_{.05}$ in Table 4.

10. The F distribution with 4 and 4 degrees of freedom is given by

$$f(F) = \begin{cases} 6F(1 + F)^{-4} & F > 0 \\ 0 & F \leq 0 \end{cases}$$

If random samples of size 5 are taken from two normal populations having the same variance, find the probability that the ratio of the larger to the smaller sample variance exceeds 3.

7

Inferences Concerning Means

7.1 POINT ESTIMATION

On page 163 we stated that the purpose of most statistical investigations is to generalize from information contained in random samples about the populations from which the samples were obtained, namely, to make a statistical inference. In the classical approach to statistical inference, these methods are divided into the major areas of *estimation* and *tests of hypotheses*, with estimation subdivided further into *point estimation* and *interval estimation*. More recently, the methods of statistical inference have been unified under the general concepts of *decision theory*, that is, under the general concepts of decision making in the face of uncertainty. In this section we shall discuss the general problem of point estimation, applying it to the estimation of a population mean, μ, with the mean of a random sample.

Basically, point estimation concerns the choosing of a *statistic*, a single number calculated from sample data (and perhaps other information), for which we have some expectation, or assurance, that it is "reasonably close" to the parameter it is supposed to estimate. To explain what we mean here by "reasonably close" is not an easy task; first, the value of the parameter is unknown, and second, the value of the statistic is unknown until after the sample has been obtained. Thus, we can only ask whether, upon repeated sampling, the distribution of the statistic has certain desirable properties akin to "closeness." For instance, we know from Theorem 6.1 that the sampling distribution of $\bar{x}$ has the same mean as the population from which the sample is obtained; hence, we can expect that the means of repeated random samples from a given population will center on the mean of this population and not about some other value. To formulate this property more generally, let us now make the following definition:

> *A statistic $\hat{\theta}$ is said to be an unbiased estimate of the parameter θ if and only if the mean of its sampling distribution is equal to θ.*

Thus, we call a statistic unbiased if "on the average" its values can be expected to equal the parameter it is supposed to estimate.

Generally speaking, the property of unbiasedness is one of the more desirable properties in point estimation, although it is by no means essential and it is sometimes outweighed by other factors. One shortcoming of the criterion of unbiasedness is that it will generally not provide a *unique* statistic for a given problem of estimation.

Example. For instance, it can be shown that for random samples of size 2 the mean $\dfrac{x_1 + x_2}{2}$ as well as the weighted mean $\dfrac{ax_1 + bx_2}{a + b}$, where a and b are positive constants, are unbiased estimates of the mean of the population. For samples of any size, so are the sample median and the midrange (the mean of the largest value and the smallest) if we assume, furthermore, that the population is symmetric.

This suggests that we must seek a further criterion for deciding which of several unbiased estimates is "best" for estimating a given parameter. Such a criterion becomes evident when we compare the sampling distributions of the mean and the median for random samples of size n taken from the same *normal* population. Although these two distributions have the same mean, namely, the population mean μ, and although they are both symmetrical and bell-shaped, *their variances differ.* In Theorem 6.1 we proved that for infinite populations the variance of the sampling distribution of the mean is σ^2/n, and it can be shown for the corresponding sampling distribution of the median that its variance is approximately $1.57 \cdot \sigma^2/n$. Thus, it is more likely for any given sample that the mean will be closer to μ than the median. Note that this does not imply that the sample mean is necessarily always closer than the median; in fact, in any given problem we have no way of knowing which of the two is actually closer.

This important property, in which we compare the variances of the sampling distributions of statistics, is formalized by means of the following definition:

A statistic $\hat{\theta}_1$ is said to be a more efficient unbiased estimate of the parameter θ than the statistic $\hat{\theta}_2$, if:

1. $\hat{\theta}_1$ and $\hat{\theta}_2$ are both unbiased estimates of θ.
2. The variance of the sampling distribution of $\hat{\theta}_1$ is less than the variance of the sampling distribution of $\hat{\theta}_2$.

We have, thus, seen that for random samples from normal populations the mean is more efficient than the median as an estimate of μ; in fact, it can be shown that in most situations met in actual practice the variance of the sampling distribution of no other statistic is less than that of the sampling distribution of the mean. In other words, in most practical situations the sample mean is an acceptable statistic for estimating a population mean μ. (There exist several other criteria for assessing the "goodness" of methods of point estimation, but we shall not discuss them in this book.)

When we use a sample mean to estimate the mean of a population, we know that although we are using a method of estimation which has certain desirable properties, the chances are slim, virtually nonexistent, that the estimate is *exactly* equal to μ. Hence, it would seem desirable to accompany such a point estimate of μ with some statement as to how close we might reasonably expect the estimate to be. The error, $\bar{x} - \mu$, is the difference between the estimate and the quantity it is supposed to estimate. To examine this error, let us make use of the fact that for large n

$$\frac{\bar{x} - \mu}{\sigma/\sqrt{n}}$$

is a value of a random variable having approximately the standard normal distribution. Consequently, we can assert with a probability of $1 - \alpha$ that

$$-z_{\alpha/2} < \frac{\bar{x} - \mu}{\sigma/\sqrt{n}} < z_{\alpha/2}$$

or that

$$\frac{|\bar{x} - \mu|}{\sigma/\sqrt{n}} < z_{\alpha/2}$$

where $z_{\alpha/2}$ is such that the normal curve area to its right equals $\alpha/2$. If we now let E stand for $|\bar{x} - \mu|$, the magnitude of the error of estimate, we have

$$E < z_{\alpha/2} \cdot \frac{\sigma}{\sqrt{n}}$$

with a probability of $1 - \alpha$. In other words, if we estimate μ by means of a random sample of size n, we can assert with a probability of $1 - \alpha$ that the error, $|\bar{x} - \mu|$, is less than $z_{\alpha/2} \cdot \sigma/\sqrt{n}$, at least for sufficiently large values of n.

Example. To illustrate, suppose that the mean of the sulfur oxides emission data on page 137, namely, $\bar{x} = 18.85$ tons, is to be used to estimate μ, the plant's *actual average daily emission* of these pollutants. To see what we can assert with a probability of 0.95 about the possible size of the error of this estimate, let us approximate σ with the value obtained for s on page 155, namely, $\sqrt{30.77} = 5.55$. This is justifiable since the sample was large. Thus, substituting $s = 5.55$, $n = 80$, and $z_{\alpha/2} = 1.96$ into the formula for E, we find that we can assert with a probability of 0.95 that the error of the estimate is less than

$$1.96 \cdot \frac{5.55}{\sqrt{80}} = 1.22 \text{ tons}$$

Of course, the error of the estimate is less than 1.22 or it is not, *and we really don't know which happens to be right*, but if we had to bet, then 95 to 5 (or 19 to 1) would be *fair odds* that the error is less than 1.22 tons.

If we take the above inequality for E and solve it for n, we obtain

$$n < \left[\frac{z_{\alpha/2} \cdot \sigma}{E}\right]^2$$

and it follows that if we select the sample size n so that

$$n = \left[\frac{z_{\alpha/2} \cdot \sigma}{E}\right]^2$$

we can assert with a probability of $1 - \alpha$ that the error of estimating μ by means of $\bar{x}$ will be less than E. To be able to use this formula for calculating the sample size needed to estimate μ in a given situation, it is necessary that we specify α, σ, and E. Thus, we must give not only the maximum tolerable error E and the population standard deviation σ, but also the probability $1 - \alpha$ with which we want to assert that the maximum error will be less than E. The population standard deviation is usually estimated with prior data of a similar kind, and sometimes a good guess will have to do.

Example. An efficiency expert has to determine the average time it takes a mechanic to rotate the tires of a car, and he wants to be able to assert with a probability of 0.99 that the error of his estimate, the mean of an appropriate sample, will be off by less than half a minute. If it is known from past experience that it is reasonable to let $\sigma = 1.6$ minutes, substitution into the formula yields

$$n = \left[\frac{2.58 \cdot 1.6}{0.5}\right]^2 = 68.2$$

or 69 rounded *up* to the nearest integer. Thus, the efficiency expert will have to take a sample of size 69 (that is, time 69 mechanics while they are rotating the tires of a car).

The methods we have discussed require that σ be known or that it can be approximated with the sample standard deviation s, thus requiring that n be large. However, if it is reasonable to assume that we are sampling from a normal population, we can base our argument on Theorem 6.3 instead of Theorem 6.2, namely, on the fact that

$$t = \frac{\bar{x} - \mu}{s/\sqrt{n}}$$

is a value of a random variable having the Student-t distribution with $n - 1$ degrees of freedom. Duplicating the steps on page 183, we thus arrive at the result that we can assert with a probability of $1 - \alpha$ that

$$E < t_{\alpha/2} \cdot \frac{s}{\sqrt{n}}$$

where E is the error we make in using $\bar{x}$ to estimate μ, and $t_{\alpha/2}$ is as defined on page 174.

Example. In 6 determinations of the melting point of tin, a chemist obtained a mean of 232.26 degrees centigrade with a standard deviation of 0.14 degrees. If he uses this mean as *the* melting point of tin, let us see what we can say with a probability of 0.98 about the possible size of his error. Substituting $n = 6$, $s = 0.14$ and 3.365 (the value of $t_{0.01}$ for 5 degrees of freedom) into the inequality for E, we get

$$E < 3.365 \cdot \frac{0.14}{\sqrt{6}} \simeq 0.19$$

Thus, we can assert with a probability of about 0.98 that he is off by at most 0.19 degree.

7.2   INTERVAL ESTIMATION

Since point estimates cannot really be expected to coincide with the quantities they are intended to estimate, it is sometimes preferable to replace them with *interval estimates*, that is, intervals for which we can assert with a reasonable degree of certainty that they contain the parameter under consideration. To illustrate the construction of such an interval, suppose that we have a random sample of size n, where n is large, from a population having the unknown mean μ and the *known* variance σ^2. Referring to the double inequality on page 183, namely,

$$-z_{\alpha/2} < \frac{\bar{x} - \mu}{\sigma/\sqrt{n}} < z_{\alpha/2}$$

which we asserted with a probability of $1 - \alpha$, we can now apply simple algebra and rewrite it as

$$\bar{x} - z_{\alpha/2} \cdot \frac{\sigma}{\sqrt{n}} < \mu < \bar{x} + z_{\alpha/2} \cdot \frac{\sigma}{\sqrt{n}}$$

Thus, we can claim with a probability of $1 - \alpha$ that the interval from $\bar{x} - z_{\alpha/2} \cdot \frac{\sigma}{\sqrt{n}}$ to $\bar{x} + z_{\alpha/2} \cdot \frac{\sigma}{\sqrt{n}}$ contains μ. It is customary to refer to an interval of this kind as a *confidence interval for* μ having the *degree of confidence* $1 - \alpha$.

Example. At this point it is well to reconsider just what is meant by the statement "We can claim with a probability of $1 - \alpha$ that such an interval contains μ." If a random sample of size $n = 100$ is taken from a population having $\sigma = 5.1$ and we obtain a sample mean of 21.6, then a 0.95 confidence interval for μ is given by

$$21.6 \pm 1.96 \cdot \frac{5.1}{\sqrt{100}},$$

namely, the interval from 20.6 to 22.6. Since the mean of the given population either is contained or is not contained in this interval, it would hardly seem reasonable to speak of the probability of such an event. Indeed, what we really mean when we claim that the interval from 20.6 to 22.6 is a 0.95 confidence interval for the mean of the population is that *in repeated sampling* 95 percent of the confidence intervals obtained with the above formula contain the means of the respective populations. Thus, although we shall never know whether the population mean is really contained in the interval from 20.6 to 22.6 in the given example, *we do have the assurance that the method used to obtain the interval is 95 percent reliable, that is, it can be expected to work 95 percent of the time.*

The formula obtained for a $1 - \alpha$ confidence interval for μ has the unfortunate feature that it requires knowledge of the population standard deviation. It is "unfortunate" because in most practical problems σ is unknown; therefore, we generally have little recourse but to replace σ with an estimate in the hope that, at least for large samples, the resulting confidence interval will be a close approximation. Substituting for σ the sample standard deviation s, which has desirable properties as a point estimate of σ (see Chapter 8), we use the interval

$$\bar{x} - z_{\alpha/2} \cdot \frac{s}{\sqrt{n}} < \mu < \bar{x} + z_{\alpha/2} \cdot \frac{s}{\sqrt{n}}$$

as an approximate *large-sample confidence interval for* μ having the degree of confidence $1 - \alpha$.

Example. To illustrate the use of this last formula, let us refer again to the sulfur oxides emission data on page 137, for which $n = 80$, $\bar{x} = 18.85$ tons, and $s = 5.55$ tons. Substituting all these values together with $z_{\alpha/2} = 1.96$ into the formula, we find that a 0.95 confidence interval for μ, the plant's actual average daily emission of sulfur oxides, is given by $18.85 \pm 1.96 \cdot \dfrac{5.55}{\sqrt{80}}$ or 18.85 ± 1.22, which can be written as

$$17.63 < \mu < 20.07$$

Observe the close relationship between the width of the confidence interval obtained here and the maximum error asserted with a probability of 0.95 on page 183. This kind of relationship will always exist so long as the sampling distribution of the statistic on which the methods are based is *symmetrical*.

When the sample size n is not large enough to approximate σ with s, we proceed as on page 185 *provided it is reasonable to assume that we are sampling from a normal population*. Thus, with $t_{\alpha/2}$ defined as on page 174, we obtain the $1 - \alpha$ confidence interval

$$\bar{x} - t_{\alpha/2} \cdot \frac{s}{\sqrt{n}} < \mu < \bar{x} + t_{\alpha/2} \cdot \frac{s}{\sqrt{n}}$$

It applies to samples from normal populations, but in accordance with the discussion on page 175, it may be used so long as the sample does not exhibit any pronounced departures from normality.

Example. If the mean weight loss of $n = 16$ grinding balls after a certain length of time in mill slurry is 3.42 grams with a standard deviation of 0.68 grams, a 0.99 confidence interval for μ is given by $3.42 \pm 2.947 \cdot \dfrac{0.68}{\sqrt{16}}$; in other words, $2.92 < \mu < 3.92$.

7.3 BAYESIAN ESTIMATION

In recent years, there has been mounting interest in methods of inference which look upon parameters as random variables. This idea is not really new, but these *Bayesian methods*, as they are called, have received considerable impetus and much wider applicability through the concept of personal, or subjective, probability. In fact, this is why supporters of the personal concept of probability refer to themselves as *Bayesians*, or *Bayesian statisticians*.

In this section we shall present a Bayesian method of estimating the mean μ of a population. Proponents of the subjective, or personal, point of view in probability look upon μ as a random variable whose distribution is indicative of *how strongly a person feels about the various values which μ can take on*. In other words, they suggest that in any problem where we want to estimate the mean of a population, a person will feel most strongly about some particular value of μ, and that this enthusiasm will diminish for values of μ which are further and further away from the one he likes the most. Like any distribution we use in actual practice, this kind of subjective *prior distribution* for the possible values of μ has a mean which we shall denote μ_0 and a standard deviation which we shall denote σ_0.

Example. To illustrate the concept of a prior distribution of the mean, let us refer again to the air pollution example on page 137, and let us suppose that *before the sample data were actually obtained* the chief engineer of the plant said that his feelings about the true average daily emission of sulfur oxides can best be described by a normal distribution with the mean $\mu_0 = 17.50$ tons and the standard deviation $\sigma_0 = 2.50$ tons. Given this information, we could ask, for example, how he felt about the possibility that the plant's true average daily emission of these pollutants is somewhere between 18.00 and 19.00 tons. The answer to this question is given by the area of the shaded region of the *first* diagram of Figure 7.1, and, as can easily be verified (see Exercise 20 on page 194), he would have assigned a probability of about 0.15 to the mean actually lying between 18.00 and 19.00 tons. Note that the interval from 18.00 tons to 19.00 tons is *not* a confidence interval which the chief engineer could have asserted with a degree of confidence of 0.15. No, *based on prior considerations, he would be making a direct probability statement about the mean of the population*, namely, about the plant's true average daily emission of sulfur oxides.

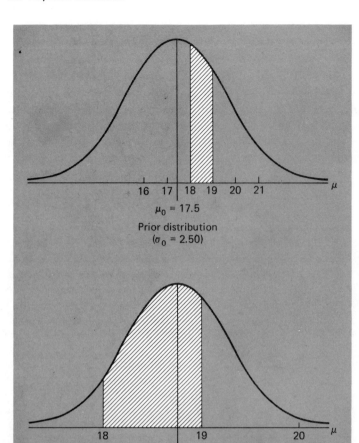

Figure 7.1. Prior and posterior distribution of the mean.

In *Bayesian estimation*, prior information about the possible values of the population mean, as expressed in terms of a prior distribution, is combined with *direct sample evidence* consisting of a random sample of size n, its mean $\bar{x}$, and its standard deviation s (which may have to serve as an estimate of σ, the standard deviation of the population). Treating the sample mean as well as the population mean as random variables and using the notation introduced on page 129, this can be accomplished by referring to the *conditional distribution* of the population mean *given* $\bar{x}$, whose values are written $g_1(\mu \,|\, \bar{x})$. Using the result of Exercise 10 on page 131, $g_1(\mu \mid \bar{x})$ can be expressed in terms of $f_1(\mu)$, $f_2(\bar{x})$, and $g_2(\bar{x} \mid \mu)$, where $f_1(\mu)$ is a value of the *prior distribution* of the mean of population and $g_2(\bar{x} \mid \mu)$ is a value of the sampling distribution of the mean for a *given* population mean μ. If, as in our example, the prior distribution of the population mean is a normal distribution with the parameters μ_0 and σ_0, and n is large enough to treat the sampling distribution of the mean as a normal distribution with the parameters μ and $\sigma/\sqrt{n}$, it can be shown that the *posterior distribution* of the population mean (namely, the *conditional distribution* of the population mean given $\bar{x}$) is also a normal distribution and that its mean and standard deviation are

$$\mu_1 = \frac{n\bar{x}\sigma_0^2 + \mu_0\sigma^2}{n\sigma_0^2 + \sigma^2} \quad \text{and} \quad \sigma_1 = \sqrt{\frac{\sigma^2\sigma_0^2}{n\sigma_0^2 + \sigma^2}}$$

Derivations of these results may be found in several of the textbooks on theoretical statistics listed in the Bibliography at the end of this book.

Continuation of Example. Let us now combine the prior information given on page 188, that based on the feelings of the chief engineer, with the fact that a random sample of size $n = 80$ yielded a mean of $\bar{x} = 18.85$ tons and a standard deviation of $s = 5.55$ tons. Using s as an estimate of σ and substituting these values together with $\mu_0 = 17.50$ and $\sigma_0 = 2.5$ into the above formulas for μ_1 and σ_1, we obtain

$$\mu_1 = \frac{80(18.85)(2.5)^2 + 17.50(5.55)^2}{80(2.5)^2 + (5.55)^2} = 18.77$$

and

$$\sigma_1 = \sqrt{\frac{(5.55)^2(2.5)^2}{80(2.5)^2 + (5.55)^2}} = 0.60$$

Note that the *mean of the posterior distribution*, namely, $\mu_1 = 18.77$, provides us with a *point estimate* of the mean of the population. It is based on direct evidence as well as prior information; in fact, it is a *weighted mean* (see page 158 and also Exercise 17 on page 193) of $\bar{x}$ and μ_0.

To continue with this example, let us now find the *posterior probability* that the population mean lies on the interval from 18.00 to 19.00 tons. This time the probability is given by the area of the shaded region of the *second* diagram of Figure 7.1, and, as the reader will be asked to verify in Exercise 20 on page 194, the answer is 0.55. Note that knowledge about the sample with $\bar{x} = 18.85$ increased the probability associated with the interval from 18.00 to 19.00 tons from 0.15 to 0.55.

EXERCISES

1. To illustrate that the mean of a random sample is an *unbiased* estimate of the mean of the population, consider five slips of paper numbered, respectively, 3, 6, 9, 15, and 27.
   (a) List all possible samples of size 3 that can be taken *without* replacement from this finite population.
   (b) Calculate the mean of each of the samples listed in (a), and, assigning each sample a probability of $\frac{1}{10}$, verify that the mean of these $\bar{x}$'s equals 12, namely, the mean of the population.

2. To verify the claim that the mean is generally *more efficient* than the median (namely, that it is subject to smaller chance fluctuations), a student conducted an experiment consisting of 12 tosses of three dice. The following are his results: 2, 4, and 6; 5, 3, and 5; 4, 5, and 3; 5, 2, and 3; 6, 1, and 5; 3, 2, and 1; 3, 1, and 4; 5, 5, and 2; 3, 3, and 4; 1, 6, and 2; 3, 3, and 3; 4, 5, and 3.
   (a) Calculate the twelve medians and the twelve means.
   (b) Group the medians and the means obtained in (a) into separate distributions having the classes 1.5–2.5, 2.5–3.5, 3.5–4.5, and 4.5–5.5.
   (c) Draw histograms of the two distributions obtained in (b) and explain how they illustrate the claim that the mean is generally more efficient than the median.

3. While performing a certain task under simulated weightlessness, the pulse rate of 32 astronaut trainees increased on the average by 26.4 beats per minute with a standard deviation of 4.28 beats per minute. What can one assert with a probability of 0.95 about the possible size of the error, if this sample mean is used to estimate the true average increase in the pulse rate of astronaut trainees performing the given task?

4. Use the data of Exercise 3 to construct a 0.95 confidence interval for the true average increase in the pulse rate of astronaut trainees performing the given task.

5. To estimate the average time it takes to assemble a certain computer component, the efficiency expert of an electronics firm timed 40 technicians in the performance of this task, getting a mean of 12.73 minutes and a standard deviation of 2.06 minutes.
   (a) What can we say with a probability of 0.99 about the possible size of the error if $\bar{x} = 12.73$ minutes is used as an estimate of the actual average time required to do the job?

    (b) Use the data to construct a 0.98 confidence interval for the true average time it takes to assemble the computer component.

6. Referring to Exercise 5, with what probability can we assert that the sample mean is within 30 seconds of the true mean?

7. Find the largest error one can expect to make with a probability of 0.90 when using the mean of a random sample of size $n = 64$ to estimate the mean of a population having a variance of 2.56.

8. In a study of automobile collision insurance costs, a random sample of 80 body repair costs on a particular kind of damage had a mean of $472.36 and a standard deviation of $62.35. What is the probability that an error of no more than $10 is made when estimating the true average repair cost of this kind of damage as $472.36?

9. If we wanted to determine the average mechanical aptitude of a large group of workers, how large a random sample would we need to be able to assert with a probability of 0.95 that the sample mean will be within 3 points of the true mean? Assume that it is known from past experience with similar data that $\sigma = 20$.

10. Referring to Exercise 8 and using $62.35 as an estimate of σ, how large a sample would be needed to be able to assert with a probability of 0.99 that the sample mean will be within $10 of the true mean?

11. In the example on page 174, twenty fuses were subjected to a 20 percent overload, and the times it took them to blow had a mean of 10.63 minutes and a standard deviation of 2.48 minutes. Based on these data, what can we say with a probability of 0.95 about the possible size of our error, if we use 10.63 minutes as an estimate of the true average time it takes such fuses to blow with a 20 percent overload?

12. Given that 10 bearings made by a certain process have a mean diameter of 0.5060 cm and a standard deviation of 0.0040 cm (see Exercise 3 on page 179), construct a 0.99 confidence interval for the actual average diameter of bearings made by this process.

13. Inspecting ceramic tiles prior to their shipment, a quality control engineer detects 2, 3, 6, 0, 4, and 9 defectives in six cartons, each containing 144 tiles. What can we assert with a probability of 0.99 about the possible size of his error, if he uses the mean of this sample to estimate the true average number of defective tiles per carton?

14. In an air pollution study, the following amounts of suspended benzene-soluble organic matter (in micrograms per cubic meter) were obtained at an experiment station for eight different samples of air: 2.2, 1.8, 3.1, 2.0, 2.4, 2.0, 2.1, and 1.2. Construct a 0.95 confidence interval for the corresponding true mean.

15. In order to test the durability of a new paint, a highway department had test strips painted across heavily traveled roads in 15 different locations. If on

the average the test strips disappeared after they had been crossed by 146,692 cars and the standard deviation is 14,380 cars, construct a 0.99 confidence interval for the true average number of cars it will take to wear off this paint.

16. Referring to Exercise 15 and using 14,380 as an estimate of σ, find the sample size that would have been necessary to be able to assert with a probability of 0.95 that the sample mean will not be off by more than 10,000. [*Hint*: First, estimate n_1 by using $z = 1.96$, then use $t_{0.025}$ with $n_1 - 1$ degrees of freedom to obtain a second estimate n_2; repeat this procedure until the last two values of n thus obtained are equal.]

17. Show that the expression for the mean of the posterior distribution on page 190 can be written as

$$\mu_1 = \frac{\dfrac{n}{\sigma^2} \cdot \bar{x} + \dfrac{1}{\sigma_0^2} \cdot \mu_0}{\dfrac{n}{\sigma^2} + \dfrac{1}{\sigma_0^2}}$$

namely, as a *weighted mean* of $\bar{x}$ and μ_0, with their respective weights being $\dfrac{n}{\sigma^2}$ and $\dfrac{1}{\sigma_0^2}$. Indicate how these weights depend

(a) on the sample size n;
(b) on the variability of the population;
(c) on the variability of the prior distribution.

18. A sales manager's feelings about the average monthly demand for one of his company's products may be described by means of a normal distribution with $\mu_0 = 3,800$ units and $\sigma_0 = 260$ units.

(a) What probability does he, thus, assign to the true average (monthly demand) being somewhere on the interval from 3,500 to 4,000 units?

(b) If data for nine months show an average demand for 3,702 units with a standard deviation of 390 units, how would the original point estimate of $\mu_0 = 3,800$ units be modified in the light of this information?

(c) How would the probability of part (a) be modified in the light of the information given in part (b)?

19. A distributor of soft-drink vending machines knows that in a supermarket one of his machines will sell on the average 835 drinks per week. Of course, this *mean* will vary somewhat from market to market, and this variation is measured by a standard deviation of 12.4. So far as a machine placed in a particular market is concerned, the number of drinks sold will vary from week to week, and this variation is measured by a standard deviation of 43.6.

(a) If this distributor plans to put one of his soft-drink vending machines into a brand new supermarket, what estimate would he use for the number of drinks he can expect to sell per week?

(b) How would he modify his estimate if during 10 weeks the machine sells on the average 592 drinks per week?

(c) How would he modify his original estimate if during 50 weeks the machine sells on the average 917 drinks per week?

20. Verify that the probabilities represented by the shaded regions of the two diagrams of Figure 7.1 are, respectively, 0.15 and 0.55.

7.4 TESTS OF HYPOTHESES

There are many problems in which we are not directly concerned with the *actual* value of a parameter; instead we want to know whether its value exceeds a given number, is less than a given number, falls into a given interval, and so on. Rather than estimate the value of a parameter, we thus want to *decide* whether a statement (or statements) concerning the parameter is true or false; that is, we want to test a hypothesis about the parameter. For example, in quality-control work a random sample may be taken to determine whether the "process mean" (for a given kind of measurement) has remained unchanged or whether it has changed to such an extent that the process has gone "out of control" and adjustments will have to be made.

Example. To illustrate the general concepts involved in this kind of decision problem, suppose that a paint manufacturer claims that the average drying time of his new "fast-drying" paint is 20 minutes, and that a government agency wants to test the validity of this claim. Suppose, furthermore, that 36 boards painted, respectively, with paint from 36 different one-gallon cans of this paint dried on the average in 20.75 minutes. This figure exceeds the average of 20 minutes claimed by the manufacturer, but *is it large enough to reject the claim*; that is, is this sufficient evidence to take appropriate action against the paint manufacturer? *Is it not possible that the discrepancy is due entirely to chance and that the paint is as good as claimed even though the sample mean turned out to be high?*

In testing the paint manufacturer's claim, the government agency is faced with the problem of establishing a criterion which will enable it to take suitable actions. Surely, if the *true* average drying time of the "fast-drying" paint is 20 minutes or less, there is no cause for any kind of regulatory action. On the other hand, how much leeway should be allowed to the manufacturer if his product is not quite as good as claimed? Should some action be taken if the true mean (drying time) μ is greater than 20.5 minutes, 21 minutes, 21.5 minutes, and so forth? Suppose, for the sake of argument, it is decided that a practical threat to construction planning exists if the average drying time exceeds 21 minutes. We can then think of the set of all possible values of μ (the set of the positive real numbers) as being divided into the

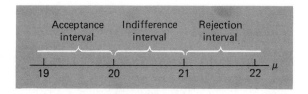

Figure 7.2. Decision intervals.

three regions shown in Figure 7.2. If μ lies in the rejection interval, some regulatory action should be taken and it would be a serious error not to reject the paint manufacturer's claim; if μ lies in the acceptance interval, there is certainly no cause for any action and it would be a serious error if, by chance, the agency did reject the manufacturer's claim; if μ lies in the indifference interval, it is difficult to argue whether or not regulatory action should be taken and no serious error is made in either case.

If μ could actually be known, the decision-making problem faced by the government agency would be resolved by simply referring to the criterion of Figure 7.2. Since this is not the case in the given problem (or any similar practical situation), the decision will have to be based instead on the results of a random sample, usually on the value obtained for the sample mean. Thus, it is important to realize that $\bar{x}$ might be less than 20 even though μ exceeds 21, and that $\bar{x}$ might exceed 21 even though μ is less than 20.

This last argument illustrates the fact that *errors are unavoidable* when decisions are based on the results of random samples and, furthermore, that these errors are of two different kinds. The sample data can lead to the rejection of the claim even though μ lies in the acceptance interval, and they can lead to the acceptance of the claim even though μ lies in the rejection interval. Schematically, the situation may be described by means of the following table:

	μ lies in acceptance interval	*μ lies in rejection interval*
Accept the claim	No error	Type II error
Reject the claim	Type I error	No error

Note that if μ lies in the indifference interval, no *serious* error is committed whatever decision is made.

To judge the merits of any decision criterion, it is essential to know the probabilities that it will lead to Type I and Type II errors, namely, the probabilities of *erroneously rejecting* or *erroneously accepting* a hypothesis (assumption or claim). Denoting these probabilities by α and β, that is,

$$\alpha = P \text{ (Type I error)}$$

$$\beta = P \text{ (Type II error)}$$

it is important not only to determine the values of α and β for given criteria, but to construct criteria for which α and β take on preassigned values.

> **Continuation of Example.** To illustrate the first of these problems, suppose that the government agency proceeds as indicated on page 194 (namely, uses the paint from 36 different one-gallon cans on 36 boards) and decides to reject the paint manufacturer's claim if $\bar{x}$, the mean drying time in this sample of size $n = 36$, exceeds 20.75 minutes. To simplify the discussion, we shall assume also that the standard deviation of such drying times can be expected to equal $\sigma = 2.4$ minutes. Now α is the probability of a *false rejection*, namely, the probability that $\bar{x} > 20.75$ when actually $\mu \leq 20$. Thus, α cannot be determined unless some value of μ (less than or equal to 20) is specified, but it is easy to see that the probability of committing a Type I error with the given criterion is *greatest* when $\mu = 20$. In fact, we could think of the acceptance interval as consisting only of the point $\mu = 20$, in which case there would be only one possible value of α, namely, the probability of getting an $\bar{x}$ greater than 20.75 when $\mu = 20$.
>
> To calculate this value of α, we approximate the sampling distribution of $\bar{x}$ with a normal distribution having a mean of 20 and a standard deviation of $\dfrac{2.4}{\sqrt{36}} = 0.4$. Thus, α is given by the area under the standard normal curve to the right of
>
> $$z = \frac{20.75 - 20}{0.4} = 1.875$$
>
> and, as can be seen from Table 3, it equals approximately 0.03.
>
> So far as Type II errors are concerned, we can argue analogously that β is a maximum for $\mu = 21$; that is, for all values of the rejection interval ($\mu \geq 21$) the probability of a Type II error is greatest for

$\mu = 21$. Thinking of the rejection interval as consisting only of the point $\mu = 21$, we can find β by approximating the sampling distribution of $\bar{x}$ with a normal distribution having a mean of 21 and a standard deviation of $\frac{2.4}{\sqrt{36}} = 0.4$. Thus, β is given by the area under the standard normal curve to the left of

$$z = \frac{20.75 - 21}{0.4} = -0.625$$

and, as can be seen from Table 3, it equals approximately 0.27.

Using the method illustrated in this example, we can always calculate α and β for testing an hypothesis H_0 of the form $\mu = \mu_0$ against an alternative hypothesis H_1 of the form $\mu = \mu_1$ on the basis of a given criterion, so long as σ is known and n is large enough to use the normal distribution to approximate the sampling distribution of the mean. Now let us turn to the other kind of problem mentioned on page 196, namely, that of constructing criteria for which α and β take on preassigned values. Actually, this involves determining the sample size n and also the "dividing line" of the criterion, which was 20.75 minutes in our example. If we denote this dividing line by the letter C, it can be seen from Figure 7.3 (where we arbitrarily let $\mu_1 > \mu_0$) that

$$z_\alpha = \frac{C - \mu_0}{\sigma/\sqrt{n}} \quad \text{and} \quad -z_\beta = \frac{C - \mu_1}{\sigma/\sqrt{n}}$$

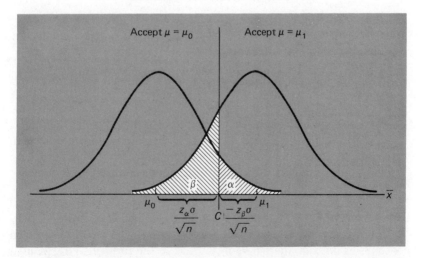

Figure 7.3. Determination of decision criterion.

where z_α and z_β are such that the normal curve areas to their right equal α and β, respectively. Solving the first of these two equations for C, we get

$$C = \mu_0 + z_\alpha \cdot \frac{\sigma}{\sqrt{n}}$$

for the dividing line of the criterion; that is, we reject the hypothesis that $\mu = \mu_0$ and accept the alternative hypothesis that $\mu = \mu_1$ if the mean of a random sample of size n exceeds $\mu_0 + z_\alpha \cdot \dfrac{\sigma}{\sqrt{n}}$. To determine n, we eliminate C from the equations for z_α and $-z_\beta$, getting

$$n = \frac{\sigma^2(z_\alpha + z_\beta)^2}{(\mu_1 - \mu_0)^2}$$

Continuation of Example. With reference to the drying-time example, suppose that the government agency is willing to risk Type I and Type II errors with respective probabilities of $\alpha = 0.05$ and $\beta = 0.10$. Since $z_{0.05} = 1.645$ and $z_{0.10} = 1.28$ according to Table 3, we find that the required sample size is

$$n = \frac{(2.4)^2(1.645 + 1.28)^2}{(21 - 20)^2}$$

or $n = 50$ rounded *up* to the nearest integer. Correspondingly, the dividing line of the criterion is

$$C = 20 + 1.645 \cdot \frac{2.4}{\sqrt{50}} = 20.56$$

Thus, the government agency can accomplish its goal by taking a random sample of size 50, accepting the paint manufacturer's claim so long as the mean of the sample does not exceed 20.56 minutes, and rejecting the claim if the mean of the sample exceeds 20.56 minutes.

In Section 7.5 we shall demonstrate another method, based on special tables, for handling problems of this kind.

In testing an hypothesis of the form $\mu = \mu_0$, where μ_0 is a specified constant, against an alternative hypothesis of the form $\mu > \mu_0$, we used the test criterion: *reject the hypothesis* $\mu = \mu_0$ *if* $\bar{x} > C$, *where C is usually determined so that the probability of committing a Type I error equals some preassigned value* α. The set of values of $\bar{x}$ which, thus, leads to the rejection of the hypothesis $\mu = \mu_0$ (and the acceptance of the alternative) is called the *critical region*; in our example it was the set of all real numbers greater than C. If the alternative hypothesis is of the form $\mu < \mu_0$, the general procedure is the same, but all inequalities are reversed, and the critical region becomes $\bar{x} < K$, where K must again be chosen so that the probability of committing a Type I error equals some preassigned value α. This kind of test would arise, for example, if we wanted to investigate the claim that a certain kind of car will average 15 miles per gallon in city traffic, and we are concerned mostly about the possibility that the actual figure is less.

So far, the tests we have discussed have been *one-tail tests*; that is, the hypothesis H_0 that $\mu = \mu_0$ has been rejected for values of $\bar{x}$ falling into one "tail" of its sampling distribution. If we now consider the alternative $\mu \neq \mu_0$, we would want to reject H_0 for values of $\bar{x}$ that are appreciably less than or greater than μ_0, and the resulting critical region would be of the form $\bar{x} < C_1$ *or* $\bar{x} > C_2$. This kind of test would arise, for example, if we wanted to check whether the mean diameter of a certain kind of steel pipe is 3.0 inches, as specified. Clearly, there could be complications if the mean diameter is either too large or too small.

The constants C_1 and C_2 for such a *two-tail test* are usually chosen so that the two "tails" are equal (that is, alternative values of μ at the same distance on either side of μ_0 have the same chance of being accepted) and the probability of committing a Type I error equals some preassigned constant α. As can be seen with the aid of Figure 7.4, the constants C_1 and C_2 may thus be obtained by solving the equations

$$\frac{C_1 - \mu_0}{\sigma/\sqrt{n}} = -z_{\alpha/2} \quad \text{and} \quad \frac{C_2 - \mu_0}{\sigma/\sqrt{n}} = z_{\alpha/2}$$

and the results are

$$C_1 = \mu_0 - z_{\alpha/2} \cdot \frac{\sigma}{\sqrt{n}} \quad \text{and} \quad C_2 = \mu_0 + z_{\alpha/2} \cdot \frac{\sigma}{\sqrt{n}}$$

Example. To illustrate, suppose that a process for making steel pipe is *under control* if the diameter of the pipe has a mean of 3.0000

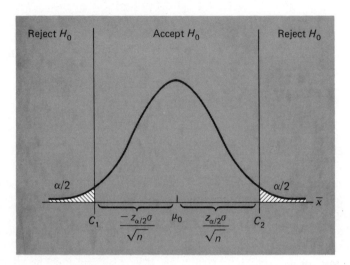

Figure 7.4. Two-tail test criterion.

inches with a standard deviation of 0.0250 inch. To check whether the process is under control, a random sample of size 30 is taken each day and the criterion is such that the probability of committing a Type I error is $\alpha = 0.05$. To find C_1 and C_2 we substitute the given values together with $z_{0.025} = 1.96$ into the formulas given above, and we get $s = \frac{\sigma}{\sqrt{n}}$

$$C_1 = 3.0000 - 1.96 \cdot \frac{0.0250}{\sqrt{30}} = 2.9911$$

and

$$C_2 = 3.0000 + 1.96 \cdot \frac{0.0250}{\sqrt{30}} = 3.0089$$

Thus, the process is regarded to be out of control if $\bar{x}$ is less than 2.9911 inches or greater than 3.0089 inches.

The calculation of β, the probability of a Type II error, is the same as before, and we shall illustrate it here for the case where $\mu_1 = 3.0050$. Referring to the sampling distribution of $\bar{x}$ with the mean $\mu_1 = 3.0050$ and the standard deviation $\frac{0.0250}{\sqrt{30}} = 0.0046$ (see Figure 7.5), we shall have to determine the area under the curve between $C_1 = 2.9911$ and $C_2 = 3.0089$, and it will be left to the reader to verify in Exercise 3 on page 208 that the answer is approximately 0.80. Thus, if the true mean diameter of the steel pipe is 3.0050 inches, the probability is about 0.80 that the given criterion will *erroneously* accept the hypothesis H_0 that the mean diameter is 3.0000 inches.

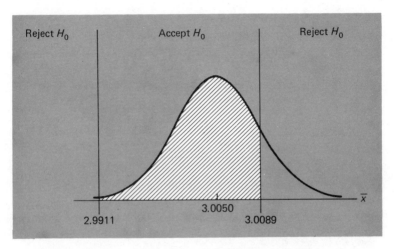

Figure 7.5. Determination of probability of Type II error.

7.5 OPERATING CHARACTERISTIC CURVES

In the drying-time example on page 194, the choice of H_0, namely, the choice of the hypothesis that $\mu = 20$ minutes, was dictated by the paint manufacturer's claim; on the other hand, the choice of the alternative hypothesis $\mu = 21$, for which we calculated the probability of a Type II error, was essentially arbitrary. Thus, it might be of interest to see what happens for other values of μ. To this end, we shall investigate the probability of *accepting* the hypothesis H_0 for any given value of μ, denoted $L(\mu)$, and it should be observed that *this is the probability of committing a Type II error for all values of μ which should be rejected, and the probability of not committing a Type I error for all values of μ which should be accepted.*

Example. Continuing with the drying-time problem on page 196, where we had $\mu_0 = 20$, $\sigma = 2.4$, $n = 36$, and the dividing line of the criterion was $C = 20.75$, it can easily be verified that the methods used on page 197 yield the results shown in the following table and in

Figure 7.6 (see also Exercise 5 on page 208):

Value of μ	Probability of accepting H_0
19.50	0.999
19.75	0.99
20.00	0.97
20.25	0.89
20.50	0.73
20.75	0.50
21.00	0.27
21.25	0.11
21.50	0.03
21.75	0.01
22.00	0.001

Note that the probability of committing a Type II error diminishes when μ is increased, and that the probability of *not* committing a Type I error approaches 1 when μ becomes smaller and smaller than 20.00.

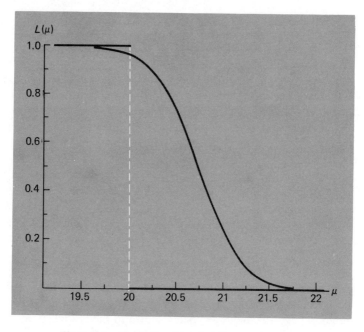

Figure 7.6. Operating characteristic curve.

The graph of $L(\mu)$ for various values of μ, shown in Figure 7.6, is called the *operating characteristic curve* or, simply, the *OC curve* of the test criterion. Ideally, we should want to reject the hypothesis H_0 that $\mu = \mu_0$ in favor of the alternative hypothesis H_1 that $\mu = \mu_1 > \mu_0$ when actually μ exceeds μ_0, and to accept it when actually μ is less than or equal to μ_0. Thus, the "ideal" *OC* curve for our example would be given by the heavy lines of Figure 7.6. In actual practice, *OC* curves can only approximate such an "ideal" curve, with the approximation becoming better as the sample size is increased. To illustrate this point, the reader will be asked to indicate in Exercise 6 on page 208 how the *OC* curve of the drying-time example would have been "improved" if the sample size had been $n = 64$, while keeping the values of μ_0, σ, and α fixed.

Figure 7.6 presents the picture of a typical *OC* curve for the case where the alternative hypothesis is $\mu_1 > \mu_0$. When the alternative hypothesis is $\mu_1 < \mu_0$, the *OC* curve becomes the *mirror image* of that of Figure 7.6, reflected about the (dashed) vertical line through μ_0. For a two-tail test, when the alternative hypothesis is $\mu \neq \mu_0$, the *OC* curve will be as shown in Figure 7.7, and it should be observed that $L(\mu)$ now is the probability

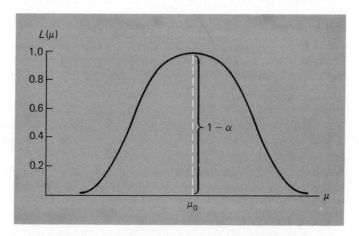

Figure 7.7. Operating characteristic curve for two-tail test.

of a Type II error for all values of μ except μ_0; at μ_0 it is still the probability of *not* committing a Type I error. Due to the symmetry of this kind of *OC* curve about the (dashed) vertical line through μ_0, all problems relating to Type II errors can be answered on the basis of the values of $L(\mu)$ for $\mu > \mu_0$.

In many practical situations, the probabilities of Type II errors can be determined directly from charts like those given in Table 8 at the end of the book. They enable us to read off the value of β which corresponds to given values of μ_0, μ_1, σ, n, and α, and they are based on the assumption that the sampling distribution of $\bar{x}$ is a normal distribution; hence, *they can be used when n is large or the population from which we are sampling has roughly the shape of a normal distribution.* Charts (a) and (b) apply to *one-tail tests* with $\alpha = 0.05$ or $\alpha = 0.01$, which are the two values most often used in actual practice. They apply regardless of whether the alternative hypothesis is $\mu > \mu_0$ or $\mu < \mu_0$, since the quantity which we plot on the horizontal scale is not μ (as in Figure 7.6) but

$$d = \frac{|\mu - \mu_0|}{\sigma}$$

namely, the *absolute value* of the difference between μ and μ_0 divided by σ. The following are some examples which illustrate this method; note that we could not have used it for the problem dealing with the drying times since α equalled 0.03.

Example. Suppose we want to investigate the claim that the sound intensity of certain vacuum cleaners may be looked upon as a random variable having a normal distribution with a mean of 75.20 decibels and a standard deviation of 3.6 decibels. Specifically, we shall want to test the hypothesis $\mu_0 = 75.20$ against the alternative hypothesis $\mu > 75.20$ on the basis of measurements of the sound intensity of $n = 15$ of these machines; the probability of a Type I error is to be $\alpha = 0.05$. Using the formula on page 198, we find that the hypothesis $\mu_0 = 75.20$ will be *accepted* so long as the mean of the sample does not exceed

$$C = \mu_0 + z_\alpha \cdot \frac{\sigma}{\sqrt{n}} = 75.20 + 1.645 \cdot \frac{3.6}{\sqrt{15}} = 76.73$$

and it may be of interest to know the probability that we would, thus, commit a Type II error when actually the mean sound intensity of the vacuum cleaners is 77.00 decibels. Since

$$d = \frac{|77.00 - 75.20|}{3.6} = 0.50$$

we mark this point on the horizontal scale of chart (a) of Table 8, and, going up vertically until we come to the point where the line is crossed by the *OC* curve for $n = 15$, we find that the corresponding probability of a Type II error is about 0.39. Proceeding in the same way, it can also be shown that when $\mu = 76.00$ decibels the probability of a Type II error is about 0.76, and when $\mu = 78.00$ decibels the probability of a Type II error is about 0.09.

Example. To give an example where the alternative hypothesis is $\mu < \mu_0$, suppose we want to check the claim that in city traffic the mileage a person will get per tankful with a certain kind of car may be looked upon as a random variable having a normal distribution with a mean of 15.0 miles per gallon and a standard deviation of 0.8 mile per gallon. Specifically, we shall want to test the hypothesis $\mu_0 = 15.0$ against the alternative hypothesis $\mu < 15.0$ on the basis of the mileage obtained in city traffic with a tankful of gas by $n = 10$ of these cars; the probability of a Type I error is to be $\alpha = 0.01$. Using the formula for C on page 198 with $-z_\alpha$ substituted for z_α, we find that the hypothesis $\mu_0 = 15.0$ will be *accepted* so long as the mean of the sample is not less than

$$C' = 15.0 - 2.33 \cdot \frac{0.8}{\sqrt{10}} = 14.41$$

and it may be of interest to know the probability that we would, thus, commit a Type II error when actually the average mileage per tankful is 14.0 miles per gallon. Since

$$d = \frac{|14.0 - 15.0|}{0.8} = 1.25$$

we mark this point on the horizontal scale of chart (b) of Table 8, and, going up vertically until we come to the point where the line is crossed by the OC curve for $n = 10$, we find that the corresponding probability of a Type II error is about 0.06. Proceeding in the same way, it can also be shown, for example, that when $\mu = 14.5$ miles per gallon, the probability of a Type II error is about 0.64.

Table 8 can also be used to solve problems like the one posed on page 197, namely, that of determining the sample size which is required to test the hypothesis $\mu = \mu_0$ against the alternative hypothesis $\mu = \mu_1$ while α and β take on preassigned values.

Example. Referring again to the example on page 198, where we had $\mu_0 = 20$, $\mu_1 = 21$, $\sigma = 2.4$, $\alpha = 0.05$, and $\beta = 0.10$, we first calculate

$$d = \frac{|21 - 20|}{2.4} = 0.42$$

Then, locating the point corresponding to $d = 0.42$ and $\beta = 0.10$ on chart (a) of Table 8, we find that it lies barely above the OC curve for $n = 50$. Thus, a sample of size $n = 50$ will serve the stated purpose, and this agrees with the result obtained on page 198.

For two-tail tests, that is, for tests of the hypothesis $\mu = \mu_0$ against the alternative hypothesis $\mu \neq \mu_0$, the procedure is the same as before, with the exception that we must now refer to charts (c) and (d) of Table 8. Note that these charts contain only the right-hand halves of the respective OC curves (as shown in Figure 7.7 on page 203), since they are symmetrical about μ_0.

Example. To illustrate, let us verify the value of β obtained in the example on page 200 (dealing with the diameters of the steel pipe), where we wanted to test the hypothesis $\mu = 3.0000$ inches against the alternative hypothesis $\mu \neq 3.0000$ inches, with $\sigma = 0.0250$, $n = 30$, and $\alpha = 0.05$. To determine the probability of a Type II error when $\mu = 3.0050$ inches, we first calculate

$$d = \frac{|3.0000 - 3.0050|}{0.0250} = 0.20$$

and mark this point on the horizontal scale of chart (c) of Table 8. Then, going up vertically until we come to the point where the line is crossed by the OC curve for $n = 30$, we find that the corresponding probability of a Type II error is just about 0.80.

To continue with this example, suppose we want to know by how much we would have to increase the sample size so that β (for $\mu = 3.0050$) would be reduced to 0.50. Locating the point corresponding to $d = 0.20$ and $\beta = 0.50$ on chart (c) of Table 8, we find that it lies just about on the OC curve for $n = 75$. Thus, it would suffice to increase the sample size from $n = 30$ to $n = 75$.

The purpose of the discussion of this and the preceding sections has been to introduce some of the basic problems connected with the testing of statistical hypotheses. Although the methods we have presented are objective—that is, two experimenters analyzing the same data under the same conditions would arrive at the identical results—their use does entail some arbitrary, or subjective, considerations. For instance, in the example on page 194 it was partially a subjective decision to "draw the line" between satisfactory and unsatisfactory values of μ at 21 minutes. It was also partially a subjective decision to use a sample of 36 one-gallon cans of the paint, and to reject the manufacturer's claim for values of

$\bar{x}$ exceeding 20.75 minutes. Approaching the problem differently, the government agency investigating the paint manufacturer's claim could have specified values of α and β, thus controlling the risks to which they are willing to be exposed. The choice of α, the probability of a Type I error, could have been based on the consequences of making that kind of error, namely, the manufacturer's cost of having a good product condemned, the possible cost of subsequent litigation, the manufacturer's cost of unnecessarily adjusting his machinery, the cost to the public of not having the product available when needed, and so forth. The choice of β, the probability of a Type II error, could similarly have been based on the consequences of making that kind of error, namely, the cost to the public of buying an inferior product, the manufacturer's savings in using inferior ingredients but loss in good will, again the cost of possible litigation, and so forth. It should be obvious that it would be extremely difficult to put "cash values" on all these eventualities, but they must nevertheless be considered, at least indirectly, in choosing suitable criteria for testing statistical hypotheses.

In recent years, attempts have been made to incorporate all these matters within a formal theory called *decision theory*. Although many important advances have been made, it should be recognized that such a theory does not eliminate the arbitrariness, or subjectiveness, discussed above; it merely incorporates these matters within the theory. This means that the use of decision theory requires that we actually put cash values on all possible consequences of our decisions. Although this has the advantage that it makes the experimenter (the engineer) more "cost conscious," it also has the disadvantage of requiring information which very often cannot be obtained.

In this text we shall discuss mainly the Neyman-Pearson theory, also called the classical theory of testing hypotheses. This means that we shall consider cost factors and other considerations that are partly arbitrary and partly subjective only insofar as they will affect the choice of a sample size, the choice of an alternative hypothesis, the choice of α and β, and so forth.

EXERCISES

1. If the criterion on page 196 is modified so that the paint manufacturer's claim is rejected for $\bar{x} > 20.50$ minutes, find
(a) the probability of a Type I error;
(b) the probability of a Type II error when $\mu = 21$ minutes.

2. Suppose that in the gasoline mileage problem on page 205, where we had $\mu_0 = 15.0$ miles per gallon, the alternative hypothesis $\mu < 15.0$, $\sigma = 0.8$, and $n = 10$, the hypothesis $\mu = 15.0$ is rejected for $\bar{x} < 14.2$. Find
(a) the probability of a Type I error;
(b) the probability of a Type II error when $\mu = 14.0$.

3. Verify that the area of the shaded region of Figure 7.5 is approximately 0.80.

4. Suppose that in the problem on page 199 dealing with the manufacture of steel pipe, where we had $\mu_0 = 3.0000$ inches, the alternative hypothesis $\mu \neq 3.0000$ inches, $\sigma = 0.0250$ inch, and $n = 30$, the hypothesis $\mu_0 = 3.0000$ is rejected if $\bar{x}$ is less than $C_1 = 2.9930$ or greater than $C_2 = 3.0070$. Find
(a) the probability of a Type I error;
(b) the probability of a Type II error when $\mu = 3.0050$.

5. With reference to the drying-time example, verify the values of the OC curve given in the table on page 202.

6. Suppose that in the drying-time example on page 196, n is changed from 36 to 64, while the other quantities remain $\mu_0 = 20$, $\sigma = 2.4$, and $\alpha = 0.03$. Find
(a) the new dividing line of the test criterion;
(b) the probabilities of Type II errors for the values of μ in the table on page 202, and plot the graph of the corresponding OC curve.
Compare the OC curve obtained here with the one given on page 202.

7. It is desired to test the hypothesis $\mu = 0$ against the alternative $\mu > 0$ on the basis of a random sample of size 9 from a normal population with the variance $\sigma^2 = 1$. If the probability of a Type I error is to be $\alpha = 0.01$,
(a) verify that the region of rejection is $\bar{x} > 0.78$;
(b) calculate β for $\mu = 0.5$, 1.0, and 1.5, and draw a rough sketch of the OC curve.
Also, use Table 8 to verify the values obtained for β in part (b).

8. It is desired to test the hypothesis $\mu_0 = 40$ against the alternative hypothesis $\mu_1 = 42$ on the basis of a random sample from a normal population with the standard deviation $\sigma = 4$. If the probability of a Type I error is to be 0.05 and the probability of a Type II error is to be 0.24, find the required size of the sample by using
(a) the formula on page 198;
(b) Table 8.

9. It is desired to test the hypothesis $\mu_0 = 100$ against the alternative hypothesis $\mu \neq 100$ on the basis of a random sample from a normal population with the standard deviation $\sigma = 16$. If the probability of a Type I error is to be 0.01 and the probability of a Type II error is to be 0.20 when $\mu = 92$, find the required size of the sample by using
(a) the formula on page 198;
(b) Table 8.

10. If a random sample of size $n = 6$ is used to test the hypothesis that the mean of a normal population with $\sigma = 4$ is $\mu_0 = 50$ against the alternative hypothesis that $\mu > 50$, and $\alpha = 0.01$, use Table 8 to determine the probability of committing a Type II error when
    (a) $\mu = 52$;   (b) $\mu = 53$;
    (c) $\mu = 54$;   (d) $\mu = 55$;
    (e) $\mu = 56$;   (f) $\mu = 57$.

11. Repeat Exercise 10, using the alternative hypothesis $\mu \neq 50$ instead of the alternative hypothesis $\mu > 50$.

12. If a random sample of size $n = 8$ is used to test the hypothesis that the mean of a normal population with $\sigma = 20$ is $\mu_0 = 200$ against the alternative hypothesis $\mu \neq 200$, and $\alpha = 0.05$, use Table 8 to determine the probability of committing a Type II error when
    (a) $\mu = 190$;   (b) $\mu = 185$;
    (c) $\mu = 180$;   (d) $\mu = 175$;
    (e) $\mu = 170$.

13. Repeat Exercise 12, using the alternative hypothesis $\mu < 200$ instead of the alternative hypothesis $\mu \neq 200$.

7.6   NULL HYPOTHESES AND SIGNIFICANCE TESTS

There are many situations in which the hypothesis $\mu = \mu_0$ is chosen in such a way that we are willing to "reserve judgment" about its validity *unless there is clear evidence that leads to its rejection.* Thus, set up as a "straw man" with the objective of determining whether or not it can be rejected, we refer to such an hypothesis H_0 as a *null hypothesis.*

The idea of setting up a null hypothesis is not an uncommon one, even in nonstatistical thinking. In fact, this is exactly what is done in an American court of law, where an accused is assumed to be innocent unless he is proven guilty "beyond a reasonable doubt." The null hypothesis states that the accused is *not guilty,* and the probability expressed subjectively by the phrase "beyond a reasonable doubt" reflects the probability α of risking a Type I error. Thus, the "burden of proof" is always on the prosecution in the sense that the accused is found not guilty unless the null hypothesis of innocence is clearly disproved. Note that this does not imply that the defendant has been proved innocent if found not guilty; it implies only that he has not been proved guilty. Of course, since we cannot legally "reserve judgment" if proof of guilt is not established, the accused is freed and we act as if the null hypothesis of innocence were accepted. Note that this is precisely what we may have to do in tests of statistical hypotheses, when we cannot afford the luxury of reserving judgment.

To establish a parallel between this argument and the kind of practical problem in which tests of null hypotheses are ordinarily applied, let us consider the following examples:

Example. Suppose that a decision has to be made whether to buy a new automatic stamping machine, and it is felt that the machine would be economical only if μ, the average number of acceptable pieces produced per hour, were greater than 1,000. Thus, we have a situation which calls for a test of the null hypothesis $\mu = 1,000$, but we cannot tell immediately whether to use the alternative hypothesis $\mu < 1,000$ or the alternative hypothesis $\mu > 1,000$. If the burden of proof is to be placed on the rather expensive machine, it would be appropriate to use the alternative hypothesis $\mu > 1,000$, *installing the machine only if the null hypothesis can be rejected on the basis of sample data.* Note that if $\alpha = 0.05$ for this test, there is only a 5 percent chance of *erroneously* rejecting the null hypothesis and installing the new stamping machine. On the other hand, if the burden of proof is to be placed on the existing machine (which, for some other reason, is looked upon with disfavor), it would be appropriate to use the alternative hypothesis $\mu < 1,000$, *installing the machine so long as the null hypothesis cannot be rejected on the basis of sample data.* In this case, $\alpha = 0.05$ would be the probability of *erroneously* rejecting the null hypothesis and *not* installing the new machine. The choice of the appropriate alternative hypothesis in this and in similar situations is a practical rather than a statistical problem; as we have indicated, it depends on where we wish to place the burden of proof.

Example. To give an example in which we would use the alternative hypothesis $\mu \neq \mu_0$, where μ_0 is the value assumed under a null hypothesis, suppose that a food processor wants to check whether the correct amount of instant coffee goes into his 4-ounce jars. Since the label reads "4 ounces," the food processor cannot afford to put much less than 4 ounces into each jar for fear of losing customer acceptance or running afoul of the law, nor can he afford to put much more than 4 ounces in each jar for fear of losing part of his profit. Thus, the food processor is concerned with the alternative hypothesis $\mu \neq 4$, and the production process will be left undisturbed unless the null hypothesis $\mu = 4$ can be rejected on the basis of sample data.

Tests of the null hypothesis $\mu = \mu_0$ may reasonably be described as tests of whether $\bar{x}$ is *significantly less* than μ_0, *significantly greater* than

μ_0, or *significantly different* from μ_0, depending on the alternative hypothesis with which we are concerned. By "significantly less," "significantly greater," and "significantly different" we mean that the discrepancy between $\bar{x}$ and μ_0 (namely, the difference between what we *get* and what we *expect*) is too large to be reasonably attributed to chance. Thus, we refer to a test of a null hypothesis as a *test of significance*, or a *significance test*.

Example. Suppose we look at the example on page 204 as a significance test of the null hypothesis that the average sound intensity of the vacuum cleaners is $\mu = 75.20$ decibels, and that the alternative hypothesis is again $\mu > 75.20$. Then, if a random sample of size 15 has the mean $\bar{x} = 77.12$, which *exceeds* $C = 76.73$ (the "dividing line" obtained on page 204), we say that the difference between $\bar{x} = 77.12$ and $\mu_0 = 75.20$ is *significant* when $\alpha = 0.05$; in other words, we conclude that the null hypothesis $\mu_0 = 75.20$ will have to be rejected. Note, however, that if we had been willing to risk a Type I error with a probability of $\alpha = 0.01$ instead of 0.05, the dividing line of the criterion would have been

$$C = 75.20 + 2.33 \cdot \frac{3.6}{\sqrt{15}} = 77.37$$

and we would have had to conclude that the difference between $\bar{x} = 77.12$ and $\mu_0 = 75.20$ is *not significant* when $\alpha = 0.01$.

This illustrates the fact that the significance of a difference between $\bar{x}$ and μ_0 may well depend on α, the probability of a Type I error, which in this connection is referred to as the *level of significance*. Thus, in the above example, the null hypothesis $\mu_0 = 75.20$ decibels can be rejected at the level of significance $\alpha = 0.05$, but *not* at the level of significance $\alpha = 0.01$. *Whether we choose $\alpha = 0.05, 0.01$, or some other value, in any given problem, will have to depend on the risks, or consequences of committing a Type I error.*

7.7 HYPOTHESES CONCERNING ONE MEAN

Let us now examine the general problem of testing the null hypothesis $\mu = \mu_0$ against one of the *one-sided alternatives* $\mu < \mu_0$ or $\mu > \mu_0$, or against the *two-sided alternative* $\mu \neq \mu_0$. Considering first the case where we are sampling from a normal population and σ is known, or the sample is large, so that the sampling distribution of $\bar{x}$ is approximately a normal distribution and σ can be estimated in terms of the sample standard deviation, we find that this problem has already been solved in Section 7.4. Given μ_0, σ, n, and α, the critical region is $\bar{x} > C$ when the alternative hypothesis is $\mu > \mu_0$, $\bar{x} < C'$ when the alternative hypothesis is $\mu < \mu_0$, and $\bar{x} < C_1$ *or* $\bar{x} > C_2$ when the alternative hypothesis is $\mu \neq \mu_0$, where the formula for C is on page 198, the formula for C' is like the one for C with $-z_\alpha$ substituted for z_α, and the formulas for C_1 and C_2 are on page 199. An equivalent, but simpler, method of specifying the critical region is to base it on the statistic

$$z = \frac{\bar{x} - \mu_0}{\sigma/\sqrt{n}}$$

instead of $\bar{x}$. If z_α is, as before, such that the area under the standard normal curve to its right equals α, the critical regions for testing the null hypothesis H_0, namely, the hypothesis $\mu = \mu_0$, can be expressed as in the following table:

CRITICAL REGIONS FOR TESTING H_0: $\mu = \mu_0$
(Normal Population and σ Known, or Large Sample)

Alternative hypothesis	Reject H_0 if
$\mu < \mu_0$	$z < -z_\alpha$
$\mu > \mu_0$	$z > z_\alpha$
$\mu \neq \mu_0$	$z < -z_{\alpha/2}$ or $z > z_{\alpha/2}$

Example. To illustrate, let us return to the problem concerning the automatic stamping machine. The null hypothesis is $\mu = 1,000$ and, putting the burden of proof on the new machine, we shall use the alternative hypothesis $\mu > 1,000$. Suppose, furthermore, that the decision is to be based on a sample of $n = 64$ independent trials (each of which contains one hour's production), and that the mean and the standard deviation of the number of acceptable pieces per trial are, respectively, $\bar{x} = 1,028$ and $s = 126$. Although σ is not actually known, the sample is large enough to approximate it with $s = 126$, and we thus obtain

$$z = \frac{1{,}028 - 1{,}000}{126/\sqrt{64}} = 1.78$$

If the level of significance is to be $\alpha = 0.05$, we find from Table 3 that the dividing line of the criterion, the *critical value*, is $z_{0.05} = 1.645$, and since $z = 1.78$ is greater than 1.645, we find that the null hypothesis can be rejected at the 0.05 level. Thus, we would decide that the new machine is to be installed. In Exercise 4 on page 222 the reader will be asked to determine what action would be taken if we had used $\alpha = 0.01$ instead, and the probability of the Type II error we would be risking with this criterion when actually $\mu = 1{,}050$ for the automatic stamping machine.

If the sample size is small and σ is unknown, the tests just described cannot be used. However, if the sample comes from a normal population (to within a reasonable degree of approximation), we can make use of the theory discussed in Section 6.3 and base the test of the null hypothesis $\mu = \mu_0$ on the statistic

$$t = \frac{\bar{x} - \mu_0}{s/\sqrt{n}}$$

The resulting critical regions are as shown in the following table, where t_α is as defined on page 174; that is, the area to its right under the t distribution with $n - 1$ degrees of freedom is equal to α:

CRITICAL REGIONS FOR TESTING H_0: $\mu = \mu_0$
(Normal Population, σ Unknown)

Alternative hypothesis	Reject H_0 if for $n - 1$ degrees of freedom
$\mu < \mu_0$	$t < -t_\alpha$
$\mu > \mu_0$	$t > t_\alpha$
$\mu \neq \mu_0$	$t < -t_{\alpha/2}$ or $t > t_{\alpha/2}$

Example. To illustrate, let us reconsider the problem of deciding whether changes have to be made in the machinery that fills the jars of instant coffee (see page 210), namely, the problem of testing the null hypothesis $\mu = 4$ ounces against the alternative hypothesis $\mu \neq 4$ ounces. Suppose that the level of significance is to be $\alpha = 0.01$, and that the net weights of the contents of a sample of 25 jars have a mean of $\bar{x} = 3.97$ and a standard deviation of $s = 0.04$ ounce. To decide whether the machinery will have to be adjusted, we calculate

$$t = \frac{3.97 - 4}{0.04/\sqrt{25}} = -3.75$$

and since this is less than -2.797, the value of $-t_{0.005}$ with 24 degrees of freedom (see Table 4), *the null hypothesis will have to be rejected.* Note that in spite of this result, the food processor may not wish to adjust his machinery, since the actual loss due to underfilling the jars by a very small amount may be less than the cost of experimenting with adjustments. This illustrates the important point that a result which is *statistically significant* may not be *commercially significant.* Such factors are considered in *decision theory* (see page 207), and under the circumstances it might be more appropriate to test the null hypothesis $\mu = 4$ against an alternative such as $\mu < 3.95$ or $\mu > 4.05$, if it is felt that in either case there is cause for an adjustment of the machinery.

Had the sample data yielded $\bar{x} = 4.02$ and $s = 0.04$ in this example, t would have equalled 2.50, the null hypothesis could *not* have been rejected, and the food processor would have risked committing a Type II error by not adjusting his machinery. The probabilities of committing Type II errors when using the t statistic are difficult to calculate, but there exist special charts, analogous to those of Table 8, which give the *OC* curves for various values of n; they may be found, for example, in *National Bureau of Standards Handbook 91*, listed in the Bibliography at the end of this book.

7.8 HYPOTHESES CONCERNING TWO
MEANS

When dealing with population means, we are frequently faced with the problem of making decisions about the relative values of two or more means. Leaving the general problem until Chapter 12, we shall devote this section to tests concerning the difference between *two* means. For example, if two methods of welding are being considered for use with railroad rails, we may take samples and decide which is better by comparing their mean strengths; also, if a licensing examination is given to engineers who graduated from two different colleges, we may want to decide whether any observed difference between the means of the scores of the students from the two colleges is significant or whether it may be attributed to chance.

Formulating the problem more generally, we shall consider two populations having the means μ_1 and μ_2 and the variances σ_1^2 and σ_2^2, and we shall want to test the null hypothesis $\mu_1 - \mu_2 = \delta$, where δ is a specified constant, on the basis of independent random samples of size n_1 and n_2. Analogous to the tests concerning one mean, we shall consider tests of this null hypothesis against each of the alternatives $\mu_1 - \mu_2 < \delta$, $\mu_1 - \mu_2 > \delta$, and $\mu_1 - \mu_2 \neq \delta$. The test, itself, will depend on the difference between the sample means, $\bar{x}_1 - \bar{x}_2$, and if both samples come from normal populations with known variances (or both samples are large), it can be based on the statistic

$$z = \frac{(\bar{x}_1 - \bar{x}_2) - \delta}{\sigma_{\bar{x}_1 - \bar{x}_2}}$$

whose sampling distribution is (or is approximately) the standard normal distribution. Here $\sigma_{\bar{x}_1 - \bar{x}_2}$ is the standard deviation of the sampling distribution of the difference between the sample means, and its value for random samples from infinite populations may be obtained with the use of the following theorem, which we shall state without proof:

THEOREM 7.1. *If the distributions of two independent random variables have the means μ_1 and μ_2 and the variances σ_1^2 and σ_2^2, then the distribution of their* sum *(or* difference*) has the mean $\mu_1 + \mu_2$ (or $\mu_1 - \mu_2$) and the variance $\sigma_1^2 + \sigma_2^2$.*

To find the variance of the difference between the means of two independent random samples of size n_1 and n_2 from infinite populations, note first that the variances of the two means, themselves, are

$$\sigma_{\bar{x}_1}^2 = \frac{\sigma_1^2}{n_1} \quad \text{and} \quad \sigma_{\bar{x}_2}^2 = \frac{\sigma_2^2}{n_2}$$

where σ_1^2 and σ_2^2 are the variances of the respective populations. Thus, by

Theorem 7.1

$$\sigma^2_{\bar{x}_1 - \bar{x}_2} = \frac{\sigma^2_1}{n_1} + \frac{\sigma^2_2}{n_2}$$

and the test statistic can be written as

$$z = \frac{(\bar{x}_1 - \bar{x}_2) - \delta}{\sqrt{\dfrac{\sigma^2_1}{n_1} + \dfrac{\sigma^2_2}{n_2}}}$$

Analogous to the table on page 212, the critical regions for testing the null hypothesis $\mu_1 - \mu_2 = \delta$ are as follows:

CRITICAL REGIONS FOR TESTING H_0: $\mu_1 - \mu_2 = \delta$
(Normal Populations with σ_1 and σ_2 Known, or Large Samples)

Alternative hypothesis	Reject H_0 if
$\mu_1 - \mu_2 < \delta$	$z < -z_\alpha$
$\mu_1 - \mu_2 > \delta$	$z > z_\alpha$
$\mu_1 - \mu_2 \neq \delta$	$z < -z_{\alpha/2}$ or $z > z_{\alpha/2}$

Example. To illustrate this procedure, suppose that a company claims that its light bulbs are superior to those of its main competitor on the basis of a study which showed that a sample of $n_1 = 40$ of its bulbs had an average "lifetime" of 647 hours of continuous use with a standard deviation of 27 hours, while a sample of $n_2 = 40$ bulbs made by its main competitor had an average "lifetime" of 638 hours of continuous use with a standard deviation of 31 hours. If we wish to test at the 0.05 level of significance whether the observed difference of 9 hours between the two sample means is significant or whether it can be attributed to chance, placing the burden of proof on the claim, the appropriate null and alternative hypotheses are, respectively, $\mu_1 - \mu_2 = 0$ and $\mu_1 - \mu_2 > 0$. Accordingly, we put $\delta = 0$ in the formula for z and the test statistic becomes

$$z = \frac{647 - 638}{\sqrt{\dfrac{27^2}{40} + \dfrac{31^2}{40}}} = 1.38$$

(Note that we have approximated the population variances with the variances of the respective samples, which is justifiable since both samples are fairly large.) Since the value which we obtained does *not* exceed the critical value $z_{0.05} = 1.645$, we find that the null hypothesis *cannot be rejected*; thus, we conclude that the observed difference between the two sample means is *not significant* at the level of significance $\alpha = 0.05$ or, in other words, that it can well be attributed to chance.

If we actually wanted to *accept* the null hypothesis that the average lifetimes of the two kinds of light bulbs are equal, we would be risking Type II errors, for which the probabilities would depend on the actual differences $\delta' = \mu_1 - \mu_2$ between the two population means. Fortunately, these can be determined with the use of Table 8 (so long as we are sampling from normal populations with known standard deviations or both samples are large). The quantity which we mark on the horizontal scale is

$$d = \frac{|\delta - \delta'|}{\sqrt{\sigma_1^2 + \sigma_2^2}}$$

where δ is the value of $\mu_1 - \mu_2$ assumed under the null hypothesis and δ' is the alternative value of $\mu_1 - \mu_2$ with which we are concerned. If $n_1 = n_2 = n$, the probability of a Type II error is read off the *OC* curve corresponding to this value of n; if $n_1 \neq n_2$, it is read off the *OC* curve corresponding to

$$n = \frac{\sigma_1^2 + \sigma_2^2}{\dfrac{\sigma_1^2}{n_1} + \dfrac{\sigma_2^2}{n_2}}$$

Continuation of Example. Suppose we are concerned about the possibility of *not rejecting* the null hypothesis $\delta = 0$, when actually the average lifetime of the first kind of light bulb exceeds that of the second kind of light bulb by $\delta' = 16$ hours. Substituting into the formula for d, we obtain

$$d = \frac{|0 - 16|}{\sqrt{27^2 + 31^2}} = 0.39$$

and, marking this point on the horizontal scale of chart (a) of Table 8 and, going up vertically until we come to the OC curve for $n = 40$, we find that the corresponding probability of a Type II error is approximately 0.20.

If either (or both) samples are *small* and the population variances are *unknown*, we can base tests of the null hypothesis $\mu_1 - \mu_2 = \delta$ on a suitable t statistic, provided it is reasonable to assume that *both populations are normal with* $\sigma_1 = \sigma_2$. Under these conditions it can be shown that the sampling distribution of the statistic

$$t = \frac{(\bar{x}_1 - \bar{x}_2) - \delta}{s_{\bar{x}_1 - \bar{x}_2}}$$

is the t distribution with $n_1 + n_2 - 2$ degrees of freedom. In this formula the denominator involves a "pooled estimate" of the population variance.

To clarify what we mean here by a "pooled estimate" of the population variance, let us first consider the problem of estimating the variance of the distribution of the difference between two sample means. Under the assumption that $\sigma_1^2 = \sigma_2^2 (= \sigma^2)$, this variance is given by

$$\sigma_{\bar{x}_1 - \bar{x}_2}^2 = \frac{\sigma_1^2}{n_1} + \frac{\sigma_2^2}{n_2} = \sigma^2 \left(\frac{1}{n_1} + \frac{1}{n_2} \right)$$

and we now estimate σ^2 by "pooling" the two sums of squared deviations from the respective sample means. In other words, we estimate σ^2 by means of

$$\frac{\sum (x_1 - \bar{x}_1)^2 + \sum (x_2 - \bar{x}_2)^2}{n_1 + n_2 - 2} = \frac{(n_1 - 1)s_1^2 + (n_2 - 1)s_2^2}{n_1 + n_2 - 2}$$

where $\sum (x_1 - \bar{x}_1)^2$ is the sum of the squared deviations from the mean for the first sample, while $\sum (x_2 - \bar{x}_2)^2$ is the sum of the squared deviations from the mean for the second sample. We divide by $n_1 + n_2 - 2$, since there are $n_1 - 1$ independent deviations from the mean in the first sample, $n_2 - 1$ in the second, and we thus have $n_1 + n_2 - 2$ independent deviations from the mean to estimate the population variance. Substituting this estimate of σ^2 into the above expression for $\sigma_{\bar{x}_1 - \bar{x}_2}^2$ and then substituting the square root of the result into the denominator of the formula for t above, we finally obtain

$$t = \frac{(\bar{x}_1 - \bar{x}_2) - \delta}{\sqrt{(n_1 - 1)s_1^2 + (n_2 - 1)s_2^2}} \sqrt{\frac{n_1 n_2 (n_1 + n_2 - 2)}{n_1 + n_2}}$$

for the statistic on which we shall base the test. The corresponding critical regions for testing the null hypothesis $H_0: \mu_1 - \mu_2 = \delta$ are as shown in the following table:

CRITICAL REGIONS FOR TESTING $H_0: \mu_1 - \mu_2 = \delta$
(Normal Populations, $\sigma_1 = \sigma_2 = \sigma$, σ Unknown)

Alternative hypothesis	Reject H_0 if for $n_1 + n_2 - 2$ degrees of freedom
$\mu_1 - \mu_2 < \delta$	$t < -t_\alpha$
$\mu_1 - \mu_2 > \delta$	$t > t_\alpha$
$\mu_1 - \mu_2 \neq \delta$	$t < -t_{\alpha/2}$ or $t > t_{\alpha/2}$

Example. To illustrate this kind of test, let us consider the following measurements of the heat-producing capacity of the coal produced by two mines (in millions of calories per ton):

Mine 1: 8,260, 8,130, 8,350, 8,070, 8,340
Mine 2: 7,950, 7,890, 7,900, 8,140, 7,920, 7,840

The means of these two samples are, respectively, 8,230 and 7,940, whose difference is *quite large*, but it remains to be seen whether it is *significant*, namely, whether it will enable us to reject the null hypothesis $\mu_1 - \mu_2 = 0$ against the alternative hypothesis $\mu_1 - \mu_2 \neq 0$, say, at the level of significance $\alpha = 0.05$. Calculating first the two quantities $(n_1 - 1)s_1^2$ and $(n_2 - 1)s_2^2$, we get

$$\sum (x_1 - \bar{x}_1)^2 = 30^2 + 100^2 + 120^2 + 160^2 + 110^2$$
$$= 63{,}000$$

and

$$\sum (x_2 - \bar{x}_2)^2 = 10^2 + 50^2 + 40^2 + 200^2 + 20^2 + 100^2$$
$$= 54{,}600$$

Then, substituting these two quantities together with $\bar{x}_1 = 8{,}230$, $\bar{x}_2 = 7{,}940$, $\delta = 0$, $n_1 = 5$, and $n_2 = 6$ into the formula for t, we obtain

$$t = \frac{(8{,}230 - 7{,}940)}{\sqrt{63{,}000 + 54{,}600}} \sqrt{\frac{5 \cdot 6 \cdot 9}{11}} = 4.19$$

Since this exceeds 2.262 the value of $t_{0.025}$ for 9 degrees of freedom, the null hypothesis will have to be *rejected*; in other words, we conclude that the average heat-producing capacity of the coal from the two mines is not the same.

In this last example we arbitrarily went ahead and performed a *two-sample t test*, tacitly assuming that the population variances were equal. Fortunately, the test is not overly sensitive to small differences between the population variances, and the procedure used in this instance is quite justifiable. To be on safer grounds, however, we should first have tested whether the difference between the sample variances may be attributed to chance; a procedure for performing such a test will be given in Chapter 8.

If the difference between the sample variances is large or if it is otherwise unreasonable to treat the population variances as being equal, we cannot use the two-sample *t* test just described. However, there are several alternative methods that can be used instead, which do not require the assumption of equal population variances. In one of these, the *paired-sample t test*, we work with the differences of paired observations, one coming from each sample, and the pairing is random (unless the data are actually given as pairs, say, when we are dealing with the gasoline consumption of a number of cars *before and after* the installation of certain anti-pollution devices). Of course, if $n_1 \neq n_2$, some of the observations in the larger sample will have to be discarded. Using these differences, we then perform the one-sample *t* test described in Section 7.7 to determine whether their mean is significantly different from δ.

Example. To determine the effectiveness of an industrial safety program, the following data were collected on the average weekly loss of man hours due to accidents in 12 plants during "one year before and one year after" the program was put into operation:

$$
\begin{array}{llll}
37.1 \text{ and } 28.0, & 72.5 \text{ and } 59.3, & 26.6 \text{ and } 24.7, & 125.0 \text{ and } 120.3 \\
45.8 \text{ and } 46.2, & 54.3 \text{ and } 43.6, & 13.2 \text{ and } 15.4, & 79.5 \text{ and } 75.1 \\
12.6 \text{ and } 18.3, & 34.9 \text{ and } 29.7, & 39.3 \text{ and } 35.0, & 26.6 \text{ and } 24.8
\end{array}
$$

The differences between these paired observations are 9.1, 13.2, 1.9, 4.7, -0.4, 10.7, -2.2, 4.4, -5.7, 5.2, 4.3, and 1.8, and we shall want to check whether we can reject the null hypothesis $\mu = 0$ against the alternative hypothesis $\mu > 0$ (namely, that the industrial safety program *is* effective) at $\alpha = 0.05$. Since the mean and the standard deviation of the differences are, respectively, $\bar{x} = 3.92$ and $s = 5.38$, substitution into the formula for t on page 213 with $\mu_0 = 0$ yields

$$
t = \frac{3.92 - 0}{5.38/\sqrt{12}} = 2.52
$$

Since this exceeds 1.796, the value of $t_{0.05}$ for 11 degrees of freedom, we find that the null hypothesis will have to be rejected; in other words, we conclude that the industrial safety program *is effective.* Note, however, that if we had used the level of significance $\alpha = 0.01$, we would not have been able to arrive at this conclusion, and this illustrates the important point that *the level of significance should always be chosen before the data are analyzed in any fashion.*

Although this paired-sample t test can be used when sampling from normal populations *regardless of whether the samples are independent (and the pairing is random) or the population standard deviations are equal,* it has two disadvantages. First, some of the data may have to be discarded when the sample sizes are unequal, and second, there is a serious loss of information in the sense that the test is performed as if there were only n observations instead of $2n$ observations. An alternate test which avoids these disadvantages when the samples are independent is given in Exercise 22 below.

EXERCISES

1. A city's police department is considering replacing the tires on its cars with radial tires. If μ_1 is the average number of miles they get out of their old tires and μ_2 is the average number of miles they will get out of the new tires, the null hypothesis they shall want to test is $\mu_1 - \mu_2 = 0$.
   (a) What alternative hypothesis should they use if they do not want to buy the radial tires unless they are definitely proven to give a better mileage? In other words, the burden of proof is put on the radial tires and the old tires are to be kept unless the null hypothesis can be rejected.
   (b) What alternative hypothesis should they use if they are anxious to get the new tires (which have some other nice features) unless they actually give a poorer mileage than the old tires? Note that now the burden of proof is on the old tires, which will be kept only if the null hypothesis can be rejected.

2. A producer of extruded plastic products finds that his mean daily inventory is 1,250 pieces. A new marketing policy has been put into effect and it is desired to test the null hypothesis that the mean daily inventory is still the same. What alternative hypothesis should be used if
   (a) it is desired to know whether or not the new policy changes the mean daily inventory;
   (b) it is desired to demonstrate that the new policy actually reduces the mean daily inventory;

(c) the new policy will be retained so long as it cannot be shown that it actually increases the mean daily inventory?

3. An oceanographer wants to check whether the average depth of the ocean in a certain region is 57.4 fathoms, as had previously been recorded. What can he conclude at the level of significance $\alpha = 0.05$ if soundings taken at 40 random locations in the given region yielded a mean of 59.1 fathoms with a standard deviation of 5.2 fathoms?

4. Referring to the example on page 212 and using $s = 126$ as an estimate of σ, but changing the level of significance from $\alpha = 0.05$ to $\alpha = 0.01$, what action would you now take? Find from Table 8 the probability of a Type II error when $\mu = 1,050$.

5. A random sample of 6 steel beams has a mean compressive strength of 58,392 psi (pounds per square inch) with a standard deviation of 648 psi. Use this information and the level of significance $\alpha = 0.05$ to test whether the true average compressive strength of the steel from which this sample came is 58,000 psi.

6. The specifications for a certain kind of ribbon call for a mean breaking strength of 175 pounds. If six pieces of the ribbon (randomly selected from different rolls) have a mean breaking strength of 172.8 pounds with a standard deviation of 7.9 pounds, test the null hypothesis $\mu = 175$ against the alternative hypothesis $\mu < 175$ at the level of significance $\alpha = 0.05$.

7. In a labor-management discussion it was brought up that workers (employed in a certain large plant) take on the average 53.6 minutes to get to work.
   (a) Show that this figure is substantiated (that is, the null hypothesis $\mu = 53.6$ cannot be rejected against the alternative hypothesis $\mu \neq 53.6$ at the level of significance $\alpha = 0.05$) by a survey in which a random sample of 60 workers took on the average 54.8 minutes with a standard deviation of 8.1 minutes.
   (b) Using $s = 8.1$ as an estimate of σ and referring to Table 8, find the probability that the criterion used here might lead to a Type II error when actually it takes workers on the average 57.6 minutes to get to work.

8. Test runs with six models of an experimental engine showed that they operated, respectively, for 24, 28, 21, 23, 32, and 22 minutes with a gallon of a certain kind of fuel. Is this evidence at the level of significance $\alpha = 0.01$ that *in general* this engine will operate below a desired standard (average) of 29 minutes per gallon?

9. A random sample from a company's very extensive files shows that orders for a certain piece of machinery were filled, respectively, in 10, 12, 19, 14, 15, 18, 11, and 13 days. Use the level of significance $\alpha = 0.01$ to test the claim that on the average such orders are filled in 10.5 days. Choose the alternative hypothesis so that rejection of the null hypothesis $\mu = 10.5$ implies that it takes longer than indicated.

10. Tests performed with a random sample of 40 diesel engines produced by a large manufacturer showed that they had a mean thermal efficiency of 31.4 percent with a standard deviation of 1.6 percent.
    (a) Use this information and the level of significance $\alpha = 0.01$ to test the null hypothesis $\mu = 32.3$ percent against the alternative hypothesis that the true average thermal efficiency of the manufacturer's diesel engines is less than 32.3 percent.
    (b) Using $s = 1.6$ percent as an estimate of σ and referring to Table 8, find the probability with which this test criterion would expose us to a Type II error when actually the true average thermal efficiency of the manufacturer's diesel engines is 31.7 percent.
    (c) How large a sample should have been used in the first place, so that the probability of the Type II error asked for in part (b) would have been only 0.15?

11. The diameters of rotor shafts in a lot have a mean of 0.249 inch and a standard deviation of 0.003 inch. The inner diameters of bearings in another lot have a mean of 0.255 inch and a standard deviation of 0.002 inch.
    (a) What are the mean and the standard deviation of the clearances between shafts and bearings selected from these lots?
    (b) If a shaft and a bearing are selected at random, what is the probability that the shaft will not fit inside the bearing? (Assume that both dimensions are normally distributed.)

12. An investigation of two kinds of photocopying equipment showed that 75 failures of the first kind of equipment took on the average 83.2 minutes to repair with a standard deviation of 19.3 minutes, while 75 failures of the second kind of equipment took on the average 90.8 minutes to repair with a standard deviation of 21.4 minutes.
    (a) Test the null hypothesis $\mu_1 - \mu_2 = 0$ (namely, the hypothesis that on the average it takes an equal amount of time to repair either kind of equipment) against the alternative hypothesis $\mu_1 - \mu_2 \neq 0$ at the level of significance $\alpha = 0.05$.
    (b) Using 19.3 and 21.4 as estimates of σ_1 and σ_2 and referring to Table 8, find the probability of accepting the null hypothesis $\mu_1 - \mu_2 = 0$ with the criterion of part (a) when actually $\mu_1 - \mu_2 = -12$.

13. Random samples are taken from two normal populations with $\sigma_1 = 10.8$ and $\sigma_2 = 14.4$ to test the null hypothesis $\mu_1 - \mu_2 = 53.2$ against the alternative hypothesis $\mu_1 - \mu_2 > 53.2$ at the level of significance $\alpha = 0.01$. Use Table 8 to determine the common sample size $n = n_1 = n_2$ that is required if the probability of accepting the null hypothesis is to be 0.09 when $\mu_1 - \mu_2 = 66.7$.

14. Measuring samples of nylon yarn taken from two spinning machines, it was found that 8 samples from the first machine had a mean denier of 9.67 with a standard deviation of 1.81 while 10 samples from the second machine had a mean denier of 7.43 with a standard deviation of 1.48. Use these results to

test the null hypothesis $\mu_1 - \mu_2 = 1.5$ against the alternative hypothesis $\mu_1 - \mu_2 > 1.5$ at the level of significance $\alpha = 0.05$. Assume that the populations are normal and have the same variance.

15. Measurements of the fat content of two kinds of ice cream, Brand A and Brand B, yielded the following sample data:

> *Brand A:* 13.5, 14.0, 13.6, 12.9, and 13.0 percent
> *Brand B:* 12.9, 13.0, 12.4, 13.5, and 12.7 percent

Test the null hypothesis $\mu_1 = \mu_2$ (where μ_1 and μ_2 are the respective true average fat contents of the two kinds of ice cream) against the alternative hypothesis $\mu_1 \neq \mu_2$ at the level of significance $\alpha = 0.05$.

16. Studying the flow of traffic at two busy intersections between 4 P.M. and 6 P.M. (to determine the possible need for turn signals), it was found that on 40 weekdays there were on the average 247.3 cars approaching the first intersection from the south which made left turns, while on 30 weekdays there were on the average 254.1 cars approaching the second intersection from the south which made left turns. The corresponding sample standard deviations are $s_1 = 15.2$ and $s_2 = 18.7$.
    (a) Test the null hypothesis $\mu_1 - \mu_2 = 0$ against the alternative hypothesis $\mu_1 - \mu_2 \neq 0$ at the level of significance $\alpha = 0.01$.
    (b) Using 15.2 and 18.7 as estimates of σ_1 and σ_2 and referring to Table 8, find the probability if accepting the null hypothesis $\mu_1 - \mu_2 = 0$ when actually $|\mu_1 - \mu_2| = 15.6$.

17. To test the claim that the resistance of electric wire can be reduced by at least 0.050 ohm by alloying, 25 values obtained for each, alloyed wire and standard wire, produced the following results:

	Mean	Standard deviation
Alloyed wire	0.083 ohm	0.003 ohm
Standard wire	0.136 ohm	0.002 ohm

Use the level of significance $\alpha = 0.05$ to determine whether the claim has been substantiated.

18. As part of an industrial training program, some trainees are instructed by Method A which is straight teaching-machine instruction, and some are instructed by Method B which also involves the personal attention of an instructor. If random samples of size 10 are taken from large groups of trainees instructed by each of these two methods, and the scores which they obtained in an appropriate achievement test are

> *Method A:* 71, 75, 65, 69, 73, 66, 68, 71, 74, 68
> *Method B:* 72, 77, 84, 78, 69, 70, 77, 73, 65, 75

test the claim that Method B is more effective. Use the level of significance $\alpha = 0.05$.

19. The following data were obtained in an experiment designed to check whether there is a systematic difference in the weights obtained with two different scales:

| | Weight in grams | |
	Scale I	Scale II
Rock specimen 1	11.23	11.27
Rock specimen 2	14.36	14.41
Rock specimen 3	8.33	8.35
Rock specimen 4	10.50	10.52
Rock specimen 5	23.42	23.41
Rock specimen 6	9.15	9.17
Rock specimen 7	13.47	13.52
Rock specimen 8	6.47	6.46
Rock specimen 9	12.40	12.45
Rock specimen 10	19.38	19.35

Use the level of significance $\alpha = 0.05$ to test whether the difference of the means of the weights obtained with the two scales is significant.

20. The following are the *Brinell* hardness values obtained for samples of two magnesium alloys:

Alloy A: 66.3, 60.7, 63.5, 64.9, 61.8, 64.3, 60.2, 62.0,
64.7, 73.6, 65.1, 64.5, 68.4, 63.2, 64.0
Alloy B: 71.3, 65.4, 69.6, 62.6, 63.9, 68.8, 69.2, 65.3,
70.1, 64.8, 68.9, 70.7, 71.1, 65.8, 66.2

Randomly pair each value obtained for Alloy A with a different value obtained for Alloy B, and use the method described on page 220, the *paired-sample t test*, to test the null hypothesis $\mu_A = \mu_B$ against the alternative hypothesis $\mu_A < \mu_B$ at the level of significance $\alpha = 0.05$. Explain why it would probably have been unreasonable to use the "standard" two-sample t test of Section 7.8 in this exercise.

21. The following are the number of sales which a sample of 9 salesmen of industrial chemicals in California and a sample of 6 salesmen of industrial chemicals in Oregon made over a certain fixed period of time:

California: 59, 68, 44, 71, 63, 46, 69, 54, 48
Oregon: 50, 31, 62, 52, 70, 41

Randomly pair each value of the second sample with a different value of the first sample, and use the *paired-sample t test* to test the null hypothesis $\mu_0 - \mu_1 = 0$ against the alternative hypothesis $\mu_0 - \mu_1 \neq 0$ at the level of significance $\alpha = 0.01$. Explain why it would probably have been unreasonable to use the "standard" two-sample t test of Section 7.8 in this example.

22. When dealing with two independent random samples from normal populations whose variances are not necessarily equal, the following *Smith-Satterthwaite* test can be used to test the null hypothesis $\mu_1 - \mu_2 = \delta$. The test statistic is given by

$$t' = \frac{(\bar{x}_1 - \bar{x}_2) - \delta}{\sqrt{\dfrac{s_1^2}{n_1} + \dfrac{s_2^2}{n_2}}}$$

and its sampling distribution can be approximated by the t distribution with

$$\frac{\left(\dfrac{s_1^2}{n_1} + \dfrac{s_2^2}{n_2}\right)^2}{\dfrac{(s_1^2/n_1)^2}{n_1 - 1} + \dfrac{(s_2^2/n_2)^2}{n_2 - 1}}$$

degrees of freedom. Use this test for the data of Exercise 20 and compare the answer with the one previously obtained.

23. Use the formula for t on page 218 to construct a $1 - \alpha$ confidence interval for δ, the difference between the two population means.

24. Use the formula obtained in Exercise 23 to construct a 0.95 confidence interval for the true difference between the average resistance of the alloyed and standard wires referred to in Exercise 17.

8

Inferences Concerning Variances

8.1 THE ESTIMATION OF VARIANCES

There were several instances in the preceding chapter where we estimated a population standard deviation by means of a sample standard deviation as defined by the formula

$$s = \sqrt{\frac{\sum_{i=1}^{n} (x_i - \bar{x})^2}{n - 1}}$$

We substituted s for σ in the large-sample confidence interval for μ on page 187, in the large-sample test concerning μ on page 212, and in the large-sample test concerning the difference between two means on page 216. Since there are many statistical procedures in which s is substituted for σ, or s^2 for σ^2, let us show first that s^2 is, in fact, an *unbiased estimate* of σ^2; namely, that the mean of the sampling distribution of s^2 equals σ^2.

If $f(x_1, x_2, \ldots, x_n)$ is the joint density of the sample values $x_1, x_2, \ldots,$ and x_n, it follows from the discussion on page 129 that the mean of the sampling distribution of s^2 is given by

$$\int_{-\infty}^{\infty} \int_{-\infty}^{\infty} \cdots \int_{-\infty}^{\infty} s^2 f(x_1, x_2, \ldots, x_n) \, dx_1 \, dx_2 \ldots dx_n$$

$$= \int_{-\infty}^{\infty} \int_{-\infty}^{\infty} \cdots \int_{-\infty}^{\infty} \sum_{i=1}^{n} \frac{(x_i - \bar{x})^2}{n - 1} f(x_1, x_2, \ldots, x_n) \, dx_1 \, dx_2 \ldots dx_n$$

If we now write

$$\sum_{i=1}^{n} (x_i - \bar{x})^2 = \sum_{i=1}^{n} x_i^2 - n\bar{x}^2$$

and interchange the operations of summation and integration, the expression for the mean of the distribution of s^2 becomes

$$\frac{1}{n - 1} \sum_{i=1}^{n} \int_{-\infty}^{\infty} \int_{-\infty}^{\infty} \cdots \int_{-\infty}^{\infty} x_i^2 f(x_1, x_2, \ldots, x_n) \, dx_1 \, dx_2 \ldots dx_n$$

$$- \frac{n}{n - 1} \int_{-\infty}^{\infty} \int_{-\infty}^{\infty} \cdots \int_{-\infty}^{\infty} \bar{x}^2 f(x_1, x_2, \ldots, x_n) \, dx_1 \, dx_2 \ldots dx_n$$

Assuming without loss of generality that the population mean μ is equal to zero, we find that these last two integrals have already been evaluated on pages 166 and 167, where it was shown that their respective values are σ^2 and σ^2/n. Thus, the mean of the sampling distribution of s^2 is given by

$$\frac{1}{n - 1} \sum_{i=1}^{n} \sigma^2 - \frac{n}{n - 1} \cdot \frac{\sigma^2}{n} = \frac{n\sigma^2}{n - 1} - \frac{\sigma^2}{n - 1} = \sigma^2$$

and this completes the proof of the unbiasedness of s^2 as an estimate of σ^2. (Note that, had we divided by n instead of $n - 1$ in defining s^2, the resulting estimator would have been biased; the mean of its sampling distribution would have been $\frac{n - 1}{n} \cdot \sigma^2$.)

Although the sample *variance* is an unbiased estimator of σ^2, it does not follow that the sample *standard deviation* is also an unbiased estimator of σ; in fact, it is not. However, for large samples the bias is small and it is common practice to estimate σ with s.

Besides s, population standard deviations are sometimes estimated in terms of the *sample range R*, which we defined earlier as the difference between the largest and the smallest values in a sample. Given a random sample of size n from a *normal population*, it can be shown that the sampling distribution of R has the mean $d_2\sigma$ and the standard deviation $d_3\sigma$, where d_2 and d_3 are constants which depend on the size of the sample, as shown in the following table:

n	2	3	4	5	6	7	8	9	10
d_2	1.128	1.693	2.059	2.326	2.534	2.704	2.847	2.970	3.078
d_3	0.853	0.888	0.880	0.864	0.848	0.833	0.820	0.808	0.797

Thus, the statistic R/d_2 is an unbiased estimator for σ, and the standard deviation of its sampling distribution is given by $d_3\sigma/d_2$. For very small samples ($n \leq 5$), R/d_2 provides nearly as good an estimate of σ as does s, but as the sample size increases it becomes far more efficient to use s. The range is used to estimate σ primarily in problems of quality control, where sample sizes are usually small and computational ease is an important requirement. This application will be discussed in detail in Chapter 14.

Example. To illustrate the use of R as an estimate of σ, let us refer to the example on page 219, which dealt with the heat-producing capacity of the coal from two mines. For the first mine, the smallest of the five sample values is 8,070, the largest is 8,350, so that the range is $R = 8,350 - 8,070 = 280$, and the corresponding estimate of σ is

$$\frac{R}{d_2} = \frac{280}{2.326} = 120.4$$

Note that this is close to the standard deviation of this sample, which equals

$$\sqrt{\frac{63,000}{4}} = 125.5$$

In Exercise 2 on page 232, the reader will be asked to verify that for the sample from the other mine the estimates of σ based on R and s are, respectively, 118.4 and 104.5.

Interval estimates of σ or σ^2 are almost always based on the sample variance. Dealing with random samples from *normal populations*, we make use of Theorem 6.4, according to which

$$\frac{(n-1)s^2}{\sigma^2}$$

is a value of a random variable having the chi-square distribution with

$n - 1$ degrees of freedom. Thus, if χ_1^2 and χ_2^2 cut off left- and right-hand tails of area $\alpha/2$ under the chi-square distribution with $n - 1$ degrees of freedom, we can assert with a degree of confidence of $1 - \alpha$ that

$$\chi_1^2 < \frac{(n - 1)s^2}{\sigma^2} < \chi_2^2$$

Solving this inequality for σ^2, we obtain the following $1 - \alpha$ confidence interval for σ^2:

$$\frac{(n - 1)s^2}{\chi_2^2} < \sigma^2 < \frac{(n - 1)s^2}{\chi_1^2}$$

If we now take the square root of each member of this inequality, we obtain a corresponding $1 - \alpha$ confidence interval for σ. Note that the above confidence intervals, obtained by taking "equal tails," do not actually give the *shortest* confidence intervals for σ^2 and σ, because the chi-square distribution is not symmetrical. Nevertheless, they are used in most applications in order to avoid complicated calculations.

Example. To illustrate the construction of a confidence interval for σ (or σ^2), let us return to the example on page 176, and let us suppose that the refractive indices of 20 pieces of glass (randomly selected from a large shipment purchased by an optical firm) had a variance of $1.20 \cdot 10^{-4}$. To construct a 0.95 confidence interval for σ, we find from Table 5 that for 19 degrees of freedom

$$\chi_1^2 = \chi_{.975}^2 = 8.907 \quad \text{and} \quad \chi_2^2 = \chi_{.025}^2 = 32.852$$

Substituting these values together with $n = 20$ and $s^2 = 1.20 \cdot 10^{-4}$ into the above confidence interval formula for σ^2, we get

$$\frac{(19)(1.20 \cdot 10^{-4})}{32.852} < \sigma^2 < \frac{(19)(1.20 \cdot 10^{-4})}{8.907}$$

or

$$0.000069 < \sigma^2 < 0.000256$$

and, hence,

$$0.0083 < \sigma < 0.0160$$

The method which we have discussed applies only to random samples from normal populations (or at least to random samples from populations sufficiently close to normal so that the method provides a good approximation). If the sample size is large, it can be shown that for samples from distributions that are approximately normal, the sampling distribution of s can be approximated closely with a normal distribution having the mean σ and the standard deviation $\sigma/\sqrt{2n}$. Hence,

$$z = \frac{s - \sigma}{\sigma/\sqrt{2n}}$$

is a value of a random variable having approximately the standard normal distribution, and solving the inequality

$$-z_{\alpha/2} < \frac{s - \sigma}{\sigma/\sqrt{2n}} < z_{\alpha/2}$$

for σ, we thus obtain the following $1 - \alpha$ large-sample confidence interval for σ:

$$\frac{s}{1 + \dfrac{z_{\alpha/2}}{\sqrt{2n}}} < \sigma < \frac{s}{1 - \dfrac{z_{\alpha/2}}{\sqrt{2n}}}$$

Example. To illustrate the use of this procedure, let us refer to the example on page 170, where 40 one-gallon cans of a certain kind of paint covered on the average 513.3 square feet with a standard deviation of 31.5 square feet. To construct a 0.99 confidence interval for σ, we substitute $n = 40$, $s = 31.5$, and $z_{0.005} = 2.58$ into the formula, getting

$$\frac{31.5}{1 + \dfrac{2.58}{\sqrt{80}}} < \sigma < \frac{31.5}{1 - \dfrac{2.58}{\sqrt{80}}}$$

and

$$24.5 < \sigma < 44.3$$

EXERCISES

1. Use the data of Exercise 8 on page 222 to estimate σ for the number of minutes the experimental engine will operate with the given fuel
   (a) in terms of the sample standard deviation;
   (b) in terms of the sample range.

Compare the two estimates by expressing their difference as a percentage of the first.

2. With reference to the example on page 229, show that for the sample from the second mine the estimates of σ based on R and s are, respectively, 118.4 and 104.5.

3. Use the data of Exercise 20 on page 225 to estimate σ for the *Brinell* hardness of Alloy A
   (a) in terms of the sample standard deviation;
   (b) in terms of the sample range.
   Compare the two estimates by expressing their difference as a percentage of the first.

4. Use the information given in Exercise 6 on page 222 to construct a 0.95 confidence interval for the standard deviation of the breaking strength of the given kind of ribbon.

5. With reference to Exercise 9 on page 222, construct a 0.99 confidence interval for the variance of the amount of time it takes to fill an order for a piece of the given kind of machinery.

6. While performing a strenuous task, the pulse rate of 25 workers increased on the average by 18.4 beats per minute with a standard deviation of 4.9 beats per minute. Find a 0.95 confidence interval for the corresponding population standard deviation, using
   (a) the small-sample technique based on the chi-square distribution;
   (b) the large-sample technique based on the normal distribution.
   Compare the two confidence intervals.

7. Using the data of part (a) of Exercise 7 on page 222, construct a 0.99 confidence interval for the standard deviation of the amount of time it takes workers (employed at the given plant) to get to work.

8. In the example on page 230 we indicated that the refractive indices of 20 pieces of glass had a variance of $1.20 \cdot 10^{-4}$. Using $\sqrt{1.20 \cdot 10^{-4}} = 0.011$ as a preliminary estimate of σ and the large-sample theory about the sampling distribution of s on page 231, find the minimum sample size needed to estimate σ with a maximum error of 0.0025 at a 0.95 degree of confidence.

9. Using the value of s obtained in Exercise 3, construct a 0.98 confidence interval for σ, measuring the actual variability in the hardness of Alloy A.

10. If 32 measurements of the boiling point of sulfur had a mean of 444.8°C and a standard deviation of 0.83°C, construct a 0.99 confidence interval for the true standard deviation of such measurements.

11. With reference to Exercise 10 on page 223, construct a 0.95 confidence interval for σ, measuring the true variability of the thermal efficiency of such diesel engines.

8.2 HYPOTHESES CONCERNING ONE VARIANCE

In this section we shall consider the problem of testing the null hypothesis that a population variance equals a specified constant against a suitable one-sided or two-sided alternative; that is, we shall test the null hypothesis $\sigma^2 = \sigma_0^2$ against one of the alternatives $\sigma^2 < \sigma_0^2$, $\sigma^2 > \sigma_0^2$, or $\sigma^2 \neq \sigma_0^2$. Tests like these are important whenever it is desired to control the uniformity of a product or an operation. For example, suppose that a silicon disc, or "wafer," is to be cut into small squares, or "dice," to be used in the manufacture of a semiconductor device. Since certain electrical characteristics of the finished device may depend on the thickness of the die, it is important that all dice cut from a wafer have approximately the same thickness. Thus, not only must the mean thickness of a wafer be kept within specifications, but also the variation in thickness from location to location on the wafer.

Using the same sampling theory as in the preceding section, namely, the fact that for random samples from normal population with the variance σ_0^2

$$\chi^2 = \frac{(n-1)s^2}{\sigma_0^2}$$

is a value of a random variable having the chi-square distribution with $n - 1$ degrees of freedom, we can use this χ^2 statistic to test the null hypothesis $\sigma^2 = \sigma_0^2$ as shown in the following table:

CRITICAL REGIONS FOR TESTING $H_0 : \sigma^2 = \sigma_0^2$
(Normal Population)

Alternative hypothesis	Reject H_0 if for $n - 1$ degrees of freedom
$\sigma^2 < \sigma_0^2$	$\chi^2 < \chi^2_{1-\alpha}$
$\sigma^2 > \sigma_0^2$	$\chi^2 < \chi^2_{\alpha}$
$\sigma^2 \neq \sigma_0^2$	$\chi^2 < \chi^2_{1-\alpha/2}$ or $\chi^2 > \chi^2_{\alpha/2}$

In this table χ_α^2 is as defined on page 176. Note that "equal tails" are used in performing the two-tail test, which is actually not the "best" procedure since the chi-square distribution is not symmetrical.

Example. To illustrate this type of *chi-square test*, let us assume that the thicknesses of 15 dice cut from a silicon wafer have a standard deviation of 0.64 mil and that the lapping process which ground the wafers to the proper thickness is acceptable only if σ, the population standard deviation of the dice thicknesses, is at most 0.50 mil. This means that we shall want to test the null hypothesis $H_0: \sigma = 0.50$, against the alternative $H_1: \sigma > 0.50$, and we shall do so at a level of significance of 0.05. Since this is equivalent to testing the null hypothesis $\sigma^2 = (0.50)^2 = 0.25$ against the alternative $\sigma^2 > 0.25$, the value of the test statistic becomes

$$\chi^2 = \frac{(14)(0.64)^2}{0.25} = 22.94$$

which does not exceed 23.685, the value of $\chi_{0.05}^2$ for 14 degrees of freedom. Thus, the null hypothesis *cannot be rejected;* even though the sample standard deviation exceeded 0.50, the evidence is not sufficient to arrive at the conclusion that the lapping process is unsatisfactory.

There exist tables, like Table 8, which enable us to read off the probabilities of Type II errors connected with this kind of test. As given in the *National Bureau of Standards Handbook 91* (see Bibliography at the end of the book), they contain the *OC* curves for the different one-sided and two-sided alternatives, for $\alpha = 0.05$ and $\alpha = 0.01$, and for various values of n. The quantity which we mark on the horizontal scale is the ratio $\frac{\sigma_1}{\sigma_0}$, where σ_1 is the alternative value of σ for which we want to determine the probability of erroneously accepting the null hypothesis $\sigma = \sigma_0$.

If the population from which we are sampling is not normal but the sample size is *large* ($n \geq 30$ is the usual rule of thumb), the null hypothesis $\sigma = \sigma_0$ can be tested with the use of the statistic

$$z = \frac{s - \sigma_0}{\sigma_0/\sqrt{2n}}$$

whose sampling distribution is approximately the standard normal distribution. The only difference in the tests is that z and z_α replace χ^2 and χ_α^2.

Example. The specifications for the mass production of certain bearings require, among other things, that the standard deviation of their diameters should not exceed 0.0040 cm. If a random sample of size $n = 35$, taken to test the null hypothesis $\sigma = 0.0040$ against the alternative $\sigma > 0.0040$ at the level of significance $\alpha = 0.01$, yielded $s = 0.0053$ for the diameters of the bearings, we get

$$z = \frac{0.0053 - 0.0040}{0.0040/\sqrt{70}} = 2.72$$

Since this exceeds $z_{0.01} = 2.33$, we find that the null hypothesis will have to be *rejected;* the bearings do *not* meet the specifications with regard to the variability of their diameters.

8.3   HYPOTHESES CONCERNING TWO VARIANCES

The two-sample t test for the difference between two means, described in Section 7.5, requires the assumption that the population variances are equal. Before proceeding with this test, therefore, it would be desirable to put this assumption to a test. In this section we shall describe a test of the null hypothesis $H_0 : \sigma_1^2 = \sigma_2^2$ against an appropriate alternative, which applies to independent random samples from two normal populations. As we shall discover in Chapter 12, the test has many other important applications.

If independent random samples of size n_1 and n_2 are taken from normal populations having the same variance, it follows from Theorem 6.5 that the statistic

$$F = \frac{s_1^2}{s_2^2}$$

is a value of a random variable having the F distribution with $n_1 - 1$ and $n_2 - 1$ degrees of freedom. Thus, if the null hypothesis $\sigma_1^2 = \sigma_2^2$ is true, the ratio of the sample variances s_1^2 and s_2^2 provides a statistic on which tests of the null hypothesis can be based.

The critical region for testing H_0 against the alternative hypothesis $\sigma_1^2 > \sigma_2^2$ is $F > F_\alpha$, where F_α is as defined on page 177, namely, it cuts off a right-hand tail of area α under the F distribution with $n_1 - 1$ and $n_2 - 1$

degrees of freedom. Similarly, the critical region for testing H_0 against the alternative hypothesis $\sigma_1^2 < \sigma_2^2$ is $F < F_{1-\alpha}$, and this causes some difficulties since Table 6 only contains values corresponding to right-hand tails of $\alpha = 0.05$ and $\alpha = 0.01$. As a result, we use the reciprocal of the original test statistic and make use of the relation

$$F_{1-\alpha}(v_1, v_2) = \frac{1}{F_\alpha(v_2, v_1)}$$

first given on page 178. Thus, we base the test on the statistic $F = s_2^2/s_1^2$ and the critical region for testing $H_0: \sigma_1^2 = \sigma_2^2$ against $H_1: \sigma_1^2 < \sigma_2^2$ becomes $F > F_\alpha$, where F_α is the appropriate critical value of F with $n_2 - 1$ and $n_1 - 1$ degrees of freedom.

For the two-sided alternative $\sigma_1^2 \neq \sigma_2^2$ the critical region is $F < F_{1-\alpha/2}$ or $F > F_{\alpha/2}$, where $F = s_1^2/s_2^2$ and the degrees of freedom are $n_1 - 1$ and $n_2 - 1$. In practice, we modify this test as in the preceding paragraph, so that we can again use the table of F values corresponding to right-hand tails of $\alpha = 0.05$ and $\alpha = 0.01$. To this end we let s_M^2 represent the larger of the two sample variances, s_m^2 the smaller, and we write the corresponding sample sizes as n_M and n_m. Thus, the test statistic becomes $F = s_M^2/s_m^2$ and the critical region is as shown in the following table:

CRITICAL REGIONS FOR TESTING $H_0: \sigma_1^2 = \sigma_2^2$
(Normal Populations)

Alternative hypothesis	Test statistic	Reject H_0 if
$\sigma_1^2 < \sigma_2^2$	$F = \dfrac{s_2^2}{s_1^2}$	$F > F_\alpha(n_2 - 1, n_1 - 1)$
$\sigma_1^2 > \sigma_2^2$	$F = \dfrac{s_1^2}{s_2^2}$	$F > F_\alpha(n_1 - 1, n_2 - 1)$
$\sigma_1^2 \neq \sigma_2^2$	$F = \dfrac{s_M^2}{s_m^2}$	$F > F_{\alpha/2}(n_M - 1, n_m - 1)$

The level of significance of these tests is α and the figures indicated in parentheses are the respective degrees of freedom. Note that, as in the chi-square test, "equal tails" are used in the two-tail test as a matter of mathematical convenience, even though the F distribution is not symmetrical.

Example. To give an example of a one-tail test for the equality of two variances, suppose that it is desired to determine whether Company 1 is better at plating silverware than Company 2. Assuming that there is essentially no difficulty in controlling the average thickness of the plating, we shall base our decision on the *variability* of its thickness. To set up the test so that rejection of the null hypothesis will enable us to conclude that the plating done by Company 1 is better (less variable) than that done by Company 2, we shall have to test the null hypothesis $\sigma_1^2 = \sigma_2^2$ against the alternative hypothesis $\sigma_1^2 < \sigma_2^2$, say, at the level of signficance $\alpha = 0.05$. Supposing now that random samples of size 12 of the work done by each company yields $s_1 = 0.035$ mil and $s_2 = 0.062$ mil for the thickness of the plating, we find that the appropriate test statistic has the value

$$F = \frac{(0.062)^2}{(0.035)^2} = 3.14$$

Comparing this value with 2.82, the value of $F_{0.05}(11,11)$ obtained from Table 6(a) by linear interpolation,* we conclude that the null hypothesis will have to be rejected; in other words, the data support the contention that the plating done by Company 1 is less variable than that done by Company 2.

Example. To give an example of a two-tail test for the equality of two variances, let us return to the example on page 219, where we compared the average heat-producing capacity of the coal from two mines. In that example we went ahead and performed the two-sample t test even though there was a difference between the variances of the two samples. To justify this procedure, let us now test whether $\sigma_1^2 = \sigma_2^2$, and let us use the alternative hypothesis $\sigma_1^2 \neq \sigma_2^2$, since we are concerned only with the question whether or not these two variances are equal. Since we had $n_1 = 5$, $n_2 = 6$, $s_1^2 = \dfrac{63,000}{4} = 15,750$, and $s_2^2 = \dfrac{54,600}{5} = 10,920$ on page 219, we get

$$F = \frac{s_1^2}{s_2^2} = \frac{15,750}{10,920} = 1.44$$

Using $\alpha = 0.02$, we find from Table 6(b) that $F_{0.01}(4, 5) = 11.4$, and it follows that the null hypothesis cannot be rejected. In other words, it was justifiable to use the two-sample t test in the given example.

---

*Note that it was really unnecessary to interpolate in this case since 3.14 exceeds both $F_{0.05}(10, 11)$ and $F_{0.05}(12, 11)$.

Had we wanted to use the level of significance $\alpha = 0.05$ or $\alpha = 0.01$ in this example, we would have required tables of the values of $F_{0.025}(v_1, v_2)$ or $F_{0.005}(v_1, v_2)$; such tables may be found in the *Biometrika Tables for Statisticians* listed in the Bibliography at the end of the book. Also, *OC* curves for the various F tests discussed in this section may be found in the book by A. H. Bowker and G. J. Lieberman listed in the Bibliography, while those for the one-tail tests only may be found in the *National Bureau of Standards Handbook 91*.

EXERCISES

1. Referring to Exercise 5 on page 222, test that $\sigma = 600$ psi for the compressive strength of the given kind of steel. Use the alternative hypothesis $\sigma > 600$ and the level of significance $\alpha = 0.05$.

2. If 12 determinations of the specific heat of iron have a standard deviation of 0.0086, test the null hypothesis that $\sigma = 0.010$ for such determinations. Use the alternative hypothesis $\sigma \neq 0.010$ and the level of significance $\alpha = 0.01$.

3. Referring to Exercise 12 on page 223, test the null hypothesis that for the first kind of photocopying equipment the variability of the time that is required for repairs is given by $\sigma = 15.0$ minutes. Use the alternative hypothesis $\sigma > 15.0$ and the level of significance $\alpha = 0.05$.

4. Test the null hypothesis that $\sigma = 0.015$ inch for the diameters of certain bolts, if in a random sample of size 15 the diameters of the bolts had a variance of 0.00011. Use the level of significance $\alpha = 0.01$.

5. Referring to Exercise 6 on page 232, test the null hypothesis that for such increases of the pulse rate (while performing the given task) the variance can be expected to equal $\sigma^2 = 30.0$. Use the alternative hypothesis $\sigma^2 < 30.0$.

6. Playing ten rounds of golf on his home course, a golf professional averaged 71.3 with a standard deviation of 1.32. Test the null hypothesis that the consistency of his game on his home course is actually measured by $\sigma = 1.20$, against the alternative hypothesis that he is less consistent. Use $\alpha = 0.05$.

7. Past data indicate that the variance of measurements made on sheet metal stampings by experienced quality control inspectors is 0.18 square inch. Such measurements made by an inexperienced inspector could have too large a variance (perhaps because of inability to read instruments properly) or too small a variance (perhaps because unusually high or low measurements are discarded). If a new inspector measures 100 stampings with a variance of 0.13 square inch, test at the 0.05 level of significance whether the inspector is making satisfactory measurements.

8. Justify the use of the two-sample t test in Exercise 14 on page 223 by testing the null hypothesis that the two populations have equal variances. Use the level of significance $\alpha = 0.02$.

9. Justify the use of the two-sample t test in Exercise 15 on page 224 by testing the null hypothesis that the two populations have equal variances. Use the level of significance $\alpha = 0.02$.

10. Two different lighting techniques are compared by measuring the intensity of light at selected locations in areas lighted by the two methods. If 15 measurements in the first area had a standard deviation of 2.7 foot-candles and 21 measurements in the second area had a standard deviation of 4.2 foot-candles, can it be concluded that the lighting in the second area is less uniform? Use a 0.01 level of significance. What assumptions must be made as to how the two samples are obtained?

11. In Exercise 16 on page 224 we had $n_1 = 40$, $n_2 = 30$, $s_1 = 15.2$, and $s_2 = 18.7$. Use the level of significance $\alpha = 0.05$ to test the claim that at the second intersection there is a greater variability in the number of cars which make left turns approaching from the south between 4 P.M. and 6 P.M.

12. Pull-strength tests on 10 soldered leads for a semiconductor device yield the following results in pounds force required to rupture the bond:

$$15.8, \quad 12.7, \quad 13.2, \quad 16.9, \quad 10.6, \quad 18.8, \quad 11.1, \quad 14.3, \quad 17.0, \quad 12.5$$

Another set of 8 leads was tested after encapsulation to determine whether the pull strength has been increased by encapsulation of the device, with the following results:

$$24.9, \quad 23.6, \quad 19.8, \quad 22.1, \quad 20.4, \quad 21.6, \quad 21.8, \quad 22.5$$

How would you test the implied hypothesis? Why?

13. Random samples of size n_1 and n_2, respectively, are taken from two log-normal populations, and the resulting sample means are $\bar{x}_1 = 3.74$ and $\bar{x}_2 = 13.91$. You wish to test whether the second population has a mean value four times as large as the first.
    (a) Can you directly use a two-sample test? Why?
    (b) Is there a transformation that can be made on the data that could conceivably allow the use of a two-sample test?

9

Inferences Concerning Proportions

9.1 ESTIMATION OF PROPORTIONS

Many engineering problems deal with proportions, percentages, or probabilities. In acceptance sampling we are concerned with the *proportion* of defectives in a lot, and in life testing we are concerned with the *percentage* of certain components which will perform satisfactorily during a stated period of time, or the *probability* that a given component will last at least a given number of hours. It should be clear from these examples that problems concerning proportions, percentages, or probabilities are really equivalent; a percentage is merely a proportion multiplied by 100, and a probability can be interpreted as a proportion "in the long run."

The information that is usually available for the estimation of a proportion is the number of times, x, that an appropriate event has occurred in n trials (or observations). The point estimate, itself, is usually the relative frequency x/n, namely, the proportion of the time that the event has actually occurred. If the n trials satisfy the assumptions underlying the binomial distribution listed on pages 54 and 55, we know that the mean and the standard deviation of the number of "successes" are given by np and $\sqrt{np(1-p)}$. Dividing both of these quantities by n, we find that the mean and the standard deviation of the *proportion* of successes (namely, the relative frequency x/n) are given by

$$\mu = p \quad \text{and} \quad \sigma = \sqrt{\frac{p(1-p)}{n}}$$

This shows that the sample proportion may be regarded as an *unbiased estimator* for the parameter p of a binomial distribution, namely, the "true" proportion we are trying to estimate on the basis of the sample.

In the construction of confidence intervals for the parameter p of the binomial distribution, we meet several obstacles. First, there is the fact that the binomial distribution is *discrete*, so that it may be impossible to get an interval with a degree of confidence exactly equal to $1 - \alpha$; second, the standard deviation of the sampling distribution of x (or that of the sample proportion x/n) involves the parameter p we are trying to estimate. To construct a confidence interval for p having *approximately* a degree of confidence of $1 - \alpha$, we first select for a given set of values of p the corresponding quantities x_0 and x_1, where x_0 is the *largest integer* for which

$$\sum_{k=0}^{x_0} b(k; n, p) \le \frac{\alpha}{2}$$

while x_1 is the *smallest integer* for which

$$\sum_{k=x_1}^{n} b(k; n, p) \le \frac{\alpha}{2}$$

To emphasize the fact that x_0 and x_1 depend on the value chosen for p, we shall write these quantities as $x_0(p)$ and $x_1(p)$. We can then assert with a probability of approximately $1 - \alpha$ (and *at least* $1 - \alpha$) that

$$x_0(p) < x < x_1(p)$$

for any given value of p. To change these inequalities into confidence intervals for p, we can use a simple graphical method which is illustrated in the following example:

Example. Suppose we want to find approximate confidence intervals for p for samples of size $n = 20$. Using Table 1, we first determime x_0 and x_1 for selected values of p such that x_0 is the *largest integer* for which

$$B(x_0; 20, p) \leq 0.025$$

while x_1 is the *smallest integer* for which

$$1 - B(x_1 - 1; 20, p) \leq 0.025$$

Letting p equal 0.1, 0.2, 0.3, . . . , and 0.9, we thus obtain the values shown in the following table:

p	0.1	0.2	0.3	0.4	0.5	0.6	0.7	0.8	0.9
x_0	—	0	1	3	5	7	9	11	14
x_1	6	9	11	13	15	17	19	20	—

Plotting the points $(p, x_0(p))$ and $(p, x_1(p))$ as we did in Figure 9.1, and drawing smooth curves, one through the x_0 points and one through the x_1 points, we can now "solve" for p. For any given value of x, we obtain approximate 0.95 confidence limits for p by going horizontally to the two curves and marking off the corresponding values of p (see Figure 9.1). Thus, for $x = 4$ we obtain the approximate 0.95 confidence interval $0.05 < p < 0.45$.

Graphs similar to the one shown in Figure 9.1 are given in Tables 9(a) and 9(b) at the end of the book for various values of n and the degrees of confidence 0.95 and 0.99. These tables differ from the one of Figure 9.1 in that the relative frequency x/n is used instead of x, thus making it possible to graph curves corresponding to various values of n on the same diagram. Also, for increased accuracy, Tables 9(a) and 9(b) are arranged so that values of x/n from 0 to 0.50 are marked on the bottom scale while those from 0.50 to 1.00 are marked on the top scale of the diagram. For values of x/n from 0 to 0.50 the confidence limits for p are read off the left-hand scale of the diagram, while for values of x/n from 0.50 to 1.00 they are read off the right-hand scale.

Continuation of Example. Note that for $n = 20$ and $x = 4$ we obtain the 0.95 confidence interval $0.06 < p < 0.44$ with the use of Table

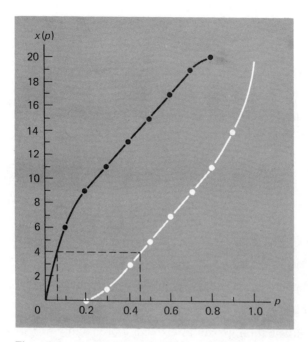

Figure 9.1. Confidence intervals for proportions ($n = 20$).

9(a). This is very close, indeed, to the confidence interval obtained earlier with the use of the somewhat cruder Figure 9.1.

When n is large and there is no reason to suspect that p is very close to either 0 or 1, we can construct approximate confidence intervals for p making use of the normal curve approximation to the binomial distribution. Since the mean and the standard deviation of the binomial distribution are np and $\sqrt{np(1-p)}$, we can assert with a probability of $1 - \alpha$ that x, the number of successes we obtain in n trials, will satisfy the inequality

$$-z_{\alpha/2} < \frac{x - np}{\sqrt{np(1-p)}} < z_{\alpha/2}$$

Solving this quadratic inequality for p, we can obtain a corresponding set of (approximate) confidence limits for p (see Exercise 10 on page 250). However, since the necessary calculations are involved, we shall use the further approximation of substituting the sample proportion x/n for p in $\sqrt{np(1-p)}$. This yields the following $1 - \alpha$ *large-sample confidence interval for p:*

$$\frac{x}{n} - z_{\alpha/2} \sqrt{\frac{\frac{x}{n}\left(1 - \frac{x}{n}\right)}{n}} < p < \frac{x}{n} + z_{\alpha/2} \sqrt{\frac{\frac{x}{n}\left(1 - \frac{x}{n}\right)}{n}}$$

Example. For instance, for $n = 100$ and $x = 36$ we obtain the 0.95 large-sample confidence interval

$$0.36 - 1.96 \sqrt{\frac{(0.36)(0.64)}{100}} < p < 0.36 + 1.96 \sqrt{\frac{(0.36)(0.64)}{100}}$$

or

$$0.266 < p < 0.454$$

Had we used Table 9(a) for this example, we would have obtained $0.27 < p < 0.46$.

The magnitude of the error we make when using x/n as an estimate of p is given by $\left|\frac{x}{n} - p\right|$, and using the normal curve approximation to the binomial distribution we can assert with a probability of $1 - \alpha$ that

$$\frac{|x - np|}{\sqrt{np(1 - p)}} < z_{\alpha/2} \quad \text{or} \quad \left|\frac{x}{n} - p\right| < z_{\alpha/2} \sqrt{\frac{p(1 - p)}{n}}$$

To be able to use this last formula, we again substitute x/n for p inside the radical and, so long as we are dealing with a large sample, we can assert with a probability of $1 - \alpha$ that *the error in using x/n as an estimate of p is less than*

$$E = z_{\alpha/2} \sqrt{\frac{\frac{x}{n}\left(1 - \frac{x}{n}\right)}{n}}$$

Example. For instance, if 136 of 400 persons interviewed as part of a poll felt their city's public transportation system is adequate and we estimate the corresponding true proportion as $\frac{136}{400} = 0.34$, we can assert with a probability of 0.95 that the error of this estimate does not exceed

$$E = 1.96 \sqrt{\frac{(0.34)(0.66)}{400}} = 0.046$$

The formula on which the above estimate of the error was based can also be used to determine the sample size that is required to attain a desired degree of precision, or reliability. Solving for n, we get

$$n = p(1 - p)\left[\frac{z_{\alpha/2}}{E}\right]^2$$

but this formula cannot be used as it stands *unless we have some information about the possible size of p* (on the basis of collateral data, say, a pilot sample). If no such information is available, we can make use of the fact that $p(1 - p)$ is *at most* equal to $\frac{1}{4}$, as can be shown by the method of elementary calculus. Thus, if

$$n = \frac{1}{4}\left[\frac{z_{\alpha/2}}{E}\right]^2$$

we can assert with a probability of *at least* $1 - \alpha$ that the error in using x/n as an estimate of p is less than E.

Example. If we want to use a sample proportion to estimate the true proportion of defectives in a very large shipment of adobe bricks, and we want to be able to assert with a degree of confidence of at least 0.95 that the size of the error is less than 0.04, we would have to take a sample of size

$$n = \frac{1}{4}\left(\frac{1.96}{0.04}\right)^2 = 600.25$$

that is, a sample of size 601 (rounded up). Had we known from experience with similar adobe brick that the proportion we are trying to estimate is in the neighborhood of 0.12, substitution into the appropriate formula would have yielded

$$n = p(1 - p)\left[\frac{z_{\alpha/2}}{E}\right]^2 \approx (0.12)(0.88)\left[\frac{1.96}{0.04}\right]^2 = 253.55$$

that is, a sample of size 254. *This serves to illustrate how some collateral information about the possible size of p can substantially reduce the size of the required sample.*

If p is very close to 0 or 1, none of the confidence intervals we have described provides a satisfactory approximation, even when n is large. Actually, in cases where p is close to 0, as might be encountered in testing for high reliability, we are really more interested in *one-sided confidence intervals* of the form $p < C$; that is, we are interested mainly in finding an *upper confidence limit* for p. (When p is close to 1, we would, correspondingly, be interested in an upper confidence limit for $1 - p$ and, hence, a *lower confidence limit* for p.) As we already pointed out on page 79, the binomial distribution is best approximated with a Poisson distribution (rather than a normal distribution) when p is small and n is large. Thus, using the relation $\lambda = np$, we can obtain an upper confidence limit for p based essentially on the Poisson distribution. Without going into any detail, let us merely state the result that this leads to the *approximate* $1 - \alpha$ confidence interval

$$p < \frac{1}{2n} \cdot \chi_\alpha^2$$

where χ_α^2 is as defined on page 176 and the number of degrees of freedom equals $2(x + 1)$. A discussion of this approximation may be found in the book by A. Hald mentioned in the Bibliography at the end of the book.

Example. For $x = 4$, $n = 2{,}000$, and $\alpha = 0.01$, for instance, we obtain the 0.99 confidence interval

$$p < \frac{1}{2(2{,}000)} \cdot 23.209 = 0.0058$$

where 23.209 is the value of $\chi_{0.01}^2$ for $2(4 + 1) = 10$ degrees of freedom. Note that if we had *erroneously* used the large-sample confidence interval formula on page 244, we would have obtained 0.0020 ± 0.0026, namely, the much *narrower* interval $p < 0.0046$.

9.2 BAYESIAN ESTIMATION

In the preceding section we looked upon the "true" proportions we tried to estimate as *unknown constants*; in Bayesian estimation these parameters are looked upon as *random variables* having *prior distributions* reflecting either the strength of one's belief about the possible values they can take on, or other indirect information. As in Section 7.3, we are thus faced with the problem of *combining prior information with direct sample evidence.*

Example. To illustrate how this might be done, suppose that a manufacturer, who regularly receives large shipments of electronic components from a vendor, knows that about 25 percent of the time 0.005 (half of one percent) of the components are defective, about 25 percent of the time 0.01 of the components are defective, and about 50 percent of the time 0.02 of the components are defective. Thus, before a shipment from this vendor is inspected, we have the following *prior distribution* for the proportion of defectives:

Value of p	Prior probability
0.005	0.25
0.01	0.25
0.02	0.50

Now suppose that 200 of these components, randomly selected from the shipment, are inspected, and only one of them is found to be defective. The probability of this happening when $p = 0.005$, $p = 0.01$, or $p = 0.02$ are, respectively, $\binom{200}{1}(0.005)^1(0.995)^{199} = 0.37$, $\binom{200}{1}(0.01)^1(0.99)^{199} = 0.27$, and $\binom{200}{1}(0.02)^1(0.98)^{199} = 0.07$, where we used the formula for the binomial distribution on page 56 and logarithms to simplify the calculations. Combining these probabilities with the prior probabilities by means of the formula for the rule of Bayes (Theorem 2.8 on page 39), we find that the *posterior probability* for $p = 0.005$ is

$$\frac{(0.25)(0.37)}{(0.25)(0.37) + (0.25)(0.27) + (0.50)(0.07)} = 0.47$$

and that the corresponding *posterior probabilities* for $p = 0.01$ and $p = 0.02$ are, respectively, 0.35 and 0.18. We have thus arrived at the following *posterior distribution* for the proportion of defective components:

Value of p	Posterior probability
0.005	0.47
0.01	0.35
0.02	0.18

Note that whereas the odds were originally 3 to 1 against $p = 0.005$, it is now almost an even bet; of course, this shift is accounted for by the fact that in the sample only $\frac{1}{200} = 0.005$ of the components inspected were defective.

It was assumed in this example that p had to be 0.005, 0.01, or 0.02, and this restriction was imposed mainly to simplify the calculations; the method would have been the same if we had considered 10 different values of p, or even 100. It would be more logical, perhaps, to let p take on any value on the continuous interval from 0 to 1, and in that case it is customary to use as the *prior distribution* the beta distribution of Section 4.8. The parameters of this distribution are α and β, and its mean, μ_0, and standard deviation, σ_0, can be expressed in terms of α and β by means of the formulas on page 120. Then, proceeding as on page 190, it can be shown that the *posterior distribution* of p, namely, the *conditional distribution* of p for a given (observed) value of x, is also a beta distribution, and that its parameters are $x + \alpha$ instead of α and $n - x + \beta$ instead of β.* Thus, it follows from the formula on page 120, that the *mean* of the posterior distribution of p is given by

$$\mu_1 = \frac{x + \alpha}{\alpha + \beta + n}$$

and we have thus arrived at a formula for a *Bayesian point estimate* of the "true" proportion with which we may be concerned in any given example.

Example. To illustrate these ideas, suppose that a person doing research for a large oil company "feels" that the proportion of customers requiring oil as well as gasoline is a random variable having as its prior distribution the beta distribution with $\alpha = 60$ and $\beta = 300$. (This "feeling" might have been based on an analysis of prior experience.) To picture this prior distribution, note that according to the formulas on page 120,

$$\mu_0 = \frac{60}{60 + 300} = 0.167$$

---

*Proofs of these results may be found in the book by J. E. Freund listed in the Bibliography at the end of the book.

and

$$\sigma_0^2 = \frac{60 \cdot 300}{(60 + 300)^2(60 + 300 + 1)} = 0.000385$$

Thus, the mean of this distribution is 0.167 and its standard deviation is about 0.020.

Now suppose that the person doing the research gets some new data, showing that in a random sample of size $n = 800$, there were only $x = 94$ who needed oil as well as gasoline. Combining these data with the prior information, we find that the mean of the posterior distribution, the point estimate of the "true" proportion, is

$$\mu_1 = \frac{94 + 60}{60 + 300 + 800} = 0.133$$

and in Exercise 17 on page 252 the reader will be asked to verify that the standard deviation of the posterior distribution is $\sigma_1 = 0.010$.

Given an appropriate table, it is possible to continue with an example like this and calculate the *posterior probabilities* that are associated with various intervals. For instance, in our example we would be able to determine the posterior probability that the true proportion of customers needing oil as well as gasoline is between 0.13 and 0.14, or, say, between 0.12 and 0.15.

EXERCISES

1. In a random sample of 200 claims filed against an insurance company writing collision insurance on cars, 84 exceeded $1,200. Use Table 9 to construct a 0.95 confidence interval for the true proportion of claims filed against this company that exceed $1,200.

2. In a sample of 100 reports of UFO sightings, 83 could easily be explained in terms of natural phenomena. Use Table 9 to construct a 0.99 confidence interval for the corresponding true proportion of reported UFO sightings that can easily be explained in terms of natural phenomena.

3. In a random sample of 300 industrial accidents, it was found that 173 were due at least partially to unsafe working conditions. Construct a 0.95 confidence interval for the corresponding true proportion
   (a) using Table 9;
   (b) using the large-sample confidence interval formula on page 244.

4. In a random sample of 160 workers exposed to a certain amount of radiation, 24 experience some ill effects. Construct a 0.99 confidence interval for the corresponding true *percentage*
   (a) using Table 9;
   (b) using the large-sample confidence interval formula on page 244.

5. In a sample survey, 200 persons with incomes of $12,000 or more were asked "Where would you most likely find out all there is to know about some news in which you are very much interested?" If 112 replied "television," construct a 0.95 confidence interval for the corresponding true proportion
   (a) using Table 9;
   (b) using the large-sample confidence interval formula on page 244.

6. In a sample survey of the "safety explosives" used in certain mining operations, explosives containing potassium nitrate were found to be used in 95 of 250 cases.
   (a) Use Table 9 to construct a 0.95 confidence interval for the corresponding true proportion.
   (b) If $\frac{95}{250} = 0.38$ is used as an estimate of the corresponding true proportion, what can we say with a probability of 0.95 about the possible size of our error?

7. In a recent study, 69 of 120 meteorites were observed to enter the earth's atmosphere with a velocity less than 26 miles per second. What can we say with a probability of 0.99 about the possible size of our error, if we estimate the corresponding true proportion as $\frac{69}{120} = 0.575$?

8. What is the size of the smallest sample required to estimate an unknown proportion to within a maximum error of 0.05 with a degree of confidence of at least 0.95. How would the required sample size be affected if it were known that the proportion to be estimated is in the neighborhood of 0.25?

9. What is the minimum sample size a public opinion poll has to use if it wants to be able to assert with a probability of at least 0.95 that its estimate of the percentage of the vote a certain candidate will get is not "off" by more than 3 percent? How would the required minimum sample size be affected if it were known that the percentage to be estimated is in the neighborhood of 60 percent?

10. Show that the double inequality on page 243 leads to the following $1 - \alpha$ confidence limits:

$$\frac{x + \frac{1}{2} z_{\alpha/2}^2 \pm z_{\alpha/2} \sqrt{\frac{x(n - x)}{n} + \frac{1}{4} z_{\alpha/2}^2}}{n + z_{\alpha/2}^2}$$

11. Use the confidence interval formula of Exercise 10 to rework Exercise 3 and compare the results.

12. In 4,000 firings of a certain kind of rocket there were ten instances in which

a rocket exploded upon ignition. Construct an upper 0.95 confidence limit for the probability that such a rocket will explode on ignition.

13. Observing the amount of pollutants in the air on 500 days, it was found that the amount of pollutants in the air in a western city exceeded 200 micrograms per cubic meter only four times. Construct an upper 0.99 confidence limit for the probability that the air pollution in this city will exceed 200 micrograms per cubic meter on any one day.

14. The head of a highway department feels that 4 out of 5 road building jobs will stay within cost estimates, while his assistant feels that it should be only 3 out of 5.

    (a) If the head of the highway department is regarded to be "three times as good" as his assistant in determining figures like these, what *prior probabilities* should we assign to their respective claims?

    (b) What *posterior probabilities* would we assign to their claims if it was found that among 12 road building jobs (randomly selected from the department's files) only two stayed within cost estimates?

    (c) Calculate the mean of the posterior distribution as a *Bayesian point estimate* of the probability that one of the department's road building jobs will stay within cost estimates, and compare it with corresponding point estimates based, respectively, on the prior information alone and on the direct sample evidence alone.

15. The purchasing agent of a firm feels that the probability is 0.80 that any one of several shipments of steel recently received will meet specifications. The head of the firm's quality control department feels that this probability is 0.90, and the chief engineer feels (somewhat more pessimistically) that it is 0.60.

    (a) If the managing director of the firm feels that in this matter the purchasing agent is 10 times as reliable as the chief engineer while the head of the quality control department is 14 times as reliable as the chief engineer, what *prior probabilities* would he assign to the respective claims?

    (b) If five of the shipments are inspected and only two meet specifications, how would this affect the prior probabilities; that is, what *posterior probabilities* should the managing director of the firm assign to the respective claims.

16. The output of a certain transistor production line is checked daily by inspecting a sample of 200 units. Over a long period of time, the process has maintained a yield of 80 percent, that is, a proportion defective of 0.20, and the variation of the proportion defective (from day to day) is measured by a standard deviation of 0.0125.

    (a) Verify that this *prior distribution* of the proportion defective can be approximated with a beta distribution having $\alpha = 200$ and $\beta = 800$.

    (b) If on a given day the sample (of 200 units) contains 86 defectives, estimate that day's proportion defective in terms of the mean of the corresponding *posterior distribution*.

17. Verify for the example on page 248 that the standard deviation of the posterior distribution is $\sigma_1 = 0.010$.

18. Records of the dean of an engineering school (collected over many years) show that on the average 75 percent of all applicants have an I.Q. of at least 115. Of course, the percentage varies somewhat from year to year and this variation is measured by a standard deviation of 2.15 percent.

    (a) Verify that the prior distribution of the *proportion* of applicants with an I.Q. of at least 115 can be approximated with a beta distribution having $\alpha = 300$ and $\beta = 100$.

    (b) If a sample check of 25 applicants in the year 1976 showed that only 12 of them have an I.Q. of at least 115, estimate that year's proportion of applicants with an I.Q. of at least 115 in terms of the mean of the corresponding *posterior distribution*.

9.3   HYPOTHESES CONCERNING ONE PROPORTION

Many of the methods used in sampling inspection, quality control, and reliability verification are based on tests of the null hypothesis that a proportion (percentage, or probability) equals some specified constant. The details of the application of such tests to quality control will be discussed in Chapter 14, where we shall also go into some problems of sampling inspection; applications to reliability and life testing will be taken up in Chapter 15.

Although there are exact tests based on the binomial distribution (which can be constructed with the use of Table 1), we shall consider here only approximate large-sample tests ($n \geq 100$) based on the normal approximation to the binomial distribution. In other words, we shall test the null hypothesis $p = p_0$ against one of the alternatives $p < p_0$, $p > p_0$, or $p \neq p_0$ with the use of the following statistic*

$$z = \frac{x - np_0}{\sqrt{np_0(1 - p_0)}}$$

If the null hypothesis is true, the sampling distribution of this statistic is approximately the standard normal distribution, and the critical regions for the test are as shown in the following table:

---

*Some authors write the numerator of this formula for z as $x \pm \frac{1}{2} - np_0$, whichever is *numerically* smaller, but there is generally no need for this *continuity correction* so long as n is large.

CRITICAL REGIONS FOR TESTING $H_0: p = p_0$
(Large Samples)

Alternative hypothesis	Reject H_0 if
$p < p_0$	$z < -z_\alpha$
$p > p_0$	$z > z_\alpha$
$p \neq p_0$	$z < -z_{\alpha/2}$ or $z > z_{\alpha/2}$

(If the sample size is not large enough to use the normal approximation of the binomial distribution, we have to use the exact test mentioned above. For instance, to test the null hypothesis $p = p_0$ against the alternative $p < p_0$, we use the critical region $x \leq k_\alpha$, where k_α is the *largest integer* such that $B(k_\alpha; n, p_0) \leq \alpha$.)

Example. To illustrate this kind of test, suppose we want to investigate whether certain detonators used with explosives in coal mining meet the requirement that at least 90 percent of the detonators will ignite the explosive when charged. Thus, we shall want to test the null hypothesis $p = 0.90$ against the alternative hypothesis $p < 0.90$, say, at the level of significance $\alpha = 0.05$. If 174 of 200 detonators function properly, we get

$$z = \frac{174 - 200(0.90)}{\sqrt{200(0.90)(0.10)}} = -1.41$$

and since this is *not* less than -1.645, the value of $-z_{0.05}$, we find that the null hypothesis cannot be rejected. In other words, there is not sufficient evidence to say that *in general* the detonators will fail to meet the required standard.

9.4 HYPOTHESES CONCERNING SEVERAL PROPORTIONS

When we compare the consumer response (percentage favorable and percentage unfavorable) to two different products, when we decide whether the proportion of defectives of a given process remains constant from day to day, when we judge whether there is a difference in political persuasion among several nationality groups, and in many similar situations, we are interested in testing whether two or more binomial populations have the same parameter p. Referring to these parameters as $p_1, p_2, \ldots,$ and p_k, we are, in fact, interested in testing the null hypothesis

$$p_1 = p_2 = \ldots = p_k$$

against the alternative that at least two of these population proportions are unequal. To perform a suitable large-sample test of this hypothesis, we require independent random samples of size $n_1, n_2, \ldots,$ and n_k from the k populations; then, if the corresponding numbers of "successes" are $x_1, x_2, \ldots,$ and x_k, the test we shall use is based on the fact that (1) for large samples the sampling distribution of

$$z_i = \frac{x_i - n_i p_i}{\sqrt{n_i p_i (1 - p_i)}}$$

is approximately the standard normal distribution, (2) the square of a random variable having the standard normal distribution is a random variable having the chi-square distribution with 1 degree of freedom, and (3) the sum of k independent random variables having chi-square distributions with 1 degree of freedom is a random variable having the chi-square distribution with k degrees of freedom. (Proofs of these last two results may be found in the book by J. E. Freund mentioned in the Bibliography.) Thus, the sampling distribution of the statistic

$$\chi^2 = \sum_{i=1}^{k} \frac{(x_i - n_i p_i)^2}{n_i p_i (1 - p_i)}$$

is approximately the chi-square distribution with k degrees of freedom. (This approximation is usually quite close so long as $n_i p_i \geq 5$ for all i.) Now, if the null hypothesis is true, and $p_1 = p_2 = \ldots = p_k = p$, the value of the above χ^2 statistic becomes

$$\chi^2 = \sum_{i=1}^{k} \frac{(x_i - n_i p)^2}{n_i p (1 - p)}$$

and in actual practice we substitute for p (which, of course, is unknown) the pooled estimate

$$\hat{p} = \frac{x_1 + x_2 + \ldots + x_k}{n_1 + n_2 + \ldots + n_k}$$

Since the null hypothesis should be rejected if the differences between the x_i and the $n_i p$ are large, the critical region is $\chi^2 \geq \chi_\alpha^2$, where χ_α^2 is as defined on page 176 and the number of degrees of freedom is $k - 1$. The loss of one degree of freedom results from the fact that p is estimated by $\hat{p}$.

We shall not illustrate this test right away, because in practice it is convenient to perform the test or express the χ^2 statistic in different (though equivalent) ways. In the special case where $k = 2$, in which we are testing the null hypothesis that two population proportions are equal, we base our decision on the statistic

$$z = \frac{\dfrac{x_1}{n_1} - \dfrac{x_2}{n_2}}{\sqrt{\hat{p}(1 - \hat{p})\left(\dfrac{1}{n_1} + \dfrac{1}{n_2}\right)}} \quad \text{with} \quad \hat{p} = \frac{x_1 + x_2}{n_1 + n_2}$$

whose sampling distribution is approximately the standard normal distribution. The test based on this statistic is equivalent to the χ^2 test described above, in the sense that the *square* of this z statistic actually equals the above χ^2 statistic with $k = 2$ (see Exercise 8 on page 259). However, it is preferable to use the z statistic, as it enables us to test the null hypothesis $p_1 = p_2$ against *one-sided as well as two-sided alternatives*; this is not the case for the χ^2 statistic, with which we can test only against the two-sided alternative $p_1 \neq p_2$. The resulting critical regions based on the above z statistic are

CRITICAL REGIONS FOR TESTING $H_0: p_1 = p_2$
(Large Samples)

Alternative hypothesis	Reject H_0 if
$p_1 < p_2$	$z < -z_\alpha$
$p_1 > p_2$	$z > z_\alpha$
$p_1 \neq p_2$	$z < -z_{\alpha/2}$ or $z > z_{\alpha/2}$

Example. To illustrate this kind of test, suppose that 16 of 200 tractors produced on one assembly line required extensive adjustments before they could be shipped, while the same was true for 14 of 400 tractors produced on another assembly line. To decide whether the difference between the corresponding proportions, namely, $\frac{16}{200} = 0.08$ and $\frac{14}{400} = 0.035$, can be attributed to chance, let us test the null hypothesis $p_1 = p_2$ against the alternative hypothesis $p_1 \neq p_2$ at the level of significance $\alpha = 0.05$. Substituting $x_1 = 16$, $n_1 = 200$, $x_2 = 14$, and $n_2 = 400$ into the formulas for $\hat{p}$ and z, we obtain

$$\hat{p} = \frac{16 + 14}{200 + 400} = 0.05$$

and

$$z = \frac{\frac{16}{200} - \frac{14}{400}}{\sqrt{(0.05)(0.95)(\frac{1}{200} + \frac{1}{400})}} = 2.38$$

Since this exceeds $z_{0.025} = 1.96$, we find that *the null hypothesis will have to be rejected*. In other words, we conclude that the true proportion of tractors requiring extensive adjustments is *not* the same for the two assembly lines.

The test we have described here applies to the null hypothesis $p_1 = p_2$, but it can easily be modified (see Exercise 14 on page 260) so that it applies also to the null hypothesis $p_1 - p_2 = \delta$, where δ is not necessarily zero.

When we apply the chi-square criterion on page 254 to the comparison of more than two sample proportions, it is convenient to look at the data as arranged in the following fashion:

	Sample 1	Sample 2	...	Sample k	Total
Successes	x_1	x_2	...	x_k	x
Failures	$n_1 - x_1$	$n_2 - x_2$	...	$n_k - x_k$	$n - x$
Total	n_1	n_2	...	n_k	n

The notation is the same as before, except for x and n, which represent, respectively, the total number of successes and the total number of trials for all samples combined. With reference to this table, the entry in the cell belonging to the ith row and jth column is called the *observed* cell frequency f_{ij}, with $i = 1, 2$ and $j = 1, 2, \ldots, k$.

Under the null hypothesis that $p_1 = p_2 = \ldots = p_k = p$, we estimate p, as before, as the total number of successes divided by the total number of trials, which we now write as $\hat{p} = x/n$. Hence, the *expected* number of successes and failures for the jth sample are estimated, respectively, as

$$e_{1j} = n_j \cdot \hat{p} = \frac{n_j \cdot x}{n} \quad \text{and} \quad e_{2j} = n_j(1 - \hat{p}) = \frac{n_j \cdot (n - x)}{n}$$

The quantities e_{1j} and e_{2j} are referred to as the *expected cell frequencies* e_{ij}, with $i = 1, 2$ and $j = 1, 2, \ldots, k$. *Thus, the expected frequency for any given cell may be obtained by multiplying the totals of the row and column to which the cell belongs, and then dividing by the grand total n.*

Using this new notation, it can be shown (see Exercise 16 on page 260) that the χ^2 statistic on page 254 can also be written in the form

$$\chi^2 = \sum_{i=1}^{2} \sum_{j=1}^{k} \frac{(f_{ij} - e_{ij})^2}{e_{ij}}$$

This formula has the advantage that it can easily be extended to the more general case, to be treated in Section 9.5, where each trial permits more than two possible outcomes, and there are, thus, more than two rows in the tabular presentation of the various frequencies.

Example. To illustrate the use of this χ^2 statistic, suppose that samples of three kinds of materials, subjected to extreme temperature changes, produced the results shown in the following table:

	Material A	Material B	Material C	Total
Crumbled	41	27	22	90
Remained intact	79	53	78	210
Total	120	80	100	300

The expected frequencies for the first two cells of the first row are

$$e_{11} = \frac{90 \cdot 120}{300} = 36 \quad \text{and} \quad e_{12} = \frac{90 \cdot 80}{300} = 24$$

and, as it can be shown that *the sum of the expected frequencies for each row or column equals that of the corresponding observed frequencies* (see Exercise 20 on page 261), we find by subtraction that e_{13} equals $90 - (36 + 24) = 30$, and that the expected frequencies for the second row equal $e_{21} = 120 - 36 = 84$, $e_{22} = 80 - 24 = 56$, and $e_{23} = 100 - 30 = 70$. Substituting these values together with the observed cell frequencies into the formula for the χ^2 statistic with $k = 3$, we obtain

$$\chi^2 = \frac{(41 - 36)^2}{36} + \frac{(27 - 24)^2}{24} + \frac{(22 - 30)^2}{30}$$

$$+ \frac{(79 - 84)^2}{84} + \frac{(53 - 56)^2}{56} + \frac{(78 - 70)^2}{70}$$

$$= 4.575$$

Since this is less than 5.991, the value of $\chi^2_{0.05}$ for $k - 1 = 2$ degrees of freedom, we find that *the null hypothesis cannot be rejected.* In other words, the data do *not* refute the claim that the probability of crumbling when subjected to extreme temperature changes is the same for the three kinds of material.

It is customary in problems of this kind to round the expected cell frequencies to the nearest integer or to one decimal. Most of the entries of Table 5 are given to three decimals, but there is seldom any need to carry more than two decimals when calculating the value of the χ^2 statistic.

EXERCISES

1. A manufacturer of submersible pumps claims that at most 30 percent of his pumps require repairs within the first five years they are in operation. What can we conclude about this claim at the level of significance $\alpha = 0.05$, if among 120 of these pumps there are 47 which require repairs within the first five years they are in operation?

2. The performance of a computer is observed over a period of two years to check the claim that the probability is 0.20 that its down time will exceed five hours in any given week. Testing the null hypothesis $p = 0.20$ against the alternative hypothesis $p \neq 0.20$, what can we conclude at the level of significance $\alpha = 0.05$, if there were only 11 weeks in which the down time of the computer exceeded five hours?

3. To check on an ambulance service's claim that at least 40 percent of its calls are life-threatening emergencies, a random sample was taken from its files, and it was found that only 49 of 150 calls were life-threatening emergencies. Test the null hypothesis $p = 0.40$ against a suitable alternative at the level of significance $\alpha = 0.01$.

4. Suppose that a nutritionist claims that at most 75 percent of the pre-school children in a certain country have protein deficient diets. What can we conclude about this claim at the level of significance $\alpha = 0.01$, if 246 of 300 pre-school children (included in a sample survey conducted in that country) are found to have protein deficient diets?

5. Suppose that 4 of 13 undergraduate engineering students state that they will go on to graduate school. Test the dean's claim that 60 percent of the undergraduate students will go on to graduate school, using the alternative hypothesis $p < 0.60$ and the level of significance $\alpha = 0.05$. [*Hint*: Use Table 1 to determine the probability of getting "at most 4 successes in 13 trials" when $p = 0.60$.]

6. Suppose we want to test the "honesty" of a coin on the basis of the number of heads we will get in 15 flips. Using Table 1, determine how few or how many heads we would have to get so that we could reject the null hypothesis $p = 0.50$ against the alternative hypothesis $p \neq 0.50$ at the level of significance $\alpha = 0.05$. What is the *actual* level of significance we would be using with this criterion?

7. It costs more to test a certain type of ammunition than to manufacture it, and, hence, only three rounds are tested from each large lot. If the lot is rejected unless all three rounds function according to specifications,
(a) sketch the *OC* curve for this test;
(b) find the actual proportion of defectives for which the test procedure will cause a lot to be rejected with a probability of 0.10.

8. Show that the *square* of the z statistic on page 255 equals the chi-square statistic on page 254 for $k = 2$.

9. A manufacturer of electronic equipment subjects samples of two competing brands of transistors to an accelerated performance test. If 45 of 180 transistors of the first kind and 34 of 120 transistors of the second kind fail the test, what can he conclude at the level of significance $\alpha = 0.05$ about the difference between the corresponding sample proportions?

10. If one method of producing rain by "seeding" clouds was successful in 16 of 50 attempts, while another method was successful in 29 of 80 attempts, can we conclude at the level of significance $\alpha = 0.05$ that the second method is better than the first?

11. Two groups of 80 patients each took part in an experiment in which one group received pills containing an anti-allergy drug, while the other group received a placebo (a pill containing no drug). If in the group given the drug 23 exhibited allergic symptoms while in the group given the placebo 41 exhibited such symptoms, is this sufficient evidence to conclude at the level of significance $\alpha = 0.01$ that the drug is effective in reducing these symptoms?

12. A study showed that 84 of 200 persons who saw a deodorant advertised during the telecast of a football game and 96 of 200 persons who saw it advertised on a variety show remembered two hours later the name of the deodor-ant. Use the level of significance $\alpha = 0.05$ to test the null hypothesis that there is no difference between the corresponding "true" proportions.

13. In a sample of the visitors to a cavern (a tourist attraction which had once served as a hide-out to a notorious outlaw), 86 of 250 men and 152 of 250 women bought souvenirs. Test whether the difference between the corre-sponding sample proportions is significant, using the level of significance $\alpha = 0.01$ and
   (a) the chi-square statistic given on page 254;
   (b) the z statistic given on page 255.
   Also verify that the *square* of the value obtained for the z statistic equals that obtained for the chi-square statistic.

14. If we want to test the null hypothesis that the difference between two popula-tion proportions equals some constant δ, not necessarily 0, we can base our decision on the statistic

$$z = \frac{\dfrac{x_1}{n_1} - \dfrac{x_2}{n_2} - \delta}{\sqrt{\dfrac{\dfrac{x_1}{n_1}\left(1 - \dfrac{x_1}{n_1}\right)}{n_1} + \dfrac{\dfrac{x_2}{n_2}\left(1 - \dfrac{x_2}{n_2}\right)}{n_2}}}$$

whose sampling distribution is approximately the standard normal distribu-tion so long as n_1 and n_2 are both large. Referring to Exercise 11, use this theory and the level of significance $\alpha = 0.05$ to test whether the percentage of patients exhibiting allergic symptoms is at least 8 percent less for those who actually receive the drug.

15. In a true-false test, a test item is considered to be *good* if it discriminates between well-prepared students and poorly-prepared students. If 205 of 250 well-prepared students and 137 of 250 poorly-prepared students answer a certain item correctly, use the theory of Exercise 14 and the level of signifi-cance $\alpha = 0.05$ to test whether for the given item the proportion of correct answers will *in general* be at least 20 percent higher among well-prepared students than among poorly-prepared students.

16. Verify that the two formulas given for the χ^2 statistic on pages 254 and 257 are equivalent.

17. If 26 of 200 tires of Brand A failed to last 20,000 miles, while the corre-sponding figures for 200 tires each of Brands B, C, and D were 23, 15, and 32, use the level of significance $\alpha = 0.05$ to test the null hypothesis that there is no difference in the quality of the four kinds of tires.

18. Tests are made on the proportion of defective castings produced by five different molds. If there were 14 defectives among 100 castings made with mold I, 33 defectives among 200 castings made with mold II, 21 defectives among 180 castings made with mold III, 17 defectives among 120 castings

made with mold IV, and 25 defectives among 150 castings made with mold V, test (at the 0.05 level of significance) whether the true proportion of defectives is the same for each mold.

19. Using a level of significance of $\alpha = 0.05$, can it be concluded from the following sample data that the proportion of employees favoring a new pension plan is *not* the same for three different government agencies:

	Agency 1	*Agency 2*	*Agency 3*
For the pension plan	67	84	109
Against the pension plan	33	66	41

20. Verify that if the expected frequencies are determined as indicated on page 257, the sum of the expected frequencies for each row or column equals the sum of the corresponding observed frequencies.

9.5   THE ANALYSIS OF *r*-by-*k* TABLES

As we suggested on page 257, the method by which we analyzed the last example of the preceding section lends itself also to the analysis of so-called *r-by-k tables*, that is, tables in which observed frequencies are arranged in *r* rows and *k* columns. Such tables arise in essentially two kinds of problems. First, we might again have samples from *k* populations, with the distinction that now each trial permits more than two possible outcomes. This might happen, for example, if persons belonging to different income groups are asked whether they favor a certain political candidate, whether they are against him, or whether they are indifferent or undecided. The other situation giving rise to an *r-by-k* table is one in which we sample from one population but classify each item with respect to two (usually qualitative) categories. This might happen, for example, if a consumer testing service rates cars as excellent, superior, average, or poor with regard to performance and also with regard to appearance. Each car tested would then fall into one of the 16 cells of a 4-by-4 table, and it is mainly in connection with problems of this kind that *r-by-k* tables are referred to as *contingency tables*.

The essential difference between the two situations giving rise to *r*-by-*k* tables is that *in the first case the column totals* (*the sample sizes*) *are fixed, while in the second case only the grand total* (*the total for the entire table*) *is fixed*. As a result, there are also differences in the null hypotheses we shall want to test. In the first case we want to test *whether the probability of obtaining an observation in the ith row is the same for each column*; symbolically, we shall want to test the null hypothesis

$$p_{i1} = p_{i2} = \ldots = p_{ik} \quad \text{for } i = 1, 2, \ldots, r$$

where p_{ij} is the probability of obtaining an observation belonging to the ith row and the jth column. The alternative hypothesis is that the p's are *not all alike at least for one row*. In the second case we shall want to test *whether the random variables represented by the two classifications are independent*; symbolically, we shall want to test the null hypothesis

$$p_{ij} = (p_{i.})(p_{.j}) \quad \text{for } \begin{cases} i = 1, 2, \ldots, r \\ j = 1, 2, \ldots, k \end{cases}$$

where $p_{i.}$ is the probability of obtaining an observation belonging to the ith row and $p_{.j}$ is the probability of obtaining an item belonging to the jth column. The alternative to this null hypothesis is that the equality does not hold for at least one pair of values of i and j.

In spite of the differences we have described, the analysis of an r-by-k table is the same for both cases. First we calculate the expected cell frequencies e_{ij} as on page 257, namely, by multiplying the totals of the respective rows and columns and then dividing by the grand total. In practice, we make use of the fact that the observed frequencies and the expected frequencies total the same for each row and column, so that only $(r - 1)(k - 1)$ of the e_{ij} have to be calculated directly, while the others can be obtained by subtraction from appropriate row or column totals. We then substitute into the formula

$$\chi^2 = \sum_{i=1}^{r} \sum_{j=1}^{k} \frac{(f_{ij} - e_{ij})^2}{e_{ij}}$$

and we reject the null hypothesis if the value of this statistic exceeds χ^2_{α} with $(r - 1)(k - 1)$ degrees of freedom. [This expression for the number of degrees of freedom is justified by the above observation that after we choose $(r - 1)(k - 1)$ of the expected cell frequencies, the others are automatically determined, that is, they may be obtained by subtraction from appropriate row or column totals.]

Example. To illustrate the general approach, suppose that the personnel manager of a large company wants to know whether there really is a relationship between an employee's performance in the company's training program and his (or her) ultimate success in the job. Suppose, furthermore, that a sample of 400 cases (taken from the company's very extensive files) yielded the results shown in the following table:

		Performance in training program			
		Below average	Average	Above average	Total
	Poor	23	60	29	112
Success in job (Employer's rating)	Average	28	79	60	167
	Very good	9	49	63	121
	Total	60	188	152	400

Calculating first the expected cell frequencies for the first two cells of the first two rows, we get

$$e_{11} = \frac{112 \cdot 60}{400} = 16.8, \qquad e_{12} = \frac{112 \cdot 188}{400} = 52.6,$$

$$e_{21} = \frac{167 \cdot 60}{400} = 25.0, \qquad e_{22} = \frac{167 \cdot 188}{400} = 78.5$$

and then, by subtraction, we find that the other expected cell frequencies are $e_{13} = 42.6$, $e_{23} = 63.5$, $e_{31} = 18.2$, $e_{32} = 56.9$, and $e_{33} = 45.9$. Thus,

$$\chi^2 = \frac{(23 - 16.8)^2}{16.8} + \frac{(60 - 52.6)^2}{52.6} + \frac{(29 - 42.6)^2}{42.6}$$

$$+ \frac{(28 - 25.0)^2}{25.0} + \frac{(79 - 78.5)^2}{78.5} + \frac{(60 - 63.5)^2}{63.5}$$

$$+ \frac{(9 - 18.2)^2}{18.2} + \frac{(49 - 56.9)^2}{56.9} + \frac{(63 - 45.9)^2}{45.9}$$

$$= 20.34$$

and since this exceeds 9.488 as well as 13.277, the values of $\chi^2_{0.05}$ and $\chi^2_{0.01}$ for $(3 - 1)(3 - 1) = 4$ degrees of freedom, we find that the null hypothesis will have to be rejected. We conclude that there is some relationship (some dependence) between an employee's performance in the training program and his (or her) success in the job.

9.6   GOODNESS OF FIT

We speak of "goodness of fit" when we try to compare an observed frequency distribution with the corresponding values of a theoretical distribution.

Example. To illustrate, suppose that during 400 five-minute intervals the air-traffic control of an airport received 0, 1, 2, . . . , and 13 radio messages with respective frequencies of 3, 15, 47, 76, 68, 74, 46, 39, 15, 9, 5, 2, 0, and 1. Suppose, furthermore, that we want to check whether these data substantiate the claim that the number of radio messages which they receive during a five-minute interval may be looked upon as a random variable having the Poisson distribution with $\lambda = 4.6$. Looking up the corresponding Poisson probabilities in Table 2 and multiplying them by 400 to get the expected frequencies, we arrive at the result shown in the following table together with the original data:

Number of radio messages	Observed frequencies	Poisson probabilities	Expected frequencies
0	3 ⎱ 18	0.010	4.0 ⎱ 22.4
1	15 ⎰	0.046	18.4 ⎰
2	47	0.107	42.8
3	76	0.163	65.2
4	68	0.187	74.8
5	74	0.173	69.2
6	46	0.132	52.8
7	39	0.087	34.8
8	15	0.050	20.0
9	9	0.025	10.0
10	5 ⎤	0.012	4.8 ⎤
11	2 ⎥ 8	0.005	2.0 ⎥ 8.0
12	0 ⎥	0.002	0.8 ⎥
13	1 ⎦	0.001	0.4 ⎦
	400		400.0

Evidently, there are some discrepancies between the observed frequencies and the expected frequencies, but it remains to be seen whether they are significant or whether they can be attributed to chance.

An appropriate test of the null hypothesis that a set of data comes from a population having a given distribution (against the alternative that the population has some other distribution) can be based on the statistic

$$\chi^2 = \sum_{i=1}^{k} \frac{(f_i - e_i)^2}{e_i}$$

where the f_i and e_i are, as before, the corresponding observed and expected frequencies. The sampling distribution of this statistic is approximately the chi-square distribution with $k - m$ degrees of freedom, where k is the number of terms in the formula for χ^2 and m is the number of quantities, obtained from the observed data, that are used in calculating the expected frequencies. Note that in our example we had to know only the total frequency, 400, to calculate the e_i, so that m equalled 1.

Continuation of Example. To satisfy the rule on page 254, according to which none of the expected frequencies in a chi-square comparison should be less than 5, we shall follow the simple practice of combining adjacent classes, the first two and the last four as indicated in the table on the preceding page. Thus, we get

$$\chi^2 = \frac{(18 - 22.4)^2}{22.4} + \frac{(47 - 42.8)^2}{42.8} + \cdots + \frac{(9 - 10.0)^2}{10.0} + \frac{(8 - 8.0)^2}{8.0}$$

$$= 6.749$$

Since this is less than 16.919, the value of $\chi^2_{0.05}$ for $10 - 1 = 9$ degrees of freedom, *the null hypothesis cannot be rejected* at the 0.05 level of significance, and we conclude that the Poisson distribution with $\lambda = 4.6$ provides a *good fit*.

EXERCISES

1. The results of polls conducted two weeks and four weeks before a gubernatorial election are shown in the following table:

	Two weeks before election	Four weeks before election
For Republican candidate	79	91
For Democratic candidate	84	66
Undecided	37	43

Use the level of significance $\alpha = 0.05$ to decide whether there has been a change in opinion during the two weeks between the two polls.

2. Suppose that in Exercise 17 on page 260 we had been interested also in how many of the tires last more than 30,000 miles, and obtained the results shown in the following table:

	Brand A	Brand B	Brand C	Brand D
Failed to last 20,000 miles	26	23	15	32
Lasted from 20,000 to 30,000 miles	118	93	116	121
Lasted more than 30,000 miles	56	84	69	47

Use the level of significance $\alpha = 0.01$ to test the null hypothesis that there is no difference in the quality of the four kinds of tires.

3. A large electronics firm which hires many handicapped workers wants to determine whether their handicaps affect such workers' performance. Use the level of significance $\alpha = 0.05$ to decide on the basis of the sample data shown in the following table whether it is reasonable to maintain that the handicaps have no effect on the workers' performance:

	Performance		
	Above average	Average	Below average
Blind	21	64	17
Deaf	16	49	14
No handicap	29	93	28

4. Tests of the fidelity and the selectivity of 190 radio receivers produced the results shown in the following table:

		Fidelity		
		Low	Average	High
Selectivity	Low	6	12	32
	Average	33	61	18
	High	13	15	0

Use the level of significance $\alpha = 0.01$ to show that there is, indeed, a relationship (dependence) between fidelity and selectivity.

5. A quality control engineer takes daily samples of four tractors coming off an assembly line and on 200 consecutive working days he obtains the data summarized in the following table:

Number requiring adjustments	Number of days
0	101
1	79
2	19
3	1

To test the claim that 10 percent of all the tractors coming off this assembly line require adjustments, look up the corresponding probabilities in Table 1, calculate the expected frequencies, and perform the chi-square test at the level of aignificance $\alpha = 0.01$.

6. With reference to Exercise 5, verify that the mean of the given distribution is 0.60, corresponding to 15 percent of the tractors requiring adjustments. Then look up the probabilities for $n = 4$ and $p = 0.15$ in Table 1, calculate the expected frequencies, and then test at the level of significance $\alpha = 0.01$ whether the binomial distribution with $n = 4$ and $p = 0.15$ provides a suitable model for this situation.

7. Suppose that in the example on page 264 we had shown first that the mean of the distribution, rounded to one decimal, is 4.5, and then tested whether the Poisson distribution with $\lambda = 4.5$ provides a good fit. What would have been the number of degrees of freedom for the appropriate chi-square criterion?

8. The following is the distribution of the daily number of power failures reported in a western city on 300 days:

Number of power failures	Number of days
0	9
1	43
2	64
3	62
4	42
5	36
6	22
7	14
8	6
9	2

Test at the level of significance $\alpha = 0.05$ whether the daily number of power failures in this city may be looked upon as a random variable having the Poisson distribution with $\lambda = 3.2$.

9. The following is the distribution of the hourly number of trucks arriving at a company's warehouse:

Trucks arriving per hour	Frequency
0	52
1	151
2	130
3	102
4	45
5	12
6	5
7	1
8	2

Find the mean of this distribution, and using its mean (rounded to one decimal) as the parameter λ, fit a Poisson distribution. Test for goodness of fit at the level of significance $\alpha = 0.05$.

10. Using any four columns of Table 7 (that is, a total of 200 random digits), construct a table showing how many times each of the digits $0, 1, \ldots$, and 9 occurred. Compare these observed frequencies with the corresponding expected frequencies (based on the assumption that the digits are randomly generated) by means of the chi-square statistic, and test at the level of significance $\alpha = 0.05$ whether the assumption of randomness is tenable.

11. The following is the distribution of the sulfur oxides emission data obtained on page 137:

Class limits (tons)	Frequency
5.0– 8.9	3
9.0–12.9	10
13.0–16.9	14
17.0–20.9	25
21.0–24.9	17
25.0–28.9	9
29.0–32.9	2
	80

As we showed on page 155, the mean of this distribution is $\bar{x} = 18.85$ and its standard deviation is $s = \sqrt{30.77} = 5.55$.

(a) Find the probabilities that a random variable having a normal distribution with $\mu = 18.85$ and $\sigma = 5.55$ takes on a value between 4.95 and 8.95, between 8.95 and 12.95, between 12.95 and 16.95, between 16.95 and 20.95, between 20.95 and 24.95, between 24.95 and 28.95, and between 28.95 and 32.95.

(b) Multiply the probabilities obtained in part (a) by the total frequency (in this case $n = 80$), thus getting the *expected normal curve frequencies* corresponding to the seven classes of the given distribution.

(c) Test the null hypothesis that the given data may be looked upon as a random sample from a normal population by comparing the observed and expected frequencies with an appropriate χ^2 statistic (using $\alpha = 0.05$). Explain why the number of degrees of freedom for this χ^2 test is given by $k - 3$, where k is the number of terms in the χ^2 statistic.

12. Among 100 vacuum tubes used in an experiment, 46 had a service life of less than 20 hours, 19 had a service life of 20 or more but less than 40 hours, 17 had a service life of 40 or more but less than 60 hours, 12 had a service life of 60 or more but less than 80 hours, and 6 had a service life of 80 hours or more. Using steps similar to those outlined in Exercise 11, test whether these lifetimes may be regarded as a sample from an exponential population with $\mu = 40$ hours. Use the level of significance $\alpha = 0.01$.

10

Nonparametric Methods

10.1 INTRODUCTION

Most of the methods of inference which we have studied so far are based on the assumption that the observations are taken from normal populations. These methods extract all the information that is available in a sample, and they usually attain the best possible precision, that is, the most reliable results. As we have pointed out earlier, the assumption that the samples are taken from normal populations is not really as stringent as it may seem. Most statistical methods based on the normal distribution are fairly *robust*, that is, they will give reasonably accurate answers even when the normality assumption is only satisfied in an approximate sense. In spite of this, there are several reasons why we may wish to use other, less precise methods—the assumption of normality may be grossly incorrect, the labor involved in carrying out the more precise methods may be excessive, or a short-cut method may be desired to determine in advance whether it is worthwhile to carry out the more detailed calculations.

Several short-cut methods have already been introduced, primarily as labor-saving devices. For example, on page 145 we introduced a quick method of estimating the mean and the standard deviation of normally distributed data, using the 50 percent and 84 percent points on a "probability" graph. On page 154 we discussed a method of coding which can materially reduce the time required to calculate the mean and the standard deviation of a set of data without any loss in precision. In Chapter 5 we observed that these calculations can further be simplified with only a minor loss in precision by grouping the observations into a frequency table.

Other short-cuts connected with point estimation were introduced in Chapters 7 and 8. We discussed the sample median, which often can be determined more easily and more rapidly than the mean, and, as was pointed out on page 181, the median gives an unbiased estimate for the mean of a symmetric population. Also, the median is superior to the mean as a measure of "location" of a highly skewed population. The sample range was introduced on page 228 as an estimator for the standard deviation of a normal population; it is obtained far more quickly than the sample standard deviation, and its precision is nearly that of s for small samples.

If there is a choice between several statistical methods which can all be used in a given situation, the criterion of *efficiency* is most commonly used as an appropriate guide. If we think of the methods based on the assumption of normality as being fully (100 percent) efficient, we can use this as a yardstick for measuring the merits of any other method. The most widely used measure of efficiency is based on the sample sizes required to give equally precise results by a given method and by the corresponding method which is fully efficient. For example, in estimating the mean of a normal population, the most efficient method involves use of the sample mean $\bar{x}$. If we wanted to use the median instead of the mean, we would have to take into account that the variance of the sampling distribution of the median is approximately $1.57 \dfrac{\sigma^2}{n}$, so that the efficiency of the median is $1/1.57$ or approximately 64 percent. In other words, the median based on a sample of size 100 gives as reliable an estimate of the mean of a normal population as the mean based on a sample of size 64. Similarly, it can be shown that the efficiency of the range estimator for the standard deviation of a normal population decreases as the sample size increases; the efficiency is 100 percent for samples of size 2, 96 percent for samples of size 5, and 81 percent for samples of size 15.

Certain methods of inference have the important advantage that they do not require the stringent assumptions of the methods based on the normal distribution, and they usually have the additional advantage that they require less burdensome calculations. They include *nonparametric methods* as well as *distribution-free methods,* where the first term applies when we are not concerned directly with the parameters of populations *of a given kind,* and the second term applies when we make no assumptions about the populations from which we are sampling, except perhaps that they must be continuous. Actually, since this distinction is rather fine, it has become the custom to refer to either kind of method simply as nonparametric.

The main *advantage* of nonparametric methods is that exact inferences can be made when the assumptions underlying so-called "standard" methods cannot all be met. The main *disadvantage* of nonparametric methods is that they may be wasteful of information, and usually have a smaller efficiency than the corresponding parametric methods *provided that the assumptions of the standard (parametric) methods can be met.* Thus, if we say that the efficiency of a certain nonparametric method is 80 percent, we may actually be understating its relative worth, for the efficiency of the corresponding "standard" method will be somewhat less than 100 percent if all assumptions are not exactly met.

In this chapter we shall outline a variety of useful methods, which can be used in place of the corresponding standard methods (described in Chapters 7 through 9) whenever the assumptions are not met or there is a need for greater ease in calculations. Certain other nonparametric methods, which can be used in place of "standard" methods not yet introduced, will be described along with the new methods in subsequent chapters.

10.2   THE SIGN TEST

In this section we shall describe nonparametric tests that are based on classifying the data according to two attributes, conveniently represented by *plus signs* and *minus signs.* For instance, in the *one-sample case* we can test the null hypothesis $\mu = \mu_0$ on the basis of a random sample of size n, by replacing each observation exceeding μ_0 with a plus sign and each observation exceeded by μ_0 with a minus sign. If the population from which we are sampling is *continuous and symmetrical,* the probability that an observation is thus replaced by a plus sign equals $\frac{1}{2}$ when H_0 is true. Consequently, the test of the null hypothesis $\mu = \mu_0$ becomes equivalent to a test of the null hypothesis $p = \frac{1}{2}$, where p is the parameter of a binomial distribution. The two-sided alternative $\mu \neq \mu_0$ is now equivalent to $p \neq \frac{1}{2}$, and the one-sided alternatives $\mu < \mu_0$ and $\mu > \mu_0$ are equivalent to $p < \frac{1}{2}$ and $p > \frac{1}{2}$, respectively, where p is the probability of getting a plus sign, namely, an observation greater than μ_0. (If a sample value happens to equal μ_0, which is possible since the values of continuous random variables are virtually always rounded, it is simply discarded.)

Example. Let us consider the following data, which are the octane ratings obtained for 15 samples of a certain kind of gasoline:

101.0	103.3	101.8	102.5	101.7	98.2	101.1	104.5
105.3	99.4	102.4	100.9	100.3	100.0	103.6	

To test the null hypothesis that the true mean octane rating of this kind of gasoline is $\mu = 100.0$ against the alternative hypothesis $\mu > 100.0$ at the level of significance $\alpha = 0.05$, we first replace each value less than 100.0 with a minus sign, each value greater than 100.0 with a plus sign, and we discard the one value which actually equalled 100.0. Thus, we get

$$+ \ + \ + \ + \ + \ - \ + \ + \ + \ - \ + \ + \ + \ +$$

and it remains to be seen whether "12 successes in 14 trials" supports the null hypothesis $p = \frac{1}{2}$ or the alternative hypothesis $p > \frac{1}{2}$. Applying the exact test criterion given on page 241 with $\alpha = 0.05$, we find from Table 1 that $k'_{0.05} = 10$; since there were 12 plus signs, it follows that the null hypothesis will have to be rejected. (More directly, we could have argued that the probability of "12 or more successes in 14 trials" is only 0.0065 according to Table 1, so that the null hypothesis must be rejected.) We conclude that the mean octane rating of the given kind of gasoline is greater than 100.0.

In this example, we had to refer directly to Table 1, but if the sample size is sufficiently large, we can use the normal curve approximation to the binomial distribution and the tests given in the table on page 212.

The sign test has important applications in problems where we are dealing with *paired* data, so that each pair can be replaced with a *plus sign* if the first value is greater than the second, a *minus sign* if the first value is smaller than the second, or be discarded if the two values are equal. As in Section 7.8 such problems arise in two kinds of situations, depending on whether the data are actually *given as pairs* as in the example on page 220, or whether they consist of independent random samples whose values are *randomly paired*.

Example. To illustrate the first case, let us refer to the example on page 220, which dealt with the effectiveness of an industrial safety program. As can easily be verified the data for the 12 plants are replaced by

$$+ \ + \ + \ + \ - \ + \ - \ + \ - \ + \ + \ +$$

and it remains to be seen whether "9 successes in 12 trials" enables us to reject the null hypothesis $p = \frac{1}{2}$ against the alternative hypothesis $p > \frac{1}{2}$ at the level of singificance $\alpha = 0.05$. Since the probability of "9 or more successes" is 0.0730 according to Table 1, it follows that *the null hypothesis cannot be rejected*. In other words, we cannot conclude that the industrial safety program is effective, and this illustrates the point made on page 272 that nonparametric methods can be wasteful of information. When we performed the t test on page 220, we were able to conclude at the level of significance $\alpha = 0.05$ that the industrial safety program *is* effective.

Example. To illustrate the second case, where we deal with two independent samples, let us consider the following data pertaining to the noise level (in decibels) recorded at two busy intersections during early morning traffic:

Intersection A: 69, 74, 77, 59, 80, 59, 71, 65, 62, 79,
76, 60, 59, 64, 71, 63, 63, 67, 62, 71

Intersection B: 59, 78, 53, 63, 67, 63, 59, 58, 64, 74,
66, 62, 68, 68, 71, 70, 56, 55, 63, 68,
55, 71

Note that there are two more values in the second sample, so that if we randomly match one value of the second sample with each value of the first sample, two values of the second sample will be left over. Actually, doing this with the use of random numbers, we obtain the pairs shown in the following table, where we indicated by means of a plus sign that the value for Intersection A is larger, and by means of a minus sign that the value for Intersection A is smaller:

Intersection A	Intersection B		Intersection A	Intersection B	
62	55	+	80	63	+
71	68	+	64	71	−
59	67	−	71	59	+
63	55	+	69	63	+
77	63	+	63	58	+
67	59	+	74	70	+
79	64	+	71	78	−
62	68	−	59	53	+
59	56	+	76	66	+
60	74	−	65	62	+

There are 15 plus signs and 5 minus signs, and we shall have to see whether "15 successes in 20 trials" supports the null hypothesis $p = \frac{1}{2}$

(namely, the null hypothesis that the true average noise level at the two intersections is the same) or the alternative hypothesis $p \neq \frac{1}{2}$. Since the probability of "15 or more successes" is 0.0207 according to Table 1, we find that the null hypothesis can be rejected at the level of significance $\alpha = 0.05$, but not at the level of significance $\alpha = 0.01$. Note that if we had used the normal approximation of the binomial distribution, substitution of $n = 20$ and $p = \frac{1}{2}$ would have yielded $\mu = 20(\frac{1}{2}) = 10$, $\sigma = \sqrt{20(\frac{1}{2})(\frac{1}{2})} = 2.236$, and

$$z = \frac{14.5 - 10}{2.236} = 2.01$$

and the result would have been exactly the same. Note that since 20 is small we made the continuity correction suggested in the footnote on page 252.

10.3 RANK-SUM TESTS

The paired-sample sign test is one of several nonparametric methods for testing the null hypothesis that two samples come from identical populations against the alternative hypothesis that the populations have unequal means. A highly efficient class of nonparametric tests of this and similar hypotheses is based on *rank sums;* that is, the observations are assigned ranks according to their order of magnitude, and the tests are performed on the basis of certain sums of these ranks. In this section we shall introduce three tests based on rank sums. The *Mann-Whitney U test* will be presented as a substitute for the two-sample t test, and it has a limiting efficiency of 95.5 percent when the assumptions underlying the corresponding t test are satisfied. A test similar to the U test, which can be used when the alternative hypothesis specifies that the two populations have unequal *dispersions*, will be taken up next. Finally we shall introduce the *Kruskal-Wallis H test*, for testing whether k samples come from identical populations against the alternative that the populations have unequal means. Like the U test, the H test also has a limiting efficiency of 95.5 percent when compared with the corresponding "standard" procedure, which will be discussed in Chapter 12.

Let us first describe the *Mann-Whitney U test*, a very popular test, by means of the following illustration:

Example. In a study of sedimentary rocks, the following diameters (in millimeters) were obtained for samples of 29 grains from two kinds of sand:

Sand I: 0.63,   0.17,   0.35,   0.49,   0.18,   0.43,   0.12,   0.20,
            0.47,   1.36,   0.51,   0.45,   0.84,   0.32,   0.40
Sand II: 1.13,   0.54,   0.96,   0.26,   0.39,   0.88,   0.92,   0.53,
            1.01,   0.48,   0.89,   1.07,   1.11,   0.58

The means of these two samples are 0.46 and 0.77, respectively, and the problem is to decide whether their difference is significant. Note that we may well be reluctant to use the standard test of Section 7.8 since the first sample shows considerably more variability than the second; its values range from 0.12 to 1.36, whereas those of the second sample range only from 0.26 to 1.13.

To perform the Mann-Whitney test, we first rank the data *jointly* in an increasing (or decreasing) order of magnitude, and for our data we, thus, obtain the following array:

0.12	0.17	0.18	0.20	0.26	0.32	0.35	0.39	0.40	0.43
I	I	I	I	II	I	I	II	I	I

0.45	0.47	0.48	0.49	0.51	0.53	0.54	0.58	0.63	0.84
I	I	II	I	I	II	II	II	I	I

0.88	0.89	0.92	0.96	1.01	1.07	1.11	1.13	1.36
II	II	II	II	II	II	II	II	I

Assigning the data *in this order* the ranks 1, 2, 3, . . . , and 30, we find that the values of the first sample (Sand I) occupy ranks 1, 2, 3, 4, 6, 7, 9, 10, 11, 12, 14, 15, 19, 20, and 29, while those of the second sample (Sand II) occupy ranks 5, 8, 13, 16, 17, 18, 21, 22, 23, 24, 25, 26, 27, and 28. There are no *ties* in this example among values belonging to different samples, but if there were, we would assign to each of the tied observations the *mean* of the ranks which they jointly occupy. (Thus, if the third and fourth values were identical, we would assign each the rank $\frac{3+4}{2} = 3.5$; if the ninth, tenth, and eleventh values were identical, we would assign each the rank $\frac{9+10+11}{3} = 10$, and so forth.)

The null hypothesis we shall want to test is that both samples come from identical populations, and it stands to reason that in that case the means of the ranks, or the sums of the ranks assigned to the values of the two samples, should be more or less the same. For our two samples, the sums of the ranks are, respectively, 162 and 273, and it remains to be seen whether their difference is large enough to reject the null hypothesis.

In the Mann-Whitney test, we actually base this decision on the statistic

$$U = n_1 n_2 + \frac{n_1(n_1 + 1)}{2} - R_1$$

where n_1 and n_2 are the respective sizes of the first and second samples, and R_1 is the sum of the ranks assigned to the values of the first sample. (In practice, we can designate the sample whose rank sum is most easily obtained as the "first sample," as it is immaterial which sample is referred to as "first.")

Under the null hypothesis that the two samples come from identical populations, it can be shown that the mean and the variance of the sampling distribution of the U statistic are given by

$$\mu_U = \frac{n_1 n_2}{2} \quad \text{and} \quad \sigma_U^2 = \frac{n_1 n_2(n_1 + n_2 + 1)}{12}$$

Note that if there are ties in rank, these formulas provide only approximations, but if the number of ties is small, such approximations will generally be good. Furthermore, if both n_1 and n_2 are *sufficiently large* (in this case, greater than 8), the sampling distribution of the U statistic can be approximated closely by a normal distribution and, hence, the test can be based on the statistic

$$z = \frac{U - \mu_U}{\sigma_U}$$

and Table 3. There also exist tables on which *exact* tests can be based when n_1 and n_2 are small, and they may be found, for example, in the book by D. B. Owen mentioned in the Bibliography.

Continuation of Example. Returning now to our numerical example, we have $n_1 = 15$, $n_2 = 14$, $R_1 = 162$, and, hence,

$$U = 15 \cdot 14 + \frac{15 \cdot 16}{2} - 162 = 168$$

$$\mu_U = \frac{15 \cdot 14}{2} = 105$$

$$\sigma_U^2 = \frac{15 \cdot 14 \cdot 30}{12} = 525$$

Thus,

$$z = \frac{168 - 105}{\sqrt{525}} = 2.74$$

and since this value exceeds $z_{0.005} = 2.58$, the upper critical value for a two-sided alternative hypothesis and $\alpha = 0.01$, we find that *the null hypothesis will have to be rejected.* We conclude that there *is* a difference between the true average grain size of the two kinds of sand.

An interesting feature of the Mann-Whitney test is that, with a slight modification, it can also be used to test the null hypothesis that two samples come from identical populations against the alternative that the two populations have *unequal dispersions*, namely, that they differ in variability or spread. As before, the values of the two samples are arranged jointly in an increasing (or decreasing) order of magnitude, but now they are ranked *from both ends toward the middle.* We assign rank 1 to the smallest value, ranks 2 and 3 to the largest and second largest values, ranks 4 and 5 to the second and third smallest, ranks 6 and 7 to the third and fourth largest, and so on. Subsequently, the calculation of U and the performance of the test are the same as before. The only difference is that with this kind of ranking a *small rank sum* tends to indicate that the population from which the sample was obtained has a *greater variation* than the other, because its values occupy the more extreme positions. We shall not attempt to illustrate this technique with reference to our grain-size example, since this test loses its *sensitivity* for detecting differences in variability when the means of the populations are not the same. The reader will be asked to use it, though, in Exercises 13 and 14 on page 281.

The *Kruskal-Wallis H test* for deciding whether k independent samples come from identical populations is conducted in a way similar to the U test. As before, the observations are ranked *jointly*, and if R_i is the sum of the ranks occupied by the n_i observations of the ith sample, the test is based on the statistic

$$H = \frac{12}{n(n+1)} \sum_{i=1}^{k} \frac{R_i^2}{n_i} - 3(n+1)$$

where $n = n_1 + n_2 + \ldots + n_k$. When $n_i > 5$ for all i and the null hypothesis holds, the distribution of the H statistic is well approximated by the chi-square distribution with $k - 1$ degrees of freedom. Special tables applying for selected small values of the n_i and k are referred to in the table by D. B. Owen mentioned in the Bibliography.

Example. To illustrate the Kruskal-Wallis H test, suppose that an experiment, designed to compare three preventive methods against corrosion, yielded the following maximum depths of pits (in thousandths of an inch) in pieces of wire subjected to the respective treatments:

Method A: 77, 54, 67, 74, 71, 66
Method B: 60, 41, 59, 65, 62, 64, 52
Method C: 49, 52, 69, 47, 56

If we rank these measurements jointly from smallest to largest, we find that those in the first sample are assigned the ranks 6, 13, 14, 16, 17, and 18, so that $R_1 = 84$; those in the second sample are assigned the ranks 1, 4.5, 8, 9, 10, 11, and 12, so that $R_2 = 55.5$; and those in the third sample are assigned the ranks 2, 3, 4.5, 7, and 15, so that $R_3 = 31.5$. Substituting into the formula for H we thus obtain

$$H = \frac{12}{18 \cdot 19}\left(\frac{84^2}{6} + \frac{55.5^2}{7} + \frac{31.5^2}{5}\right) - 3 \cdot 19$$
$$= 6.7$$

and, comparing this figure with 5.991 (the value of $\chi^2_{0.05}$ for 2 degrees of freedom), we find that *the null hypothesis will have to be rejected.* Thus, the three preventive methods against corrosion are *not* equally effective.

EXERCISES

1. With reference to the ignition times (of certain upholstery materials) of Exercise 6 on page 147, use the sign test to test the null hypothesis $\mu = 6.50$ seconds against the alternative hypothesis $\mu < 6.50$ seconds at the level of significance $\alpha = 0.01$.

2. In a laboratory experiment, 18 determinations of the coefficient of friction between leather and metal yielded the following results: 0.59, 0.56, 0.49, 0.55, 0.65, 0.55, 0.51, 0.60, 0.56, 0.47, 0.58, 0.61, 0.54, 0.68, 0.56, 0.50, 0.57, and 0.53. Use the sign test at the level of significance $\alpha = 0.05$ to test the null hypothesis $\mu = 0.55$ against the alternative hypothesis $\mu \neq 0.55$.

3. Applying the sign test to the data of Exercise 9 on page 147, the one pertaining to the boiling point of a certain silicon compound, test the null hypothesis $\mu = 158$ (degrees centigrade) against the alternative hypothesis $\mu \neq 158$. Use the level of significance $\alpha = 0.05$.

4. The quality control department of a large manufacturer obtained the following sample data (in pounds) on the breaking strength of a certain kind of

2-inch cotton ribbon: 153, 159, 144, 160, 158, 153, 171, 162, 159, 137, 159, 159, 148, 162, 154, 159, 160, 157, 140, 168, 163, 148, 151, 153, 157, 155, 148, 168, 162, and 149. Use the sign test at the level of significance $\alpha = 0.01$ to test the null hypothesis $\mu = 150$ against the alternative hypothesis $\mu > 150$.

5. With reference to the data of Exercise 18 on page 224, use the sign test at the level of significance $\alpha = 0.05$ to check the claim that Method B (of instructing the trainees) is more effective than Method A.

6. With reference to the data of Exercise 19 on page 225, use the sign test at the level of significance $\alpha = 0.10$ to check whether there is a systematic difference between weights obtained with the two scales.

7. With reference to Exercise 20 on page 225, use the sign test and the level of significance $\alpha = 0.01$ to check whether there really is a difference in the hardness of the two kinds of magnesium alloys.

8. The following are the number of speeding tickets issued by two policemen on 17 days: 7 and 10, 11 and 13, 14 and 14, 11 and 15, 12 and 9, 6 and 10, 9 and 13, 8 and 11, 10 and 11, 11 and 15, 13 and 11, 7 and 10, 8 and 8, 11 and 12, 9 and 14, 10 and 9, 13 and 16. Use the sign test at the level of significance $\alpha = 0.05$ to test the null hypothesis that on the average the two policemen issue equally many speeding tickets per day against the alternative hypothesis that on the average the second policeman issues more speeding tickets per day than the first.

9. The following are the number of minutes it took a sample of 15 men and 12 women to complete the application form for a position:

> *Men:*   16.5, 20.0, 17.0, 19.8, 18.5, 19.2, 19.0, 18.2, 20.8,
> 18.7, 16.7, 18.1, 17.9, 16.4, 18.9
> *Women:*   18.6, 17.8, 18.3, 16.6, 20.5, 16.3, 19.3, 18.4, 19.7,
> 18.8, 19.9, 17.6

Use the Mann-Whitney test at the level of significance $\alpha = 0.05$ to test the null hypothesis that the two samples come from identical populations against the alternative that the two populations have unequal means.

10. Referring to Exercise 20 on page 225, use the Mann-Whitney test at the level of significance $\alpha = 0.05$ to test whether the true average hardness of the two alloys is the same (see also Exercise 7 above).

11. Comparing two kinds of emergency flares, a consumer testing service obtained the following burning times (rounded to the nearest tenth of a minute):

> *Brand C:*   19.4, 21.5, 15.3, 17.4, 16.8, 16.6, 20.3, 22.5,
> 21.3, 23.4, 19.7, 21.0
> *Brand D:*   16.5, 15.8, 24.7, 10.2, 13.5, 15.9, 15.7, 14.0,
> 12.1, 17.4, 15.6, 15.8

Use the Mann-Whitney test and a level of significance of 0.01 to check whether it is reasonable to say that there is no difference between the true average burning times of the two kinds of flares.

12. To find the best arrangement of instruments on a control panel of an airplane, two different arrangements were compared by simulating an emergency condition and measuring the reaction time required to correct the condition. The reaction times (in *tenths* of a second) of twenty pilots (randomly assigned to the two different arrangements) were as follows:

> *Arrangement 1:*   8, 15, 10, 13, 17, 10, 9, 11, 12, 15
> *Arrangement 2:*   12, 7, 13, 8, 14, 6, 16, 7, 10, 9

Use the Mann-Whitney test at the level of significance $\alpha = 0.05$ to check the claim that the second arrangement is better.

13. With reference to Exercise 9, use the Mann-Whitney test (modified as suggested on page 278) to test the null hypothesis that the two samples come from identical populations against the alternative hypothesis that the populations have unequal dispersions. Use the level of significance $\alpha = 0.05$.

14. The following are the weekly food expenditures (in dollars) of ten families with two children chosen at random from two suburbs of a large city:

> *Suburb A:*   54.78, 62.60, 51.89, 54.50, 56.00, 59.38, 48.19,
> 70.45, 55.15, 51.95
> *Suburb B:*   50.12, 44.63, 64.91, 72.16, 74.59, 39.35, 52.76,
> 78.19, 45.75, 68.72

Use the Mann-Whitney test (modified as suggested on page 278) to test the null hypothesis that the two samples come from identical populations against the alternative that the two populations have unequal dispersions. Use the level of significance $\alpha = 0.05$.

15. The following are the number of misprints counted on pages selected at random from three Sunday editions of a newspaper:

> *April 11:*   4, 10, 2, 6, 4, 12
> *April 18:*   8, 5, 13, 8, 8, 10
> *April 25:*   7, 9, 11, 2, 14, 7

Use the Kruskal-Wallis test at the level of significance $\alpha = 0.05$ to test the null hypothesis that the three samples come from identical populations against the alternative that the compositors and/or proofreaders who worked on the three editions are not equally good.

16. The following are the miles per gallon which a test driver got for ten tankfuls each of three brands of gasoline:

> *Brand 1:*   22, 25, 32, 18, 23, 15, 30, 27, 19, 23
> *Brand 2:*   19, 22, 18, 29, 28, 32, 17, 33, 28, 20
> *Brand 3:*   30, 29, 25, 24, 15, 27, 30, 27, 18, 32

Test at the level of significance $\alpha = 0.05$ whether there is a difference in the true average performance of the three brands of gasoline.

17. So-called Franklin tests were performed to determine the insulation proper-ties of grain-oriented silicon steel specimens that were annealed in five dif-ferent atmospheres with the following results:

Atmosphere	Test results (amperes)						
1	0.58	0.61	0.69	0.79	0.61	0.59	
2	0.37	0.37	0.58	0.40	0.28	0.44	0.35
3	0.29	0.19	0.34	0.17	0.29	0.16	
4	0.81	0.69	0.75	0.72	0.68	0.85	0.57	0.77
5	0.26	0.34	0.29	0.47	0.30	0.42	

Use the Kruskal-Wallis H test and a level of significance of 0.05 to decide whether these five samples can be assumed to come from identical popu-lations.

18. A panel of seven experts was asked to rate each of five industries on the like-lihood that technological changes would produce improvement in environ-mental pollution over the next ten years. Their ratings (in the form of judg-mental probabilities) are as follows:

	Industry				
	A	B	C	D	E
Expert					
1	0.15	0.75	0.10	0.00	0.30
2	0.30	0.60	0.20	0.05	0.25
3	0.20	0.80	0.30	0.00	0.50
4	0.00	0.50	0.25	0.10	0.60
5	0.10	0.55	0.15	0.15	0.40
6	0.25	0.70	0.35	0.25	0.45
7	0.40	0.95	0.45	0.20	0.35

(a) Test whether the average scores given by the experts are significantly different (0.05 level of significance).

(b) Test whether the average scores given to the five industries are signifi-cantly different (0.01 level). What argument(s) can you make against the validity of this test?

10.4 A TEST OF RANDOMNESS

When we discussed random sampling in Chapter 6, we gave several methods which provide some assurance *in advance* that a sample taken will be random. Nevertheless, it is useful to have a technique for testing whether a sample may be looked upon as random *after it has actually been obtained*. One such technique is based on the order in which the sample values were obtained; more specifically, it is based on the number of *runs* exhibited in the sample results.

Given a sequence of two symbols, such as *H* and *T* (which might represent the occurrence of heads and tails in repeated tosses of a coin), a *run* is defined as a succession of identical symbols contained between different symbols or none at all. For example, the sequence

$$\underline{T\ T}\ \underline{H\ H}\ \underline{T\ T}\ \underline{H\ H\ H}\ \underline{T}\ \underline{H\ H\ H}\ \underline{T\ T\ T\ T}\ \underline{H\ H\ H}$$

contains 8 runs, as indicated by the underlines. The total number of runs in a sequence of *n* trials often serves as an indication that the arrangement is *not random*. For instance, if there had been only two runs consisting of ten heads followed by ten tails, we might have suspected that the probability of a success did not remain constant from trial to trial. On the other hand, had the sequence of twenty tosses consisted of alternating heads and tails, we might have suspected that the trials were not independent. In either case, there are grounds to suspect a lack of randomness. Note that our suspicion is not aroused by the *number of H's and T's*, but by the *order* in which they appeared.

If a sequence contains n_1 symbols of one kind and n_2 of another kind (and neither n_1 nor n_2 is less than 10), the sampling distribution of the *total number of runs*, *u*, can be approximated closely by a normal distribution with the mean and the standard deviation

$$\mu_u = \frac{2n_1 n_2}{n_1 + n_2} + 1 \quad \text{and} \quad \sigma_u = \sqrt{\frac{2n_1 n_2 (2n_1 n_2 - n_1 - n_2)}{(n_1 + n_2)^2 (n_1 + n_2 - 1)}}$$

Thus, the test of the null hypothesis that the arrangement of the symbols (and, hence, the sample) is random can be based on the statistic

$$z = \frac{u - \mu_u}{\sigma_u}$$

and Table 3. Special tables for performing exact tests when n_1, n_2, or both are small may be found in the tables by D. B. Owen listed in the Bibliography.

Example. To illustrate this test, let us consider the following arrangement of *defective*, *d*, and *nondefective*, *n*, pieces produced in the given order by a certain machine:

$$n\,n\,n\,n\,n\,d\,d\,d\,d\,n\,n\,n\,n\,n\,n\,n\,n\,n\,n\,d\,d\,n\,n\,d\,d\,d\,d$$

Since there are $n_1 = 10$ d's, $n_2 = 17$ n's, and $u = 6$ runs, substitution into the formulas yields

$$\mu_u = \frac{2 \cdot 10 \cdot 17}{10 + 17} + 1 = 13.59$$

$$\sigma_u = \sqrt{\frac{2 \cdot 10 \cdot 17(2 \cdot 10 \cdot 17 - 10 - 17)}{(10 + 17)^2(10 + 17 - 1)}} = 2.37$$

and

$$z = \frac{6 - 13.59}{2.37} = -3.20$$

Since this value is less than $-z_{0.005} = -2.58$, we can reject the null hypothesis of randomness at the level of significance $\alpha = 0.01$. The total number of runs is *much smaller than expected* and there is a strong indication that the defective pieces appear in clusters or groups; the reason for this will have to be uncovered by an engineer who is familiar with the process.

The run test can be used also to test the randomness of samples consisting of numerical data by counting *runs above and below the median*. Denoting an observation exceeding the median of the sample by the letter a and an observation less than the median by the letter b, we can use the resulting sequence of a's and b's to test for randomness by the method just indicated. A frequent application of this test is in quality control, where the means of successive small samples are exhibited on a graph in chronological order. The run test can then be used to check whether there might be a trend in the data, so that it is possible to adjust a machine setting or some other process variable before any serious damage occurs.

Example. To illustrate this kind of test, suppose that an engineer is concerned about the possibility that too many changes are being made in the settings of an automatic lathe. To test this hypothesis at $\alpha = 0.01$, the following mean diameters (in inches) are obtained for 40 successive shafts turned on the lathe:

0.261 0.258 0.249 0.251 0.247 0.256 0.250 0.247 0.255 0.243
0.252 0.250 0.253 0.247 0.251 0.243 0.258 0.251 0.245 0.250
0.248 0.252 0.254 0.250 0.247 0.253 0.251 0.246 0.249 0.252
0.247 0.250 0.253 0.247 0.249 0.253 0.246 0.251 0.249 0.253

The median of these measurements is 0.250, and, replacing each by the letter a if it exceeds 0.250, by the letter b if it is less than 0.250, and omitting the five which equal 0.250, we obtain the sequence

$$a\,a\,b\,a\,b\,a\,b\,a\,b\,a\,a\,b\,a\,b\,a\,a\,b\,b\,a\,a\,b\,a\,a\,b\,b\,a\,b\,a\,b\,b\,a\,b\,a\,b\,a$$

having 27 runs. Thus, $n_1 = 19, n_2 = 16, u = 27$, and we obtain

$$\mu_u = \frac{2 \cdot 19 \cdot 16}{35} + 1 = 18.37$$

$$\sigma_u = \sqrt{\frac{2 \cdot 19 \cdot 16(2 \cdot 19 \cdot 16 - 19 - 16)}{(19 + 16)^2(19 + 16 - 1)}} = 2.89$$

and

$$z = \frac{27 - 18.37}{2.89} = 2.98$$

Since this value exceeds $z_{0.01} = 2.33$, we can reject the null hypothesis that the sequence of measurements is random. Since the number of runs is larger than one might expect due to chance, it is reasonable to suppose that the lathe is being adjusted too often; it is probably adjusted after each rod is turned, to compensate for any observed discrepancy from a nominal diameter of 0.250 inch.

10.5 THE KOLMOGOROV-SMIRNOV TESTS

The Kolmogorov-Smirnov tests are nonparametric tests for differences between cumulative distributions. The *one-sample test* concerns the agreement between an observed cumulative distribution of sample values and a specified continuous distribution function; thus, it is a test of goodness of fit. The *two-sample test* concerns the agreement between two observed cumulative distributions; it tests the hypothesis whether two independent samples come from identical continuous distributions, and it is sensitive to population differences with respect to location, dispersion, or skewness.

The Kolmogorov-Smirnov one-sample test is generally more efficient than the chi-square test for goodness of fit for small samples, and it can be used for very small samples where the chi-square test does not apply. It must be remembered, however, that the chi-square test of Section 9.6 can be used in connection with discrete distributions whereas the Kolmogorov-Smirnov test cannot.

The one-sample test is based on the maximum absolute difference D between the values of the cumulative distribution of a random sample of size n and a specified theoretical distribution. To determine whether this difference is larger than can reasonably be expected, we look up the critical value of D in Table 10.

Example. To illustrate this test, suppose it is desired to check whether pinholes in electrolytic tin plate are distributed uniformly across a plated coil on the basis of the following distances (in inches) of 10 pinholes from one edge of a long strip of tin plate 30 inches wide:

$$4.8 \quad 14.8 \quad 28.2 \quad 23.1 \quad 4.4 \quad 28.7 \quad 19.5 \quad 2.4 \quad 25.0 \quad 6.2$$

Under the null hypothesis that the pinholes are uniformly distributed, the theoretical cumulative distribution with which we want to compare the observed cumulative distribution is given by

$$F(x) = \begin{cases} 0 & \text{for } x \leq 0 \\ \dfrac{x}{30} & \text{for } 0 < x < 30 \\ 1 & \text{for } x \geq 30 \end{cases}$$

The graph of this theoretical cumulative distribution, as well as that of the observed cumulative distribution, is shown in Figure 10.1. As

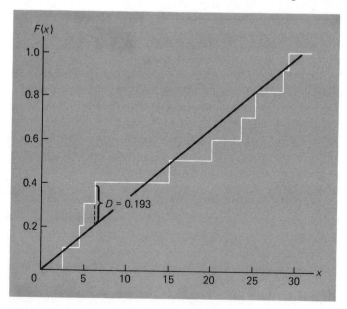

Figure 10.1. Kolmogorov-Smirnov test.

indicated in this diagram, the maximum difference between the two cumulative distributions is 0.193, and since this is less than $D_{0.05}$ = 0.410 (the critical value of D for $n = 10$ and $\alpha = 0.05$), it follows that the null hypothesis (that the pinholes are uniformly distributed) *cannot be rejected.*

The two-sample Kolmogorov-Smirnov test is based on the maximum absolute difference between the values of the two observed cumulative distributions. In principle, it is very similar to the one-sample test, and the necessary critical values can be obtained from special tables (for example, those by D. B. Owen referred to in the Bibliography).

EXERCISES

1. The following arrangement indicates whether sixty consecutive cars which went by the toll booth of a bridge had local plates, L, or out-of-state plates, O:

   $$L\,L\,O\,L\,L\,L\,L\,O\,O\,L\,L\,L\,L\,O\,L\,O\,O\,L\,L\,L\,L\,O\,L\,O\,O\,L\,L\,L\,L\,L$$
   (cont.) $$O\,L\,L\,L\,O\,L\,O\,L\,L\,L\,L\,O\,O\,L\,O\,O\,O\,O\,L\,L\,L\,L\,O\,L\,O\,O\,L\,L\,L\,O$$

   Use the level of significance $\alpha = 0.05$ to test whether this arrangement of L's and O's may be regarded as random.

2. To check the randomness of the digits in Table 7, take those in any five rows, replace the odd digits by the letter O and the even digits by the letter E, and base your decision on the total number of runs in the resulting sequence of O's and E's. Use the level of significance $\alpha = 0.05$.

3. To test whether radio signals from deep space contain a message, an interval of time could be subdivided into a number of very short intervals and it could then be determined whether the signal strength exceeded a certain level (background noise) in each short interval. Suppose that the following is part of such a record, where H denotes a high signal strength and L denotes that the signal strength does not exceed a given noise level.

   $$L\,L\,H\,L\,H\,L\,H\,L\,H\,H\,H\,L\,H\,H\,H\,L\,H\,H\,H\,L\,H\,L\,H\,L\,H\,L\,L\,H\,L$$
   $$L\,L\,H\,L\,H\,L\,H\,L\,H\,H\,H\,L\,H\,H\,H\,L\,H\,H\,H\,L\,H\,L\,H\,L\,H\,L\,L\,L\,L$$

   Test this sequence for randomness (using the 0.05 level of significance) and ascertain whether it is reasonable to assume that the signal contains a message.

4. The run test of Section 10.4 can also be used as a nonparametric alternative for testing the significance between two means. As in the Mann-Whitney

test, we rank the data belonging to the two samples jointly, write a 1 below each value belonging to the first sample, a 2 below each value belonging to the second sample, and then test for the randomness of this arrangement of 1's and 2's. If there are *too few runs*, this may well be accounted for by the fact that the two samples came from populations with unequal means.

(a) Apply this method to the example on page 275, which dealt with the grain size of two kinds of sand.

(b) Use this method to rework Exercise 14 on page 281.

5. Making use of the fact that the median is 19.0 (see Exercise 13 on page 157), test the randomness of the sulfur oxides emission data on page 137 at the level of significance $\alpha = 0.05$.

6. The following are the number of defective pieces turned out by a machine during 24 consecutive shifts: 15, 11, 17, 14, 16, 12, 19, 17, 21, 15, 17, 19, 21, 14, 22, 16, 19, 12, 16, 14, 18, 17, 24, and 13. Test for randomness at the level of significance $\alpha = 0.01$.

7. The following are 50 consecutive down times of a machine (in minutes) which were observed during a certain period of time: 22, 29, 32, 25, 33, 34, 38, 34, 29, 25, 27, 33, 34, 28, 39, 41, 24, 31, 34, 29, 34, 25, 30, 37, 40, 39, 35, 24, 32, 43, 44, 34, 40, 38, 39, 43, 46, 34, 39, 45, 42, 39, 54, 50, 38, 41, 43, 46, 52, and 55. Use the method of runs above and below the median and the level of significance $\alpha = 0.05$ to test the null hypothesis of randomness against the alternative that there is a trend.

8. Use the Kolmogorov-Smirnov test with $\alpha = 0.01$ to decide whether the boiling points of Exercise 9 on page 147 can be assumed to come from a normal population having the mean 160°C and the standard deviation 10°C. [*Hint*: Use probability graph paper.]

9. In a vibration study, certain airplane components were subjected to severe vibrations until they showed structural failures. Given the following failure times (in minutes), test whether they can be looked upon as a sample from an exponential population with the mean $\mu = 10$:

$$1.5 \quad 10.3 \quad 3.6 \quad 13.4 \quad 18.4 \quad 7.7 \quad 24.3 \quad 10.7 \quad 8.4$$
$$15.4 \quad 4.9 \quad 2.8 \quad 7.9 \quad 11.9 \quad 12.0 \quad 16.2 \quad 6.8 \quad 14.7$$

Use the level of significance $\alpha = 0.05$.

10. Survival times (days) of 8 cancer-bearing mice that have been treated with a certain anticancer drug are as follows:

$$16, \quad 11, \quad 24, \quad 18, \quad 31, \quad 15, \quad 12, \quad 21$$

Test whether these data are consistent with the assumption of a log-normal distribution of survival times. Use the level of significance $\alpha = 0.01$.

11

Curve Fitting

11.1 THE METHOD OF LEAST SQUARES

The main objective of many engineering investigations is to make predictions, preferably on the basis of mathematical equations. For instance, an engineer may wish to predict the amount of oxide that will form on the surface of a metal baked in an oven for a specified amount of time at 200°C, or the amount of deformation of a ring subjected to a compressive force of 1,000 pounds, or the time between recappings of a tire having a given tread thickness and composition. Usually, such predictions require that a formula be found which relates the dependent variable (whose value one wants to predict) to one or more independent variables. In this section we shall consider the special case where a dependent variable is to be predicted in terms of a single independent variable.

In many problems of this kind the independent variable is observed without error, or with an error which is negligible when compared with the error (chance variation) in the dependent variable. For example, in measuring the amount of oxide on the surface of a metal specimen, the baking temperature can usually be controlled with good precision, but the oxide-thickness measurement may be subject to considerable chance variation. Thus, even though the independent variable may be fixed at x, repeated measurements of the dependent variable may lead to y-values which differ considerably. Such differences among y-values can be attributed to several causes, chiefly to errors of measurement and to the existence of other, uncontrolled variables which may influence the value of y when x is fixed. Thus, measurements of the thickness of oxide layers may vary over several specimens baked for the same length of time at the same temperature because of the difficulty in measuring thickness as well as possible differences in the composition of the oven atmosphere, surface conditions of the specimens, and the like.

It should be apparent from this discussion that in this context y is the value of a random variable whose distribution depends on x. In most situations of this sort we are interested mainly in the relationship between x and the *mean* of the corresponding distribution of y's, and we refer to this relationship as the *regression curve of y on x*. (For the time being we shall assume that x is fixed, that is, not random; in Section 11.5 we shall consider the case where x and y are both values of random variables.)

Let us first treat the case where the regression curve of y on x is *linear*, that is, where, for any given x, the mean of the distribution of the y's is given by $\alpha + \beta x$. In general, an observed y will differ from this mean, and we shall denote this difference by ϵ, writing

$$y = \alpha + \beta x + \epsilon$$

Thus, ϵ is a value of a random variable and we can always choose α so that the mean of the distribution of this random variable is equal to zero. The value of ϵ for any given observation will depend on a possible error of measurement and on the values of variables other than x which might have an influence on y.

Example. To give an example where the regression curve of y on x can reasonably be assumed to be linear, suppose that a tensile ring is to be calibrated by measuring the deflection in thousandths of an inch at various loads in thousands of pounds. In the following table (giving the results of 12 measurements), the x's are the load forces in thousands of pounds and the y's are the corresponding deflections in thousandths of an inch:

x	1	2	3	4	5	6	7	8	9	10	11	12
y	16	35	45	74	86	96	106	124	134	156	164	182

It is apparent from Figure 11.1, where these data have been plotted, that it is reasonable to assume that the relationship (regression curve) is linear; that is, a straight line gives a very good approximation over the range of the available data.

With regard to this example, we face the problem of using the *data points* plotted in Figure 11.1 to estimate the parameters α and β of the regression line, and this is equivalent to finding the equation of the straight line which "best fits" the data points. In this example it may well be satisfactory to do this "by eye," and if different experimenters were to

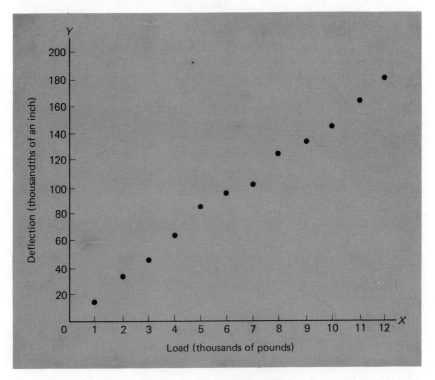

Figure 11.1. Linear regression.

fit a line in this way, they would probably all predict that at a loading of 7,500 pounds the deflection should be close to 0.115 inch. However, if we have to deal with data such as those plotted in Figure 11.2, the problem of finding a best-fitting line is not so obvious. To handle problems of this kind, we must seek a *nonsubjective* method for fitting straight lines which reflects some desirable statistical properties.

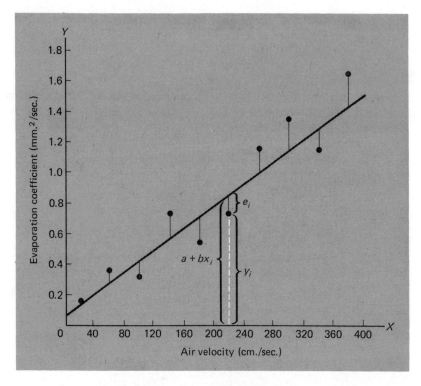

Figure 11.2. Least-squares criterion.

To state the problem formally, we have n paired observations (x_i, y_i) for which it is reasonable to assume that the regression of y on x is linear, and we want to determine the line (that is, the equation of the line) which in some sense provides the "best" fit. There are several ways in which we can interpret the word "best," and the meaning we shall give it here may be explained as follows. If we predict y by means of the equation

$$y' = a + bx$$

where a and b are constants, then e_i, the error in predicting the value of y

corresponding to the given x_i, is

$$y_i - y_i' = e_i$$

Note that the equation $y' = a + bx$ provides an *estimate* of the equation of the regression line whose actual, but unknown, equation is $y = \alpha + \beta x$. The actual error in predicting y_i is ϵ_i, and this error is estimated by the quantity $y_i - y_i' = e_i$. We shall attempt to determine a and b so that the estimated errors are in some sense as small as possible.

Since we cannot minimize each of the e_i individually, it suggests itself that we might try to make their sum $\sum_{i=1}^{n} e_i$ as close as possible to zero. However, since this sum can be made *equal* to zero by many choices of totally unsuitable straight lines for which the positive and the negative errors cancel, we shall instead minimize the sum of the *squares* of the e_i (see also definition of standard deviation). In other words, we shall choose a and b so that

$$\sum_{i=1}^{n} [y_i - (a + bx_i)]^2$$

is a minimum. Note from Figure 11.2 that this is equivalent to minimizing the sum of the squares of the vertical distances from the points to the line. This criterion, called the *criterion of least squares*, yields values for a and b (estimates for α and β) that have many desirable properties, some of which will be mentioned at the end of this section.

A necessary condition for a relative minimum is the vanishing of the partial derivatives with respect to a and b. We thus have

$$2\sum_{i=1}^{n} [y_i - (a + bx_i)](-1) = 0$$

$$2\sum_{i=1}^{n} [y_i - (a + bx_i)](-x_i) = 0$$

and rewriting these equations in a somewhat more convenient form we obtain the following equations, called the *normal equations*:

$$\sum_{i=1}^{n} y_i = an + b\sum_{i=1}^{n} x_i$$

$$\sum_{i=1}^{n} x_i y_i = a\sum_{i=1}^{n} x_i + b\sum_{i=1}^{n} x_i^2$$

The normal equations are a set of two linear equations in the unknowns a and b; their simultaneous solution gives the values of a and b for the line which, thus, provides the best fit to the given data according to the criterion of least squares. Note that they can easily be remembered as follows: we first write down the equation $y_i = a + bx_i$ and then the equation $x_iy_i = ax_i + bx_i^2$, obtained from the first by multiplying both sides by x_i. If we then sum both sides of each of these equations, we get the two normal equations (after some easy algebraic simplifications).

Example. To illustrate the method of least squares, as it is used to fit a straight line to a given set of paired data, let us apply it to the data of Figure 11.2, pertaining to the air velocities and evaporation coefficients of burning fuel droplets in an impulse engine:

Air velocity (cm/sec) x	Evaporation coefficient (mm²/sec) y
20	0.18
60	0.37
100	0.35
140	0.78
180	0.56
220	0.75
260	1.18
300	1.36
340	1.17
380	1.65

Here the x's are the air velocities in cm/sec and the y's are the evaporation coefficients in mm²/sec. Since $n = 10$,

$$\sum_{i=1}^{n} x_i = 2{,}000, \qquad \sum_{i=1}^{n} x_i^2 = 532{,}000$$

$$\sum_{i=1}^{n} y_i = 8.35, \qquad \sum_{i=1}^{n} x_iy_i = 2{,}175.40$$

the normal equations are

$$8.35 = 10a + 2{,}000b$$
$$175.40 = 2{,}000a + 532{,}000b$$

Solving these two equations, we obtain $a = 0.069$, $b = 0.0038$, and the equation of the straight line which provides the best fit in the sense of least squares is

$$y' = 0.069 + 0.0038x$$

This equation might be used, for example, to predict that the evaporation coefficient of droplets when the air volocity is 190 cm/sec will be $y' = 0.069 + (0.0038)(190) = 0.79$ mm²/sec.

It is impossible to make any exact statements about the "goodness" of this prediction unless we make some assumptions about the underlying distribution of the evaporation coefficients and about the true nature of the regression, namely, that for a given x the true mean of the y's is of the form $\alpha + \beta x$. Looking upon the values of a and b obtained from the normal equations as *estimates* of the *regression coefficients* α and β, the reader will be asked to show in Exercises 13 and 14 on page 306 that these estimates are *linear* in the observations y_i and that they are *unbiased* estimates of α and β. With these properties, we can then refer to the remarkable *Gauss-Markov theorem* which states that among all unbiased estimators for α and β which are linear in the y_i, the least-squares estimators have the smallest variance. In other words, the least-squares estimators are the most reliable in the sense that they are subject to the smallest chance variations. A proof of the Gauss-Markov theorem may be found in the book by H. Scheffe referred to in the Bibliography.

11.2 INFERENCES BASED ON THE LEAST-SQUARES ESTIMATORS

The method of the preceding section is used when the relationship between x and the mean of y is linear, or close enough to a straight line so that the least-squares line yields reasonably good predictions. In what follows we shall assume that the regression *is* linear and furthermore that the n random variables having the values y_i ($i = 1, \ldots, n$) are *independently normally distributed with the means $\alpha + \beta x_i$ and the common variance σ^2*. If we write

$$y_i = \alpha + \beta x_i + \epsilon_i$$

it follows from these assumptions that the ϵ_i are values of independent normally distributed random variables having zero means and the common variance σ^2. The various assumptions we have made here are illustrated in Figure 11.3, showing the distributions of values of y_i for several values of the x_i. Note that these additional assumptions are required to discuss the goodness of predictions based on least-squares equations, the properties of a and b as estimates of α and β, and so on; they were *not* required to obtain the original estimates based on the method of least squares.

Before we state a theorem concerning the distribution of the least-squares estimators of α and β, it will be convenient to introduce some

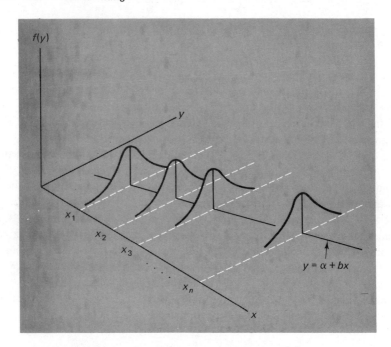

Figure 11.3. Assumptions underlying Theorem 11.1.

special notation. The following expressions pertaining to the sample values (x_i, y_i) occur so often that it is useful to write them as

$$S_{xx} = n \sum_{i=1}^{n} x_i^2 - \left(\sum_{i=1}^{n} x_i \right)^2$$

$$S_{yy} = n \sum_{i=1}^{n} y_i^2 - \left(\sum_{i=1}^{n} y_i \right)^2$$

$$S_{xy} = n \sum_{i=1}^{n} x_i y_i - \left(\sum_{i=1}^{n} x_i \right)\left(\sum_{i=1}^{n} y_i \right)$$

Using this notation, the reader will be asked to show in Exercise 15 on page 306 that the solutions of the two normal equations on page 293 can be written as

$$a = \bar{y} - b \cdot \bar{x} \quad \text{and} \quad b = \frac{S_{xy}}{S_{xx}}$$

where $\bar{x}$ and $\bar{y}$ are, respectively, the means of the x's and the y's. Note also the close relationship between S_{xx} and S_{yy} and the respective sample variances of the x's and the y's; in fact, $s_x^2 = S_{xx}/n(n - 1)$ and $s_y^2 = S_{yy}/n(n - 1)$, and we shall sometimes use this alternate notation.

The variance σ^2 defined on page 295 is usually estimated in terms of the vertical deviations of the sample points from the least-squares line. The ith such deviation is $y_i - y_i' = y_i - (a + bx_i)$ and the estimate of σ^2 is

$$s_e^2 = \frac{1}{n - 2} \sum_{i=1}^{n} [y_i - (a + bx_i)]^2$$

where s_e is, traditionally, referred to as the *standard error of estimate*. An equivalent formula for this estimate of σ^2, which is more convenient to use in actual applications, is given by

$$s_e^2 = \frac{S_{xx}S_{yy} - (S_{xy})^2}{n(n - 2)S_{xx}}$$

In these formulas the divisor $n - 2$ is used to make the resulting estimator for σ^2 *unbiased*. The "loss" of two degrees of freedom is explained by the fact that the two regression coefficients α and β had to be replaced by their least-squares estimates. It can also be shown that under the given assumptions $(n - 2)s_e^2/\sigma^2$ is a value of a random variable having the chi-square distribution with $n - 2$ degrees of freedom.

Based on the assumptions made concerning the distribution of the y's, one can prove the following theorem concerning the distributions of the least-squares estimators of the regression coefficients α and β:

THEOREM 11.1. *Under the assumptions given on page 295, the sampling distributions of the statistics*

$$t = \frac{(a - \alpha)}{s_e} \sqrt{\frac{nS_{xx}}{S_{xx} + (n\bar{x})^2}}$$

and

$$t = \frac{(b - \beta)}{s_e} \sqrt{\frac{S_{xx}}{n}}$$

are t distributions with $n - 2$ degrees of freedom.

Using the first of these statistics, we can construct a confidence interval for α, solving the double inequality $-t_{\alpha/2} < t < t_{\alpha/2}$.* The limits of the resulting confidence interval are

$$a \pm t_{\alpha/2} \cdot s_e \sqrt{\frac{S_{xx} + (n\bar{x})^2}{nS_{xx}}}$$

Similar arguments lead to the following confidence limits for β:

$$b \pm t_{\alpha/2} \cdot s_e \sqrt{\frac{n}{S_{xx}}}$$

Continuation of Example. To illustrate the construction of such confidence intervals for the regression coefficients α and β, we shall assume that the populations from which the data on evaporation coefficients on page 294 were obtained satisfy the necessary assumptions of independence, normality, and equal variances. Using the numerical results on page 294 together with

$$\sum_{i=1}^{n} y_i^2 = 9.1097$$

we first obtain

$$S_{xx} = 10(532,000) - (2,000)^2 = 1,320,000$$
$$S_{yy} = 10(9.1097) - (8.35)^2 = 21.3745$$
$$S_{xy} = 10(2,175.40) - (2,000)(8.35) = 5,054.00$$

and, hence,

$$S_e^2 = \frac{(1,320,000)(21.3745) - (5,054.00)^2}{(10)(8)(1,320,000)} = 0.0253$$

Since $t_{.025}$ equals 2.306 for $10 - 2 = 8$ degrees of freedom, we get the following 0.95 confidence intervals for α and β:

$$0.069 \pm (2.306)(0.159) \sqrt{\frac{1,320,000 + (2,000)^2}{10(1,320,000)}}$$

---

*Although the symbol α is used here both for a parameter of the regression line and for the level of confidence, the formulas are such that there should be no ambiguity.

or

$$-0.164 < \alpha < 0.302$$

and

$$0.0038 \pm (2.306)(0.159)\sqrt{\frac{10}{1,320,000}}$$

or

$$0.0028 < \beta < 0.0048$$

The results of Theorem 11.1 can also be used to establish criteria for testing hypotheses concerning α and β. In this connection, the parameter β is of special importance as it represents the *slope* of the regression line; that is, β gives the change in the mean of y corresponding to a unit increase in x. If $\beta = 0$, the regression line is horizontal and the mean of y does not depend linearly on x.

To test the null hypothesis $H_0 : \beta = \beta_0$ we can make use of the second t statistic of Theorem 11.1, namely,

$$t = \frac{(b - \beta_0)}{s_e}\sqrt{\frac{S_{xx}}{n}}$$

and the resulting critical regions are as shown in the following table:

CRITICAL REGIONS FOR TESTING $H_0 : \beta = \beta_0$

Alternative hypothesis	Reject H_0 if
$\beta < \beta_0$	$t < -t_\alpha$
$\beta > \beta_0$	$t > t_\alpha$
$\beta \neq \beta_0$	$t < -t_{\alpha/2}$ or $t > t_{\alpha/2}$

where t_α is obtained from Table 4 with $n - 2$ degrees of freedom.

Continuation of Example. To illustrate this kind of test, let us test for the significance of the linear regression (linear dependence) of the evaporation coefficient on air velocity, basing the calculations on the data given on page 294. The null hypothesis to be tested is $H_0 : \beta = 0$ and the appropriate alternative is $H_1 : \beta \neq 0$. Using the calculations on page 298, we obtain

$$t = \frac{0.0038 - 0}{0.159} \sqrt{\frac{1,320,000}{10}} = 8.36$$

and since this exceeds 2.306, the value of $t_{.025}$ for 8 degrees of freedom, we can reject the null hypothesis at the 0.05 level of significance. In other words, we conclude that there is a relationship between air velocity and the average evaporation coefficient.

Another problem, closely related to the problem of estimating the regression coefficients α and β, is that of estimating $\alpha + \beta x$, namely, the mean of the distribution of the y's for a given value of x. If x is held fixed at x_0, the quantity we want to estimate is $\alpha + \beta x_0$ and it would seem reasonable to use $a + bx_0$, where a and b are again the values obtained by the method of least squares. In fact, it can be shown that this estimator is unbiased, has the variance

$$\sigma^2 \left[\frac{1}{n} + \frac{n(x_0 - \bar{x})^2}{S_{xx}} \right]$$

and that a $1 - \alpha$ confidence interval for $\alpha + \beta x_0$ is given by

$$(a + bx_0) \pm t_{\alpha/2} \cdot s_e \sqrt{\frac{1}{n} + \frac{n(x_0 - \bar{x})^2}{S_{xx}}}$$

where $t_{\alpha/2}$ is to be obtained from Table 4 with $n - 2$ degrees of freedom.

Of even greater importance than the estimation of $\alpha + \beta x_0$ is usually the *prediction* of y' (a future value of y) when $x = x_0$. Thus, for our example we already showed on page 295 that for an air velocity of 190 cm/sec (a value not included in the experiment), the predicted evaporation coefficient is 0.79 mm²/sec. Now let us indicate a method of constructing an interval in which y can be expected to lie with a given probability when $x = x_0$. If α and β were known, we could use the fact that y is a random variable having a normal distribution with the mean $\alpha + \beta x_0$ and the variance σ^2 (or that $y - \alpha - \beta x_0$ is a value of a random variable having a normal distribution with zero mean and the variance σ^2). However, if α and β are not known, we must consider the quantity $y - a - bx_0$, where y, a, and b are all values of random variables, and the resulting theory leads to the following *limits of prediction* for y when $x = x_0$:

$$(a + bx_0) \pm t_{\alpha/2} \cdot s_e\sqrt{1 + \frac{1}{n} + \frac{n(x_0 - \bar{x})^2}{S_{xx}}}$$

where the number of degrees of freedom for $t_{\alpha/2}$ is again $n - 2$.

Continuation of Example. With reference to the air velocity and evaporation-coefficient data, let us first illustrate the construction of confidence limits for $\alpha + bx_0$ when $x_0 = 190$. Substituting the various quantities already calculated into the formula on page 300, we can write 0.95 confidence limits for $\alpha + 190\beta$ as

$$0.79 \pm (2.306)(0.159)\sqrt{\frac{1}{10} + \frac{(10)(190 - 200)^2}{1,320,000}}$$

and, hence,

$$0.67 < \alpha + 190\beta < 0.91$$

Correspondingly, 0.95 limits of prediction for a future value of y when $x = 190$ are given by

$$0.79 \pm (2.306)(0.159)\sqrt{1 + \frac{1}{10} + \frac{(10)(190 - 200)^2}{1,320,000}}$$

or 0.79 ± 0.39.

Note that although the *mean* of the distribution of y's when $x = 190$ can be estimated fairly closely, the value of a single future observation cannot be predicted with good precision. Observe from the formula for the limits of prediction that even as $n \rightarrow \infty$ the difference between the limits of prediction does *not* approach zero. The limiting width of the interval of prediction depends on s_e, which expresses the inherent variability of the data. Note, further, that if we wish to *extrapolate*, that is, to predict a future value of y corresponding to a value of x outside the range of the observations, the limits of prediction (and also the confidence limits for $\alpha + \beta x$) become increasingly wide.

Continuation of Example. For instance, 0.95 limits of prediction for y corresponding to $x = 450$ cm/sec are given by

$$1.78 \pm (2.306)(0.159)\sqrt{1 + \frac{1}{10} + \frac{10(450 - 200)^2}{1,320,000}}$$

or 1.78 ± 0.46. In other words, we can assert with a probability of 0.95 that the evaporation coefficient corresponding to an air velocity of 450 cm/sec will fall between 1.32 and 2.24 mm²/sec. Poor as it is, this prediction is also based on the assumption that the true regression continues to be linear when we go beyond the range of the observations.

EXERCISES

1. A chemical company, wishing to study the effect of extraction time on the efficiency of an extraction operation, obtained the data shown in the following table:

Extraction time (minutes) x	Extraction efficiency (percent) y
27	57
45	64
41	80
19	46
35	62
39	72
19	52
49	77
15	57
31	68

(a) Assuming the regression of y on x to be of the form $y = \alpha + \beta x$, find the normal equations for estimating the parameters α and β and solve for a and b.

(b) Graph the data together with the straight line obtained in part (a), and use it to read off the extraction efficiency one can expect when the extraction time is 35 minutes.

2. Solving the normal equations on page 293 symbolically, show that

$$a = \frac{(\sum x^2)(\sum y) - (\sum x)(\sum xy)}{n(\sum x^2) - (\sum x)^2}$$

$$b = \frac{n(\sum xy) - (\sum x)(\sum y)}{n(\sum x^2) - (\sum x)^2}$$

(The indices and limits of summation have been omitted for simplicity.) Use these formulas to verify the results obtained in part (a) of Exercise 1.

3. In the accompanying table, x is the tensile force applied to a steel specimen in thousands of pounds and y is the resulting elongation in thousandths of an inch. Assuming the regression of y on x to be linear, estimate the parameters of the regression line and construct a 0.95 confidence interval for β, the elongation of the specimen per thousand pounds of tensile stress.

x	1	2	3	4	5	6
y	14	33	40	63	76	85

4. The following data pertain to the demand for a product (in thousands of units) and its price (in cents) charged in seven different market areas:

Price x	Demand y
11	145
9	177
12	109
10	135
15	81
12	118
6	218

(a) Estimate α and β, the coefficients of the regression line of y on x.
(b) Test the null hypothesis $\beta = -20$ at a level of significance of 0.01.

5. The following show the improvement (gain in reading speed) of eight students in a speed-reading program, and the number of weeks they have been in the program:

Number of weeks	Speed gain (words per minute)
3	86
5	118
2	49
8	193
6	164
9	232
3	73
4	109

Fit a straight line to these data by the method of least squares, using number of weeks in the program as the independent variable. Also, find a 0.90 confidence interval for β.

6. Raw material used in the production of a synthetic fiber is stored in a place which has no humidity control. Measurements of the relative humidity in the storage place and the moisture content of a sample of the raw material (both in percentages) on 12 days yielded the following results:

x Humidity	y Moisture content
42	12
35	8
50	14
43	9
48	11
62	16
31	7
36	9
44	12
39	10
55	13
48	11

(a) Estimate α and β, the coefficients of the regression line of y on x.

(b) Find a 0.99 confidence interval for the mean moisture content of the raw material when the humidity of the storage place is 40 percent.

(c) Find 0.95 limits of prediction for the moisture content of the raw material when the humidity of the storage place is 40 percent.

(d) Referring to part (c), indicate to what extent the width of the interval is affected by the size of the sample and to what extent it is affected by the inherent variability of the data.

7. Referring to the data of Exercise 3:

(a) Find a 0.95 confidence interval for the mean of the distribution of the y's when $x = 3.5$.

(b) Express the 0.95 confidence limits for the mean of the distribution of the y's in terms of x_0; plot a graph of the estimated regression line and, choosing suitable values of x_0, sketch graphs of the loci of the upper and lower confidence limits on the same set of coordinate axes. (The resulting "confidence bands" do *not* give confidence intervals for the line itself; in fact, since any two confidence intervals obtained from these bands are dependent, they should be used only *once* in connection with some fixed value x_0.)

8. Referring to Exercise 1, express 0.95 limits of prediction for the extraction efficiency in terms of the extraction time x_0. Choosing suitable values of x_0, sketch graphs of the loci of the upper and lower limits of prediction on the same set of coordinate axes. (Note that, as in Exercise 7, such bands should be used only *once* in connection with some given extraction time.)

9. In Exercises 3 and 5 it would have been entirely reasonable to impose the condition $\alpha = 0$ *before* fitting a straight line by the method of least squares.

(a) Using the method of least squares, derive a formula for estimating β when the regression line is assumed to have the form $y = \beta x$.

(b) Using the data of Exercise 3 and the formula obtained in part (a), estimate β and compare the result with the estimate previously obtained without the condition that the line must pass through the origin.

(c) Using the data of Exercise 5 and the formula obtained in part (a), estimate β and compare the result with the estimate previously obtained without the condition that the line must pass through the origin.

10. The cost of manufacturing a lot of a certain product depends upon the lot size as shown in the following table:

Cost (dollars)	30	70	140	270	530	1,010	2,500	5,020
Lot size	1	5	10	25	50	100	250	500

Fit a straight line to these data by the method of least squares, using lot size as the independent variable. Also, find a 0.90 confidence interval for α, which can here be interpreted as the fixed overhead cost of manufacturing.

11. Referring to the data of Exercise 10, fit a straight line using the cost as the independent variable. Graph the resulting line together with the one found in Exercise 10 in one diagram and note that the two regression lines do *not* coincide.

12. When the sum of the x-values is equal to zero, the calculation of the coefficients of the regression line of y on x is greatly simplified; in fact, their estimates are given by

$$a = \frac{\sum y}{n} \quad \text{and} \quad b = \frac{\sum xy}{\sum x^2}$$

This simplification can also be attained when the x's are *equally spaced*, that is, when they are in arithmetic progression. We then "code" the data by substituting for the x's the values $\ldots, -2, -1, 0, 1, 2, \ldots$, when n is *odd*, or the values $\ldots, -3, -1, 1, 3, \ldots$, when n is *even*. The above formulas are then used in connection with the coded data.

(a) Use this technique to fit a least-squares trend line to the following data on the income from tourism in Arizona (in millions of dollars):

Year	Income from tourism
1965	420
1966	450
1967	480
1968	500
1969	530
1970	565
1971	600

Also use the result to predict Arizona's income from tourism in the year 1978, and calculate 0.90 limits of prediction for this estimate.

(b) Use this technique to fit a least-squares trend line to the following data on a company's annual profits in millions of dollars:

Year	Profit
1970	5.3
1971	6.1
1972	8.0
1973	8.9
1974	10.1
1975	10.7

Also use the result to predict the company's profit in 1980, and calculate 0.80 limits of prediction.

(c) What assumption are you making in connection with your predictions of tourism in 1978 and profits in 1980?

13. Using the formulas obtained in Exercise 2, show that
   (a) the expression for a is linear in the y_i;
   (b) a is an unbiased estimate of α.

14. Using the formulas obtained in Exercise 2, show that
   (a) the expression for b is linear in the y_i;
   (b) b is an unbiased estimate of β.

15. Show that the least-squares estimates of the coefficients of the regression line of y on x can be written in the form

$$a = \bar{y} - b \cdot \bar{x} \quad \text{and} \quad b = \frac{S_{xy}}{S_{xx}}$$

11.3   CURVILINEAR REGRESSION

So far we have studied only the case where the regression curve of y on x is linear; that is, where for any given x, the mean of the distribution of the y's is given by $\alpha + \beta x$. In this section, we shall first investigate cases where the regression curve is nonlinear, but where the methods of Section 11.1 can nevertheless be applied; then we shall take up the problem of polynomial regression, that is, problems where for any given x the mean of the distribution of the y's is given by $\beta_0 + \beta_1 x + \beta_2 x^2 + \ldots + \beta_p x^p$. This last technique is also used to obtain approximations when the functional form of the regression curve is unknown.

It is common practice for engineers to plot paired data on various kinds of graph paper, in order to determine whether for suitably transformed scales the points will fall close to a straight line. If that is the case, the nature of the transformation used leads to a functional form of the regression equation, and the necessary constants (parameters) can be determined by applying the method of Section 11.1 to the transformed data. If a set of paired data consisting of n points (x_i, y_i) "straightens out" when plotted on semi-log paper, for instance, this indicates that the regression curve of y on x is *exponential*, namely, that for any given x, the mean of the distribution of the y's is given by $\alpha \cdot \beta^x$. Taking logarithms to the base 10 of both sides of the *predicting equation**

$$y' = \alpha \cdot \beta^x$$

we obtain

$$\log y' = \log \alpha + x \cdot \log \beta$$

and we can now get estimates of $\log \alpha$ and $\log \beta$, and hence of α and β, by applying the method of Section 11.1 to the n pairs of values $(x_i, \log y_i)$.

Example. To illustrate the least-squares fitting of an exponential curve, let us consider the following data on the percentage of the radial tires made by a certain manufacturer that are still usable after having been driven for the given number of miles:

Miles driven (thousands) x	Percentage usable y
1	98.2
2	91.7
5	81.3
10	64.0
20	36.4
30	32.6
40	17.1
50	11.3

When plotting the data on semi-log graph paper as in Figure 11.4, we observe that the points fall fairly close to a straight line and that it is, therefore, justifiable to fit an exponential curve. Determining first the logarithms of the eight y's, we obtain, respectively, 1.9921, 1.9624, 1.9101, 1.8062, 1.5611, 1.5132, 1.2330, 1.0531, and the summations required for substitution into the normal equations are $\sum x = 158$, $\sum x^2 = 5{,}530$, $\sum \log y = 13.0312$, and $\sum x \cdot \log y = 212.1224$, where the subscripts and limits of summation have been omitted for simplicity. Again using a and b for the least-squares estimates of α and β,

---

*Any other base could have been used.

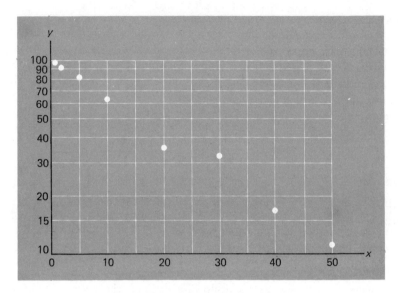

Figure 11.4.   Exponential regression.

we obtain the normal equations

$$13.0312 = 8(\log a) + 158(\log b)$$
$$212.1224 = 158(\log a) + 5{,}530(\log b)$$

and, solving this system, we get

$$\log a = 2.0002 \quad \text{and} \quad \log b = -0.0188$$

Hence, $a = 100.0$, $b = 0.96$, and the exponential curve fit by the method of least squares has the equation

$$y' = 100(0.96)^x$$

This equation can also be written as $y' = 100 \cdot 10^{-0.0188x}$ or $y' = 100 \cdot e^{-0.0433x}$, making use of the fact that $10^{-0.0188} = e^{-0.0433}$. If we want to base a prediction on an exponential equation fitted to a set of data, it is usually more convenient to use it in its logarithmic form. For instance, if we want to predict what percentage of the tires will last at least 25,000 miles, we get

$$\log y = 2.0002 + 25(-0.0188)$$
$$= 1.5302$$

and, hence, 33.9 percent.

Two other relationships, frequently arising in engineering practice, that can be fitted by the method of Section 11.1 after suitable transformations are the *reciprocal function* given by $y' = \dfrac{1}{\alpha + \beta x}$, and the *power function* $y' = \alpha \cdot x^{\beta}$. The first of these represents a linear relationship between x and $1/y'$, namely,

$$\frac{1}{y'} = \alpha + \beta x$$

and we obtain estimates of α and β by applying the method of Section 11.1 to the points $(x_i, 1/y_i)$. The second represents a linear relationship between $\log x$ and $\log y'$, namely,

$$\log y' = \log \alpha + \beta \cdot \log x$$

and we obtain estimates of $\log \alpha$ and β, and hence of α and β, by applying the method of Section 11.1 to the points $(\log x_i, \log y_i)$. Examples of other curves that can be fitted by the method of least squares after a suitable transformation are given in Exercises 6 and 7 on page 318.

If there is no clear indication about the functional form of the regression of y on x, we often assume that the underlying relationship is at least "well behaved" to the extent that it has a Taylor series expansion and that the first few terms of this expansion will yield a fairly good approximation. We thus fit to our data a polynomial, that is, a predicting equation of the form

$$y' = \beta_0 + \beta_1 x + \beta_2 x^2 + \ldots + \beta_p x^p$$

where the degree is determined by inspection of the data or by more "scientific" methods to be discussed below.

Given a set of data consisting of n points (x_i, y_i), we estimate the coefficients $\beta_0, \beta_1, \beta_2, \ldots, \beta_p$ of the pth-degree polynomial by minimizing

$$\sum_{i=1}^{n} [y_i - (\beta_0 + \beta_1 x_i + \beta_2 x_i^2 + \ldots + \beta_p x_i^p)]^2$$

In other words, we are now applying the least-squares criterion by minimizing the sum of the squares of the vertical distances from the points to the curve (see Figure 11.5). Differentiating partially with respect to $\beta_0, \beta_1, \beta_2, \ldots, \beta_p$, equating these partial derivatives to zero, rearranging some of the terms, and letting b_i be the estimate of β_i, we obtain the $p + 1$ *normal equations*

$$\sum y = nb_0 + b_1 \sum x + \ldots + b_p \sum x^p$$
$$\sum xy = b_0 \sum x + b_1 \sum x^2 + \ldots + b_p \sum x^{p+1}$$
$$\cdots\cdots\cdots\cdots\cdots\cdots\cdots\cdots\cdots\cdots\cdots\cdots\cdots$$
$$\sum x^p y = b_0 \sum x^p + b_1 \sum x^{p+1} + \ldots + b_p \sum x^{2p}$$

where the indices and limits of summation are omitted for simplicity. Note that this is a system of $p + 1$ *linear* equations in the $p + 1$ unknowns, $b_0, b_1, \ldots, b_p$. Unless the choice of the x-values is very unusual, this system of equations will have a unique solution.

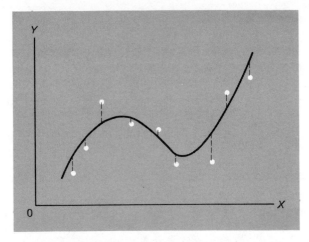

Figure 11.5. Least-squares criterion for a polynomial.

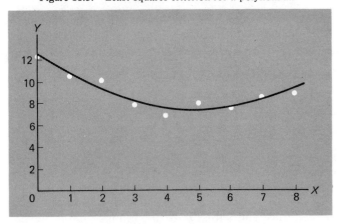

Figure 11.6. Polynomial regression curve.

Example. To illustrate the fitting of a polynomial by the method of least squares, we shall fit a quadratic (second-degree) polynomial to the following data, relating the drying time of a varnish to the amount of a certain additive:

x Amount of varnish additive (grams)	y Drying time (hours)
0	12.0
1	10.5
2	10.0
3	8.0
4	7.0
5	8.0
6	7.5
7	8.5
8	9.0

Inspection of these data plotted in Figure 11.6 indicates that a second-degree polynomial, having one relative minimum, should yield a fairly good fit. First calculating the sums required for substitution into the normal equations, we obtain

$$\sum x = 36, \qquad \sum x^2 = 204, \qquad \sum x^3 = 1296, \qquad \sum x^4 = 8772$$
$$\sum y = 80.5, \qquad \sum xy = 299.0, \qquad \sum x^2 y = 1697.0$$

and we thus have to solve the following system of three linear equations in three unknowns:

$$80.5 = 9b_0 + 36b_1 + 204b_2$$
$$299.0 = 36b_0 + 204b_1 + 1296b_2$$
$$1697.0 = 204b_0 + 1296b_1 + 8772b_2$$

Getting $b_0 = 12.2$, $b_1 = -1.85$, and $b_2 = 0.183$, we find that the equation of the least-squares polynomial is

$$y' = 12.2 - 1.85x + 0.183x^2$$

The graph of this equation is also shown in Figure 11.6.

Having obtained this equation, we might now use it to predict the drying time when the amount of additive used is 6.5 grams. Substituting $x = 6.5$ into the equation, we obtain

$$y' = 12.2 - 1.85(6.5) + 0.183(6.5)^2 = 7.9$$

that is, a predicted drying time of approximately 7.9 hours. Note that it would have been rather dangerous to use the above equation to predict, say, the drying time which corresponds to 24.5 grams of additive. The risks inherent in extrapolation discussed on page 301, in connection with fitting straight lines, are greatly increased when polynomials are used to approximate unknown regression functions.

In actual practice, it may be difficult to determine the degree of the polynomial to fit to a given set of paired data. As it is always possible to find a polynomial of degree at most $n - 1$ that will pass through each of n points corresponding to n distinct values of x, it should be clear that what we actually seek is a polynomial of *lowest possible degree* that "adequately" describes the data. As we did in our example, it is often possible to determine the degree of the polynomial by inspection of the data.

There also exists a more rigorous method for determining the degree of the polynomial to be fitted to a given set of data. Essentially, it consists of first fitting a straight line as well as a second-degree polynomial and testing the null hypothesis $\beta_2 = 0$, namely, that nothing is gained by including the quadratic term. If this null hypothesis can be rejected, we then fit a third-degree polynomial and test the hypothesis $\beta_3 = 0$, namely, that nothing is gained by including the cubic term. This procedure is continued until the null hypothesis $\beta_i = 0$ cannot be rejected in two successive steps and there is, thus, no apparent advantage to carrying the extra terms. Note that in order to perform these tests it is necessary to impose the assumptions of normality, independence, and equal variances introduced in Section 11.2. Also, these tests should never be used "blindly," that is, without inspection of the overall pattern of the data.

The use of this technique is fairly tedious and we shall not illustrate it in the text. In Exercise 11 on page 319 the reader will be asked to apply it to the varnish-additive, drying-time data in order to check whether it is worthwhile to carry the quadratic term. (If the x values are equally spaced, it is possible to simplify the necessary calculations with the use of *orthogonal polynomials*, as is explained in Volume 2 of the book by M. G. Kendall and A. Stuart referred to in the Bibliography.)

11.4 MULTIPLE REGRESSION

Before we extend the methods of the preceding sections to problems involving more than one independent variable, let us point out that the curves obtained (and the surfaces we will obtain) are not used only to make predictions. They are often used also for purposes of *optimization*, namely, to determine for what values of the independent variable (or variables) the dependent variable is a maximum or minimum. For instance, in the example on page 311 we might use the polynomial fitted to the data to conclude that the drying time is a minimum when the amount of varnish additive used is 5.1 grams (see Exercise 12 on page 320).

Statistical methods of prediction and optimization are often referred to under the general heading of *response surface analysis*. Within the scope of this text, we shall be able to introduce two further methods of response surface analysis: *multiple regression* in this section and related problems of *factorial experimentation* in Chapter 13.

In multiple regression, we deal with data consisting of n $(r + 1)$-tuples $(x_{1i}, x_{2i}, \ldots, x_{ri}, y_i)$, where the x's are again assumed to be known without error while the y's are values of random variables. Data of this kind arise, for example, in studies designed to determine the effect of various climatic conditions on a metal's resistance to corrosion; the effect of kiln temperature, humidity, and iron content on the strength of a ceramic coating; or the effect of factory production, consumption level, and stocks in storage on the price of a product.

As in the case of one independent variable, we shall first treat the problem where the regression equation is linear, namely, where for any given set of values $x_1, x_2, \ldots$, and x_r the mean of the distribution of the y's is given by

$$\beta_0 + \beta_1 x_1 + \beta_2 x_2 + \ldots + \beta_r x_r$$

For two independent variables, this is the problem of fitting a *plane* to a set of n points with coordinates (x_{1i}, x_{2i}, y_i) as is illustrated in Figure 11.7. Applying the method of least squares to obtain estimates of the coefficients β_0, β_1, and β_2, we minimize the sum of the squares of the vertical distances from the points to the plane (see Figure 11.7); symbolically, we minimize

$$\sum_{i=1}^{n} [y_i - (\beta_0 + \beta_1 x_{1i} + \beta_2 x_{2i})]^2$$

and it will be left to the reader to verify in Exercise 13 on page 320 that the resulting normal equations are

$$\sum y = nb_0 + b_1 \sum x_1 + b_2 \sum x_2$$
$$\sum x_1 y = b_0 \sum x_1 + b_1 \sum x_1^2 + b_2 \sum x_1 x_2$$
$$\sum x_2 y = b_0 \sum x_2 + b_1 \sum x_1 x_2 + b_2 \sum x_2^2$$

As before, we write the least-square estimates of β_0, β_1, and β_2 as b_0, b_1, and b_2. Note that in the abbreviated notation used $\sum x_1$ stands for $\sum_{i=1}^{n} x_{1i}$, $\sum x_1 x_2$ stands for $\sum_{i=1}^{n} x_{1i} x_{2i}$, $\sum x_1 y$ stands for $\sum_{i=1}^{n} x_{1i} y_i$, and so forth.

Example. To illustrate this method of fitting a plane to a set of data, let us consider the following data relating the number of twists required to break a forged alloy bar to the percentage of each of two alloying elements present in a metal:

y Number of twists	x_1 Percent of element A	x_2 Percent of element B
38	1	5
40	2	5
85	3	5
59	4	5
40	1	10
60	2	10
68	3	10
53	4	10
31	1	15
35	2	15
42	3	15
59	4	15
18	1	20
34	2	20
29	3	20
42	4	20

Substituting

$$\sum x_1 = 40, \quad \sum x_2 = 200, \quad \sum x_1^2 = 120, \quad \sum x_1 x_2 = 500$$
$$\sum x_2^2 = 3{,}000, \quad \sum y = 733, \quad \sum x_1 y = 1{,}989, \quad \sum x_2 y = 8{,}285$$

into the normal equations, we obtain

$$733 = 16b_0 + 40b_1 + 200b_2$$
$$1{,}989 = 40b_0 + 120b_1 + 500b_2$$
$$8{,}285 = 200b_0 + 500b_1 + 3{,}000b_2$$

The unique solution of this system of equations is $b_0 = 48.2$, $b_1 = 7.83$, $b_2 = -1.76$, and we can thus estimate (predict) the number of twists

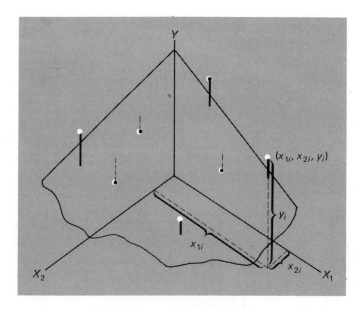

Figure 11.7. Regression plane.

required to break one of these bars by means of the equation

$$y' = 48.2 + 7.83x_1 - 1.76x_2$$

for any given pair of values of x_1 and x_2. Note that b_1 and b_2 are estimates of the average change in y resulting from a unit increase in the corresponding independent variable when the other independent variable is held fixed.

When a regression problem involves many variables and (or) the fitting of polynomials, the necessary calculations can become quite tedious. However, if the values of the independent variables can be controlled, it is possible to attain considerable simplifications by using appropriate spacings and subsequent coding (similar to that used in Exercise 12 on page 305).

Continuation of Example. Observe that in our example the x_1's as well as the x_2's are equally spaced and that, furthermore, each value (level) of x_1 occurs once in combination with each value (level) of x_2. This arrangement makes it possible to reduce the labor of obtaining and solving the normal equations. Coding the x_1 values as well as the

x_2 values -3, -1, 1, and 3,* and calling them z_1 and z_2, we can write the original data as

y	z_1	z_2
38	-3	-3
40	-1	-3
85	1	-3
59	3	-3
40	-3	-1
60	-1	-1
68	1	-1
53	3	-1
31	-3	1
35	-1	1
42	1	1
59	3	1
18	-3	3
34	-1	3
29	1	3
42	3	3

We now get $\sum z_1 = 0$, $\sum z_2 = 0$, $\sum z_1^2 = 80$, $\sum z_1 z_2 = 0$, $\sum z_2^2 = 80$, $\sum y = 733$, $\sum z_1 y = 313$, $\sum z_2 y = -351$, and the normal equations become

$$733 = 16 b_0$$
$$313 = 80 b_1$$
$$-351 = 80 b_2$$

Thus, in terms of these coded variables, we have

$$y' = 45.8 + 3.91 z_1 - 4.39 z_2$$

and the reader will be asked to verify in Exercise 16 on page 321 that (except for rounding errors) this result is equivalent to the one previously obtained.

The same coding would have produced even greater simplifications if we had wanted to fit a second-degree polynomial to our data, that is, if we had used an equation of the form

$$y' = \beta_0 + \beta_1 x_1 + \beta_2 x_2 + \beta_3 x_1 x_2 + \beta_4 x_1^2 + \beta_5 x_2^2$$

---

*Had there been five values of x_1, we would have coded them -2, -1, 0, 1, and 2. In this kind of coding the values must always be equally spaced and their sum must equal zero.

Without coding, the method of least squares would have led to a system of normal equations consisting of six simultaneous linear equations in six unknowns. As the reader will be asked to verify in Exercise 17 on page 321, the same coding as used above reduces this problem, essentially, to that of solving three simultaneous linear equations in three unknowns.

The example of this section served to illustrate how a careful choice of the values of the independent variables can simplify the calculations required in a problem of multiple regression. The principles and procedures for making this choice, whenever possible, are part of the important subject of *experimental design*. In Chapter 12 we shall consider some of the simpler, but nevertheless widely used, experimental designs. We shall focus our attention on what is called the *analysis of variance*, where tests are made concerning the significance of the effects of certain combinations of levels of the independent variables. In Chapter 13 we shall introduce the related problem of *factorial experimentation*.

EXERCISES

1. The following data pertain to the chlorine residual in a swimming pool at various times after it has been treated with chemicals:

Number of hours	Chlorine residual (parts per million)
x	y
12	1.80
24	1.47
36	1.25
48	1.10
60	1.01

Use the method of least squares to fit an exponential curve of the form $y = \alpha \cdot \beta^x$.

2. The following data pertain to the growth of a colony of bacteria in a culture medium:

Days since inoculation	Bacteria count
x	y
3	115,000
6	147,000
9	239,000
12	356,000
15	579,000
18	864,000

(a) Plot the data on semi-log graph paper and use the method of least squares to fit an exponential curve of the form $y = \alpha \cdot \beta^x$.

(b) Use the result obtained in part (a) to estimate the bacteria count at the end of the first day.

3. The following data pertain to the cosmic ray doses measured at various altitudes:

Altitude (feet) x	Dose rate (mrem/year) y
50	28
450	30
780	32
1,200	36
4,400	51
4,800	58
5,300	69

Use the method of least squares to fit an exponential curve of the form $y = \alpha \cdot e^{\beta x}$.

4. Use the method of least squares to fit a curve of the form $y = \alpha \cdot x^\beta$ to the price-demand data of Exercise 4 on page 303.

5. In an experiment designed to determine the specific heat ratio γ for a certain gas, the gas was compressed adiabatically to several predetermined volumes V, and the corresponding pressure p was measured with the following results:

p (lb/in²)	16.6	39.7	78.5	115.5	195.3	546.1
V (in³)	50	30	20	15	10	5

Assuming the ideal gas law $p \cdot V^\gamma = C$, use these data to

(a) estimate γ for this gas;

(b) construct a 0.95 confidence interval for γ. (State what assumptions will have to be made.)

6. Fit a Gompertz curve of the form

$$y = e^{e^{\alpha x + \beta}}$$

to the data of Exercise 1 and compare the fit with that of the exponential curve obtained in Exercise 1.

7. The rise of current in an inductive circuit having the time constant τ is given by

$$I = 1 - e^{-t/\tau}$$

where t is the time measured from the instant the switch is closed, and I is the ratio of the current at time t to the full value of the current given by Ohm's law. Given the measurements

I	0.073	0.220	0.301	0.370	0.418	0.467	0.517	0.578
t (sec)	0.1	0.2	0.3	0.4	0.5	0.6	0.7	0.8

estimate the time constant of this circuit from the experimental results given. [*Hint*: See Exercise 9 on page 304.]

8. The number of inches which a newly-built structure is settling into the ground is given by

$$y = 3 - 3e^{-\alpha x}$$

where x is its age in months.

x	2	4	6	12	18	24
y	1.07	1.88	2.26	2.78	2.97	2.99

Use the method of least squares to estimate α.

9. The following data pertain to the amount of hydrogen present (y, in parts per million) in core drillings made at one-foot intervals along the length of a vacuum-cast ingot (x, core location in feet from base):

x	1	2	3	4	5	6	7	8	9	10
y	1.28	1.53	1.03	0.81	0.74	0.65	0.87	0.81	1.10	1.03

Fit a parabola by the method of least squares.

10. Suitably coding the years, fit a parabola to the following data on the production of television sets (in thousands) in Japan for the years 1963 through 1969: 4,916, 5,273, 4,190, 5,652, 6,963, 9,001, and 12,118.

11. When fitting a polynomial to a set of paired data, we usually begin by fitting a straight line and using the method on page 299 to test the null hypothesis $\beta_1 = 0$. Then we fit a second-degree polynomial and test whether it is worthwhile to carry the quadratic term by comparing $\hat{\sigma}_1^2$, the *residual variance* after fitting the straight line, with $\hat{\sigma}_2^2$, the *residual variance* after fitting the second-degree polynomial. Each of these residual variances is given by the formula

$$\frac{\sum (y - y')^2}{\text{degrees of freedom}}$$

with y' computed, respectively, from the equation of the line and the equation of the second-degree polynomial. The decision whether to carry the quadratic term is based on the statistic

$$F = \frac{\hat{\sigma}_1^2 - \hat{\sigma}_2^2}{\hat{\sigma}_2^2}$$

which (under the assumptions of Section 11.2) is a value of a random variable having the F distribution with 1 and $n - 3$ degrees of freedom.

(a) Fit a straight line to the varnish-additive and drying-time data on page 311, test the null hypothesis $\beta_1 = 0$ at the 0.05 level of significance, and calculate $\hat{\sigma}_1^2$.

(b) Using the result on page 311 calculate $\hat{\sigma}_2^2$ for the given data and test at the 0.05 level whether we should carry the quadratic term. (Note that we could continue this procedure and test whether to carry a cubic term by means of a corresponding comparison of residual variances. Then we could test whether to carry a fourth-degree term, and so on. It is customary to terminate this procedure after two successive steps have not produced significant results.)

12. Referring to the example on page 311, verify that the drying time is a minimum when the amount of varnish additive used is 5.1 grams.

13. Verify that the method of least squares leads to the system of normal equations on page 313.

14. Twelve specimens of cold-reduced sheet steel, having different copper contents and annealing temperatures, are measured for hardness with the following results:

Hardness (Rockwell 30-T)	Copper content (percent)	Annealing temperature (degrees F)
78.9	0.02	1000
65.1	0.02	1100
55.2	0.02	1200
56.4	0.02	1300
80.9	0.10	1000
69.7	0.10	1100
57.4	0.10	1200
55.4	0.10	1300
85.3	0.18	1000
71.8	0.18	1100
60.7	0.18	1200
58.9	0.18	1300

Fit an equation of the form $y = \beta_0 + \beta_1 x_1 + \beta_2 x_2$, where x_1 represents the copper content, x_2 represents the annealing temperature, and y represents the hardness.

15. The following are data on the ages and incomes of five executives working for the same company, and the number of years they went to college:

Age x_1	Years college x_2	Income (dollars) y
37	4	31,200
45	0	26,800
38	5	35,000
42	2	30,300
31	4	25,400

Fit an equation of the form $y = b_0 + b_1x_1 + b_2x_2$ to the given data, and use it to estimate how much *on the average* an executive working for this company will make if he is 40 years old and has had 4 years of college.

16. Referring to the example involving the number of twists required to break a forged bar, verify that the result given on page 316, obtained by coding, is equivalent to the original result shown on page 315.

17. Verify that if the x's are coded as on page 316, the normal equations for estimating the coefficients of the second-degree polynomial

$$y' = \beta_0 + \beta_1 x_1 + \beta_2 x_2 + \beta_3 x_1 x_2 + \beta_4 x_1^2 + \beta_5 x_2^2$$

become

$$733 = 16b_0 \qquad\qquad + 80b_4 + 80b_5$$
$$315 = \qquad 80b_1$$
$$351 = \qquad\qquad 80b_2$$
$$-79 = \qquad\qquad\qquad 400b_3$$
$$3453 = 80b_0 \qquad\qquad + 656b_4 + 400b_5$$
$$3493 = 80b_0 \qquad\qquad + 400b_4 + 656b_5$$

18. Code the values of x_1 in Exercise 14 as $z_1 = -1, 0, +1$, and the values of x_2 as $z_2 = -3, -1, +1, +3$. Then write down the resulting normal equations, solve, and show that the resulting regression equation is equivalent to the one obtained in Exercise 14.

11.5 CORRELATION

So far in this chapter, we have studied problems where the independent variable (or variables) was assumed to be known without error. Although this applies to many experimental situations, there are also problems where the x's as well as the y's are values assumed by random variables. This would be the case, for instance, if we studied the relationship between rainfall and the yield of a certain crop, the relationship between the tensile strength and the hardness of aluminum, or the relationship between impurities in the air and the incidence of a certain disease. Problems like these are referred to as problems of *correlation analysis*, where it is assumed that the data points (x_i, y_i) for $i = 1, 2, \ldots, n$ are values of a pair of random variables whose joint density is given by $f(x, y)$.

The bivariate density which is most commonly used in problems of correlation analysis is the *bivariate normal distribution*. We shall introduce it here in terms of the *conditional density* $g_2(y \mid x)$ and the *marginal density* $f_1(x)$, as defined in Section 4.10. So far as $g_2(y \mid x)$ is concerned, the conditions we shall impose are practically identical with the ones we used in connection with the sampling theory of Section 11.2. For any given x, it will be assumed that $g_2(y \mid x)$ is a normal distribution with the mean $\alpha + \beta x$ and the variance σ^2. Thus, the regression of y on x is linear and the variance of the conditional density does not depend on x. Furthermore, we shall assume that the marginal density $f_1(x)$ is normal with the mean μ_1 and the variance σ_1^2. Making use of the relationship $f(x, y) = f_1(x) \cdot g_2(y \mid x)$ given on page 129, we thus obtain

$$f(x, y) = \frac{1}{\sqrt{2\pi}\,\sigma_1} e^{-(x - \mu_1)^2/2\sigma_1^2} \cdot \frac{1}{\sqrt{2\pi}\,\sigma} e^{-[y - (\alpha + \beta x)]^2/2\sigma^2}$$

$$= \frac{1}{2\pi \cdot \sigma \cdot \sigma_1} e^{-\{[y - (\alpha + \beta x)]^2/2\sigma^2 + (x - \mu_1)^2/2\sigma_1^2\}}$$

for $-\infty < x < \infty$ and $-\infty < y < \infty$. Note that this joint distribution involves the *five* parameters μ_1, σ_1, α, β, and σ.

For reasons of symmetry and other considerations to be explained below, it is customary to express the bivariate normal density in terms of the parameters $\mu_1, \sigma_1, \mu_2, \sigma_2,$ and ρ. Here μ_2 and σ_2^2 are the mean and the variance of the marginal distribution $f_2(y)$, while ρ, called the *correlation coefficient*, is defined by

$$\rho^2 = 1 - \frac{\sigma^2}{\sigma_2^2}$$

with ρ taken to be positive when β is positive (and negative when β is negative). Leaving it to the reader to show in Exercise 4 on page 329 that

$$\mu_2 = \alpha + \beta \mu_1 \quad \text{and} \quad \sigma_2^2 = \sigma^2 + \beta^2 \sigma_1^2$$

we then substitute into the above expression for $f(x, y)$ and finally obtain the following form of the bivariate normal distribution:

$$f(x,y) = \frac{1}{2\pi \cdot \sigma_1 \sigma_2 \sqrt{1 - \rho^2}} e^{-[(\frac{x-\mu_1}{\sigma_1})^2 - 2\rho(\frac{x-\mu_1}{\sigma_1})(\frac{y-\mu_2}{\sigma_2}) + (\frac{y-\mu_2}{\sigma_2})^2]/2(1-\rho^2)}$$

for $-\infty < x < \infty$ and $-\infty < y < \infty$ (see Exercise 5 on page 329).

Concerning the correlation coefficient ρ, note that $-1 \le \rho \le +1$ since $\sigma_2^2 = \sigma^2 + \beta^2 \sigma_1^2$ and, hence, $\sigma_2^2 \ge \sigma^2$. Furthermore, ρ can equal -1 or $+1$ only when $\sigma^2 = 0$, which represents the *degenerate case* where all the probability is concentrated along the line $y = \alpha + \beta x$ and there is, thus, a perfect linear relationship between the two random variables. (That is, for a given value of x, y *must* equal $\alpha + \beta x$.) The correlation coefficient is equal to zero if and only if $\sigma^2 = \sigma_2^2$, and it follows from the identity $\sigma_2^2 = \sigma^2 + \beta^2 \sigma_1^2$ that this can happen only when $\beta = 0$. Thus, $\rho = 0$ implies that the regression line of y on x is a horizontal line and, hence, that knowledge of x does not help in the prediction of y. Thus, when $\rho = \pm 1$ we say that there is a perfect linear correlation (relationship, or association) between the two random variables; when $\rho = 0$ we say that there is no correlation (relationship, or association) between the two random variables. In fact, $\rho = 0$ implies for the bivariate normal density that the two random variables are *independent* (see Exercise 6 on page 329).

For values between 0 and $+1$ or 0 and -1, we interpret ρ by referring back to the identity

$$\rho^2 = 1 - \frac{\sigma^2}{\sigma_2^2} = \frac{\sigma_2^2 - \sigma^2}{\sigma_2^2}$$

on page 322. Since σ^2 is a measure of the variation of the y's when x is *known* while σ_2^2 is a measure of the variation of the y's when x is *unknown*, $\sigma_2^2 - \sigma^2$ measures the variation of the y's that is accounted for by the linear relationship with x. Hence, ρ^2 tells us what *proportion* of the variation of the y's can be attributed to the linear relationship with x.

Given a random sample of size n—that is, n pairs of values (x_i, y_i)—it is customary to estimate ρ by means of the *sample correlation coefficient*

$$r = \frac{S_{xy}}{\sqrt{S_{xx} \cdot S_{yy}}}$$

where S_{xy}, S_{xx}, and S_{yy} are as defined on page 296. This estimator is not unbiased and, except for the factor $\sqrt{\dfrac{n-1}{n-2}}$, it may be obtained by substituting for σ_2^2 the sample variance of the y's and for σ^2 the estimator on page 297, namely, s_e^2 (see also Exercise 7 on page 329). Note that the sample correlation coefficient r is often used to measure the strength of a linear relationship exhibited by sample data even if the data do not necessarily come from a bivariate normal population.

Example. To illustrate the calculation of r, consider the following paired observations of the number of minutes it takes 10 mechanics to assemble a piece of machinery in the morning, x, and in the late afternoon, y:

x	y
11.1	10.9
10.3	14.2
12.0	13.8
15.1	21.5
13.7	13.2
18.5	21.1
17.3	16.4
14.2	19.3
14.8	17.4
15.3	19.0

The calculations required to find r are as follows, where the subscripts and limits of summation are omitted for the sake of simplicity. First we determine the necessary sums, getting $\sum x = 142.3$, $\sum y = 166.8$, $\sum x^2 = 2{,}085.31$, $\sum xy = 2{,}434.69$, $\sum y^2 = 2{,}897.80$, and then, substituting into the formulas on page 296, we obtain

$$S_{xx} = 10(2{,}085.31) - (142.3)^2 = 603.81$$
$$S_{xy} = 10(2{,}434.69) - (142.3)(166.8) = 611.26$$
$$S_{yy} = 10(2{,}897.80) - (166.8)^2 = 1{,}155.76$$

Thus,

$$r = \frac{611.26}{\sqrt{(603.81)(1{,}155.76)}} = 0.73$$

which means that $100r^2 = 53$ percent of the variation among the afternoon times is explained by (is accounted for or may be attributed to) corresponding differences among the morning times.

Whenever a value of r is based on sample data, we can perform a test of significance (a test of the null hypothesis $\rho = \rho_0$) or construct a confidence interval for ρ. For random samples from a bivariate normal population, this is usually based on the *Fisher Z transformation*, namely, the transformation from r to

$$Z = \frac{1}{2} \ln \frac{1 + r}{1 - r}$$

as it can be shown that the sampling distribution of this statistic is approximately normal with the mean $\mu_Z = \frac{1}{2} \ln \frac{1 + \rho}{1 - \rho}$ and the variance $\frac{1}{n - 3}$. Thus, we can test the null hypothesis $\rho = \rho_0$ with the statistic

$$z = \frac{Z - \mu_Z}{1/\sqrt{n - 3}} = \frac{\sqrt{n - 3}}{2} \cdot \ln \frac{(1 + r)(1 - \rho_0)}{(1 - r)(1 + \rho_0)}$$

which has approximately the standard normal distribution. In particular, we can test the null hypothesis $\rho = 0$ with the statistic

$$z = \sqrt{n - 3} \cdot Z = \frac{\sqrt{n - 3}}{2} \cdot \ln \frac{1 + r}{1 - r}$$

These tests are facilitated by the use of Table 11, which gives the values of Z corresponding to $r = 0.00, 0.01, 0.02, \ldots$, and 0.99. When r is negative, we look up the value of Z corresponding to $-r$, and then take $-Z$.

Continuation of Example. Returning to the example where we had $r = 0.73$, let us test the null hypothesis $\rho = 0$ at the level of significance $\alpha = 0.05$. Since the value of Z corresponding to $r = 0.73$ is 0.929, the test statistic becomes

$$z = \sqrt{n - 3}(0.929) = \sqrt{10 - 3}(0.929) = 2.46$$

which exceeds $z_{0.025} = 1.96$. Thus, the null hypothesis $\rho = 0$ can be rejected and we conclude that there *is* a relationship between the amount of time a mechanic takes to assemble the piece of machinery in the morning and in the late afternoon.

To construct a confidence interval for p, we first construct a confidence interval for μ_z, the mean of the sampling distribution of Z, and then convert to r and p by means of Table 11. Making use of the theory on page 325, we can write the first of these confidence intervals as

$$Z - \frac{z_{\alpha/2}}{\sqrt{n - 3}} < \mu_z < Z + \frac{z_{\alpha/2}}{\sqrt{n - 3}}$$

Example. Suppose that for a given set of paired data (pertaining, say, to the mathematics and physics grades of 30 students) we obtained $r = 0.70$. Looking up the value of Z which corresponds to $r = 0.70$ in Table 11, we get 0.867, and substituting this value together with $n = 30$ and $z_{\alpha/2} = 1.96$ into the above formula, we get the 0.95 confidence interval

$$0.867 - \frac{1.96}{\sqrt{27}} < \mu_z < 0.867 + \frac{1.96}{\sqrt{27}}$$

or

$$0.490 < \mu_z < 1.244$$

Then, looking up the values of r for which Z is closest to 0.490 and 1.244, respectively, we get the 0.95 confidence interval

$$0.45 < p < 0.85$$

for the "true" strength of the linear relationship between the grades of students in the two given subjects.

Note that the confidence interval obtained in this example is fairly wide, which illustrates the fact that *correlation coefficients based on relatively small samples are generally not very reliable.*

Example. To give an example involving negative values of r and Z, suppose that $r = 0.20$ and $n = 40$ for a given set of paired data. Since the value of Z which corresponds to $r = 0.20$ is 0.203 according to Table 11, we first get the 0.95 confidence interval

$$0.203 - \frac{1.96}{\sqrt{37}} < \mu_z < 0.203 + \frac{1.96}{\sqrt{37}}$$

or

$$-0.119 < \mu_z < 0.525$$

Then, looking up the values of r for which Z is closest to 0.119 and 0.525, respectively, we get the 0.95 confidence interval

$$-0.12 < \rho < 0.48$$

There are several serious pitfalls in the interpretation of the coefficient of correlation. First, it must be emphasized that r is an estimate of the strength of the *linear* association between the random variables; thus, as illustrated in Figure 11.8, r may be close to zero when there is actually a strong (but nonlinear) relationship between the two random variables.

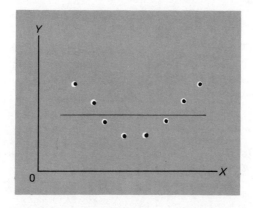

Figure 11.8. Nonlinear relationship where correlation coefficient is zero.

Second, and perhaps of greatest importance, a significant correlation does not necessarily imply a *causal* relationship between the two random variables. Although it would not be surprising, for example, to obtain a high positive correlation between the annual sales of chewing gum and the incidence of crime in the United States, one cannot conclude that crime might be reduced by prohibiting the sale of chewing gum. Both variables depend upon the size of the population, and it is this mutual relationship with a third variable (population size) which produces the positive correlation.

EXERCISES

1. The following data pertain to the resistance (ohms) and the failure time (minutes) of certain overloaded resistors:

Resistance	Failure time	Resistance	Failure time
43	32	36	36
29	20	39	33
44	45	36	21
33	35	47	44
33	22	28	26
47	46	40	45
34	28	42	39
31	26	33	25
48	37	46	36
34	33	28	25
46	47	48	45
37	30	45	36

(a) Calculate the coefficient of correlation.

(b) Test the null hypothesis $\rho = 0$ at the level of significance $\alpha = 0.05$.

2. Calculate r for the air velocities and evaporation coefficients of the example on page 294. Assuming that the necessary assumptions can be met, test for significance at $\alpha = 0.01$.

3. The following are measurements of the carbon content and the permeability index of 22 sinter mixtures:

Carbon content (%)	Permeability index	Carbon content (%)	Permeability index
4.4	12	4.1	13
5.5	14	4.9	19
4.2	18	4.7	22
3.0	35	5.0	20
4.5	23	4.6	16
4.9	29	3.6	27
4.6	16	4.9	21
5.0	12	5.1	13
4.7	18	4.8	18
5.1	21	5.2	17
4.4	27	5.2	11

(a) Calculate the coefficient of correlation.

(b) Find 99 percent confidence limits for ρ.

4. Evaluating the necessary integrals, verify the identities

$$\mu_2 = \alpha + \beta\mu_1 \quad \text{and} \quad \sigma_2^2 = \sigma^2 + \beta^2\sigma_1^2$$

on page 322.

5. Substitute $\mu_2 = \alpha + \beta\mu_1$ and $\sigma_2^2 = \sigma^2 + \beta^2\sigma_1^2$ into the formula for the bivariate density given on page 322, and show that this gives the final form shown on page 323.

6. Show that for the bivariate normal distribution (a) independence implies zero correlation, and (b) zero correlation implies indepedence.

7. By substituting the sample variance of the y's for σ_2^2 and s_e^2 for σ^2 in the formula $\rho^2 = 1 - \sigma^2/\sigma_2^2$, verify the formula for r given on page 323 (except for a multiplicative constant).

8. If data on the ages and prices of 25 pieces of equipment yielded $r = -0.57$, test the null hypothesis $\rho = 0$ at the level of significance $\alpha = 0.01$.

9. If for certain paired data $n = 18$ and $r = 0.44$, test the null hypothesis $\rho = 0.30$ against the alternative hypothesis $\rho > 0.30$ at the level of significance $\alpha = 0.05$.

10. Construct a 0.95 confidence interval for ρ when the necessary assumptions are met and
(a) $r = 0.75$ and $n = 15$;   (b) $r = 0.16$ and $n = 25$;
(c) $r = -0.57$ and $n = 32$.

11. Construct a 0.99 confidence interval for ρ when the necessary assumptions are met and
(a) $r = 0.82$ and $n = 19$;   (b) $r = 0.12$ and $n = 35$;
(c) $r = -0.20$ and $n = 14$.

12. Instead of using the computing formula on page 323, the correlation coefficient r may be obtained from the formula

$$r = \pm\sqrt{1 - \frac{\sum(y - y')^2}{\sum(y - \bar{y})^2}}$$

which is analogous to the formula used to define ρ. Although the computations required by the use of this formula are tedious, the formula has the advantage that it can be used also to measure the strength of nonlinear relationships or relationships in several variables. For instance, in the multiple linear regression example on page 314, one could calculate the predicted values by means of the equation

$$y' = 48.2 + 7.83x_1 - 1.76x_2$$

and then determine r as a measure of how strongly y, the twist required to break one of the forged alloy bars, depends on *both* percentages of alloying elements present.

(a) Using the data on page 314, find $\sum (y - \bar{y})^2$ by means of the formula $\sum (y - \bar{y})^2 = \sum y^2 - n\bar{y}^2$.

(b) Using the regression equation obtained on page 315, calculate y' for the sixteen points and then determine $\sum (y - y')^2$.

(c) Substitute the results obtained in (a) and (b) into the above formula for r. The result is the so-called *multiple correlation coefficient*.

13. When we substitute for the actual values of paired data the *ranks* which the values occupy in the respective samples, the correlation coefficient can be written in the form

$$r' = 1 - \frac{6 \cdot \sum d_i^2}{n(n^2 - 1)}$$

in which it is called the *rank correlation coefficient*. In this formula, d_i is the difference between the ranks of the paired observations (x_i, y_i) and n is the number of pairs in the sample. (In case of ties, we substitute for each of the tied observations the mean of the ranks which they jointly occupy, as in Section 10.3.)

(a) Calculate r' for the example on page 324, which deals with the time it takes mechanics to assemble a piece of machinery in the morning and in the late afternoon. Compare with the result obtained for r.

(b) Calculate r' for the data of Exercise 1 and compare with the value of r previously obtained.

(c) Calculate r' for the air velocities and evaporation coefficients of the example on page 294 and compare with the result obtained for r in Exercise 2.

14. Since the assumptions on which the significance test of Section 11.5 are based are rather stringent, it is often desirable to use a *nonparametric* alternative based on the rank correlation coefficient of Exercise 13. It is based on the theory that if there is no relationship between the x's and the y's (in fact, if they are *randomly matched*), the sampling distribution of r' can be approximated closely with a normal distribution having the mean 0 and the standard deviation

$$\sigma_{r'} = \frac{1}{\sqrt{n - 1}}$$

Use the level of significance $\alpha = 0.01$ to test the significance of the values of r' obtained in the three parts of Exercise 13.

12

Analysis
of Variance

12.1 INTRODUCTION

Some of the examples of Chapter 11 have already taught us that considerable economies in calculation can result by appropriately planning an
experiment in advance. What is even more important, proper experimental planning can give a reasonable assurance that the results of an experiment will provide clear-cut answers to questions under investigation. While
it is impossible to give in this chapter a complete discussion of *experimental
design*, including the many pitfalls to which the experimenter is exposed,
we shall begin by presenting in this section some of the general principles
of experimental design. Several of the designs most frequently used in
engineering and other applied research will be given in subsequent
sections.

Example. To illustrate some of the most important aspects of experimental design, let us consider the following situation. Suppose a steel mill supplies tin plate to three can manufacturers, the major specification being that the tin-coating weight should be at least 0.25 pound per base box. The mill and each can manufacturer has a laboratory where measurements are made of the tin-coating weights of samples taken from each shipment. Suppose, also, that some disagreement has arisen about the actual tin-coating weights of the tin plate being shipped, and it is decided to plan an experiment to determine whether the four laboratories are making consistent measurements. A complicating factor is that part of the measuring process consists of the chemical removal of the tin from the surface of the base metal; thus, it is impossible to have the identical sample measured by each laboratory to determine how closely the measurements correspond.

One possibility is to send several samples (in the form of circular discs having equal areas) to each of the laboratories. Although these discs may not actually have identical tin-coating weights, it is hoped that such differences will be small and that they will more or less "average out." In other words, it will be assumed that whatever differences there may be among the means of the four samples can be attributed to no other causes but systematic differences in measuring techniques and chance variability. This would make it possible to determine whether the results produced by the laboratories are consistent by comparing the variability of the four sample means with an appropriate measure of chance variation.

Now there remains the problem of deciding how many discs are to be sent to each laboratory and how the discs are actually to be selected. The question of sample size can be answered in many different ways, one of which is to use the formula on page 216 for the standard deviation of the sampling distribution of the difference between two means. Substituting known values of σ_1 and σ_2 and specifying what differences between the true means of any two of the laboratories should be detected with a probability of at least 0.95 (or 0.98, or 0.99), it is possible to determine $n_1 = n_2 = n$ (see Exercise 10 on page 345). Suppose that this method and, perhaps, also considerations of cost and availability of the necessary specimens lead to the decision to send a sample of 12 discs to each laboratory.

The problem of selecting the required 48 discs and allocating 12 to each laboratory is not as straightforward as it may seem at first. To begin with, suppose that a sheet of tin plate, sufficiently long and sufficiently wide, is selected and that the 48 discs are cut as shown in Figure 12.1. The twelve discs cut from strip 1 are sent to the first laboratory, the

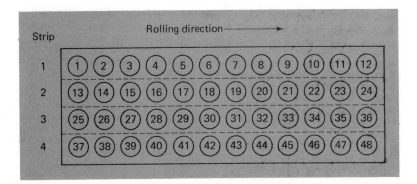

Figure 12.1. Numbering of tin-plate samples.

twelve discs from strip 2 are sent to the second laboratory, and so forth. If the four mean coating weights subsequently obtained were then found to differ significantly, would this allow us to conclude that these differences can be attributed to lack of consistency in the measuring techniques? Suppose, for instance, that additional investigation shows that the amount of tin deposited electrolytically on a long sheet of steel has a distinct and repeated pattern of variation perpendicular to the direction in which it is rolled. (Such a pattern might be caused by the placement of electrodes, "edge effects," and so forth.) Thus, even if all four laboratories measured the amount of tin consistently and without error, there could be cause for differences in the tin-coating weight determinations. The allocation of an entire strip of discs to each laboratory is such that inconsistencies among the laboratories' measuring techniques are inseparable from (or *confounded* with) whatever differences there may be in the actual amount of tin deposited perpendicular to the direction in which the sheet of steel is rolled.

One way to avoid this kind of confounding is to number the discs and allocate them to the four laboratories *at random*, such as in the following arrangement, which was obtained with the aid of a table of random numbers:

> *Laboratory A:* 3, 38, 17, 32, 24, 30, 48, 19, 11, 31, 22, 41
> *Laboratory B:* 44, 20, 15, 25, 45, 4, 14, 5, 39, 7, 40, 34
> *Laboratory C:* 12, 21, 42, 8, 27, 16, 47, 46, 18, 43, 35, 26
> *Laboratory D:* 9, 2, 28, 23, 37, 1, 10, 6, 29, 36, 33, 13

If there were any actual pattern of tin-coating thickness on the sheet of tin plate, it would be "broken up" by the *randomization*.

Although we have identified and counteracted one possible systematic pattern of variation, there is no assurance that there can be no others. For instance, there may be systematic differences in the areas of the discs caused by progressive wear of the cutting instrument, or there may be scratches or other imperfections on one part of the sheet which could affect the measurements. Thus, there is always the possibility that differences in means attributed to inconsistencies among the laboratories are actually caused by some other uncontrolled variable, and it is the purpose of randomization to avoid confounding the variable under investigation with such other variables.

By distributing the 48 discs among the four laboratories entirely at random, we have no choice but to include whatever variation may be attributable to extraneous causes under the heading of "chance variation." This may give us an excessively large estimate of chance variation which, in turn, may make it difficult to detect differences between the true laboratory means. In order to avoid this, we could, perhaps, use only discs cut from the same strip (or from an otherwise homogeneous region). Unfortunately, this kind of *controlled experimentation* presents us with new complications. Of what use would it be, for example, to perform an experiment which allows us to conclude that the laboratories are consistent (or inconsistent), *if such a conclusion is limited to measurements made at a fixed distance from one edge of a sheet*? To consider a more poignant example, suppose that a manufacturer of plumbing materials wishes to compare the performance of several kinds of material to be used in underground water pipes. If such conditions as soil acidity, depth of pipe, and mineral content of water were all held fixed, any conclusions as to which material is best would be valid only for the given set of conditions. What the manufacturer really wants to know is which material is best over a fairly wide variety of conditions, and in designing a suitable experiment it would be advisable (indeed, necessary) to specify that pipe of each material be buried at each of several depths in each of several kinds of soil, and in locations where the water varies in hardness.

The foregoing example serves to illustrate that it is seldom desirable to hold all or most extraneous factors fixed throughout an experiment in order to obtain an estimate of chance variation that is not "inflated" by variations due to other causes. (In fact, it is rarely, if ever, possible to exercise such strict control, that is, to hold *all* extraneous variables fixed.) In actual practice, experiments should be planned so that known sources of variability are deliberately varied over as wide a range as necessary; furthermore, they should be varied in such a way that their variability can be eliminated from the estimate of chance variation. One way to accomplish this is to repeat the experiment in several *blocks*, where known sources of variability (that is, extraneous variables) are held fixed in each block, but vary from block to block.

Continuation of Example. In the tin-plating problem we might thus account for variations across the sheet of steel by randomly allocating three discs from each strip to each of the laboratories as in the following arrangement:

	Strip 1	Strip 2	Strip 3	Strip 4
Laboratory A:	8, 4, 10	23, 24, 19	26, 29, 35	37, 44, 48
Laboratory B:	2, 6, 12	21, 15, 22	34, 33, 32	45, 43, 46
Laboratory C:	1, 5, 11	16, 20, 13	36, 29, 30	41, 38, 47
Laboratory D:	7, 3, 9	17, 18, 14	28, 31, 25	39, 40, 42

In this experimental layout, the strips form the blocks, and if we base our estimate of chance variation on the variability *within* each of the sixteen sets of three discs, this estimate will not be inflated by the extraneous variable, that is, differences among the strips. (Note also that, with this arrangement, differences among the means obtained for the four laboratories cannot be attributed to differences among the strips. This cannot be said for the arrangement on page 333.)

The analysis of experiments in which blocking is used to eliminate one source of variability will be discussed in Section 12.3. The analysis of experiments in which two or three sources of variability are thus eliminated will be treated in Section 12.5.

12.2 COMPLETELY RANDOMIZED DESIGNS

In this section we shall consider, in general, the statistical analysis of the *completely randomized design*, or *one-way classification*. We shall suppose that the experimenter has available the results of k independent random samples, each of size n, from k different populations (that is, data concerning k treatments, k groups, k methods of production, etc.); and he is concerned with testing the hypothesis that the means of these k populations are all equal. An example of such an experiment, with $k = 4$, is given by the layout on page 333. If we denote the jth observation in the ith sample by y_{ij}, the general schema for a one-way classification is as follows:

		Means
Sample 1:	$y_{11}, y_{12}, \ldots, y_{1j}, \ldots, y_{1n}$	$\bar{y}_1$
Sample 2:	$y_{21}, y_{22}, \ldots, y_{2j}, \ldots, y_{2n}$	$\bar{y}_2$
	$\cdot \quad \cdot \quad \cdot \quad \cdot \quad \cdot \quad \cdot$	$\cdot$
Sample i:	$y_{i1}, y_{i2}, \ldots, y_{ij}, \ldots, y_{in}$	$\bar{y}_i$
	$\cdot \quad \cdot \quad \cdot \quad \cdot \quad \cdot \quad \cdot$	$\cdot$
Sample k:	$y_{k1}, y_{k2}, \ldots, y_{kj}, \ldots, y_{kn}$	$\bar{y}_k$
		$\bar{y}.$

With reference to the experimental layout on page 333, y_{ij} ($i = 1, 2, 3, 4$; $j = 1, 2, \ldots, 12$) is the jth tin-coating weight measured by the ith laboratory, $\bar{y}_i$ is the mean of the measurements obtained by the ith laboratory, and $\bar{y}.$ is the overall mean (or *grand mean*) of all 48 observations.

To be able to test the hypothesis that the samples were obtained from k populations with equal means, we shall make several assumptions. Specifically, it will be assumed that we are dealing with *normal populations* having *equal variances*. There exist methods for testing the reasonableness of this last assumption (see the book by A. M. Mood and F. A. Graybill mentioned in the Bibliography), but the methods we shall develop in this chapter are fairly *robust*; that is, they are relatively insensitive to violations of the assumption of normality as well as the assumption of equal variances.

If μ_i denotes the mean of the ith population and σ^2 denotes the common variance of the k populations, we can express each observation y_{ij} as μ_i plus the value of a random component; that is, we can write

$$y_{ij} = \mu_i + \epsilon_{ij} \qquad \text{for } i = 1, 2, \ldots, k; j = 1, 2, \ldots, n$$

In accordance with the above assumptions, the ϵ_{ij} are values of independent, normally distributed random variables with zero means and the common variance σ^2.*

To attain uniformity with corresponding equations for more complicated kinds of designs, it is customary to replace μ_i by $\mu + \alpha_i$, where μ is the mean of the μ_i and, hence, $\sum_{i=1}^{k} \alpha_i = 0$ (see Exercise 11 on page

---

*Note that this equation, or *model*, can be regarded as a multiple regression equation; introducing the variables x_{il} which equal 0 or 1 depending on whether the two subscripts are unequal or equal, we can write

$$y_{ij} = \mu_1 x_{i1} + \mu_2 x_{i2} + \ldots + \mu_k x_{ik} + \epsilon_{ij}$$

The parameters μ_i can now be interpreted as regression coefficients, and they can be estimated by the least-squares methods of Chapter 11.

345). Using these new parameters, we can write the model equation for the one-way classification as

$$y_{ij} = \mu + \alpha_i + \epsilon_{ij} \quad \text{for } i = 1, 2, \ldots, k; j = 1, 2, \ldots, n$$

and the null hypothesis that the k population means are all equal can be replaced by the null hypothesis that $\alpha_1 = \alpha_2 = \ldots = \alpha_k = 0$. The alternative hypothesis that at least two of the population means are unequal is equivalent to the alternative hypothesis that $\alpha_i \neq 0$ for some i.

To test the null hypothesis that the k population means are all equal, we shall compare two estimates of σ^2—one based on the variation among the sample means, and one based on the variation within the samples. Since, by assumption, each sample comes from a population having the variance σ^2, this variance can be estimated by any one of the sample variances

$$s_i^2 = \sum_{j=1}^{n} \frac{(y_{ij} - \bar{y}_i)^2}{n - 1}$$

and, hence, also by their mean

$$\hat{\sigma}_W^2 = \sum_{i=1}^{k} \frac{s_i^2}{k} = \sum_{i=1}^{k} \sum_{j=1}^{n} \frac{(y_{ij} - \bar{y}_i)^2}{k(n - 1)}$$

Note that each of the sample variances s_i^2 is based on $n - 1$ degrees of freedom ($n - 1$ independent deviations from $\bar{y}_i$) and, hence, $\hat{\sigma}_W^2$ is based on $k(n - 1)$ degrees of freedom. Now, the variance of the k sample means is given by

$$s_{\bar{x}}^2 = \sum_{i=1}^{k} \frac{(\bar{y}_i - \bar{y}_.)^2}{k - 1}$$

and *if the null hypothesis is true* it estimates σ^2/n. Thus, an estimate of σ^2 based on the differences among the sample means is given by

$$\hat{\sigma}_B^2 = n \cdot s_{\bar{x}}^2 = n \cdot \sum_{i=1}^{k} \frac{(\bar{y}_i - \bar{y}_.)^2}{k - 1}$$

and it is based on $k - 1$ degrees of freedom.

If the null hypothesis is true, it can be shown that $\hat{\sigma}_W^2$ and $\hat{\sigma}_B^2$ are independent estimates of σ^2, and it follows that

$$F = \frac{\hat{\sigma}_B^2}{\hat{\sigma}_W^2}$$

is a value of a random variable having the F distribution with $k - 1$ and $k(n - 1)$ degrees of freedom. Since the *between sample variance*, $\hat{\sigma}_B^2$, can be expected to exceed the *within sample variance*, $\hat{\sigma}_W^2$, when the null hypothesis is *false*, the null hypothesis will be rejected if F exceeds F_α, where F_α is obtained from Table 6 with $k - 1$ and $k(n - 1)$ degrees of freedom.

The preceding argument has shown how the test of the equality of k means can be based on the comparison of two variance estimates. More remarkable, perhaps, is the fact that the two estimates in question [except for the divisors $k - 1$ and $k(n - 1)$] can be obtained by "breaking up" or analyzing the total variance of all nk observations into two parts. The sample variance of all nk observations is given by

$$s^2 = \sum_{i=1}^{k} \sum_{j=1}^{n} \frac{(y_{ij} - \bar{y}_{..})^2}{nk - 1}$$

and with reference to its numerator, called the *total sum of squares*, we shall now prove the following theorem.

THEOREM 12.1

$$\sum_{i=1}^{k} \sum_{j=1}^{n} (y_{ij} - \bar{y}_{..})^2 = \sum_{i=1}^{k} \sum_{j=1}^{n} (y_{ij} - \bar{y}_{i.})^2 + n \cdot \sum_{i=1}^{k} (\bar{y}_{i.} - \bar{y}_{..})^2$$

The proof of this theorem is based on the identity

$$y_{ij} - \bar{y}_{..} = (y_{ij} - \bar{y}_{i.}) + (\bar{y}_{i.} - \bar{y}_{..})$$

Squaring both sides and summing on i and j, we obtain

$$\sum_{i=1}^{k} \sum_{j=1}^{n} (y_{ij} - \bar{y}_{..})^2 = \sum_{i=1}^{k} \sum_{j=1}^{n} (y_{ij} - \bar{y}_{i.})^2 + \sum_{i=1}^{k} \sum_{j=1}^{n} (\bar{y}_{i.} - \bar{y}_{..})^2$$
$$+ 2 \sum_{i=1}^{k} \sum_{j=1}^{n} (y_{ij} - \bar{y}_{i.})(\bar{y}_{i.} - \bar{y}_{..})$$

Next, we observe that

$$\sum_{i=1}^{k} \sum_{j=1}^{n} (y_{ij} - \bar{y}_{i.})(\bar{y}_{i.} - \bar{y}_{..}) = \sum_{i=1}^{k} (\bar{y}_{i.} - \bar{y}_{..}) \sum_{j=1}^{n} (y_{ij} - \bar{y}_{i.}) = 0$$

since $\bar{y}_{i.}$ is the mean of the ith sample and, hence, $\sum_{j=1}^{n} (y_{ij} - \bar{y}_{i.}) = 0$ for all i. To complete the proof of Theorem 12.1, we have only to observe that the summand of the second sum on the right-hand side of the original identity does not involve the subscript j and that, consequently,

$$\sum_{i=1}^{k} \sum_{j=1}^{n} (\bar{y}_i - \bar{y}.)^2 = n \cdot \sum_{i=1}^{k} (\bar{y}_i - \bar{y}.)^2$$

It is customary to denote the *total sum of squares*, the left-hand member of the identity of Theorem 12.1, by *SST*. The first term on the right-hand side is $\hat{\sigma}_W^2$ times its degrees of freedom, and we refer to this sum as the *error sum of squares*, *SSE*. The term "error sum of squares" expresses the idea that the quantity estimates *random* (or *chance*) *error*. The second term on the right-hand side of the identity of Theorem 12.1 is $\hat{\sigma}_B^2$ times its degrees of freedom, and we refer to it as the *between-samples sum of squares* or the *treatment sum of squares*, *SS(Tr)*. (Most of the early applications of this kind of analysis were in the field of agriculture, where the k populations represented different treatments, such as fertilizers, applied to agricultural plots.) Note that in this notation the F ratio on page 337 can be written

$$F = \frac{SS(Tr)/(k-1)}{SSE/k(n-1)}$$

The sums of squares required for substitution into this last formula are usually obtained by means of the following short-cut formulas, which the reader will be asked to verify in Exercise 12 on page 345. We first calculate *SST* and *SS(Tr)* by means of the formulas

$$SST = \sum_{i=1}^{k} \sum_{j=1}^{n} y_{ij}^2 - C$$

$$SS(Tr) = \frac{\sum_{i=1}^{k} T_i^2}{n} - C$$

where C, called the correction term, *is given by*

$$C = \frac{T_{..}^2}{kn}$$

In these formulas, T_i is the total of the n observations in the ith sample while $T_{.}$ is the grand total of all kn observations. The error sum of squares,

SSE, is then obtained by subtraction; according to Theorem 12.1 we can write

$$SSE = SST - SS(Tr)$$

The results obtained in analyzing the total sum of squares into its components are conveniently summarized by means of the following kind of *analysis of variance table*:

Source of variation	Degrees of freedom	Sum of squares	Mean square	F
Treatments	$k - 1$	$SS(Tr)$	$MS(Tr)$ $= SS(Tr)/(k - 1)$	$\dfrac{MS(Tr)}{MSE}$
Error	$k(n - 1)$	SSE	MSE $= SSE/k(n - 1)$	
Total	$nk - 1$	SST		

Note that each *mean square* is obtained by dividing the corresponding sum of squares by its degrees of freedom.

Example. To illustrate the *analysis of variance* (as this technique is appropriately called) for a one-way classification, suppose that in accordance with the layout on page 333 each laboratory measures the tin-coating weights of 12 discs and that the results are as follows:

Laboratory A	Laboratory B	Laboratory C	Laboratory D
0.25	0.18	0.19	0.23
0.27	0.28	0.25	0.30
0.22	0.21	0.27	0.28
0.30	0.23	0.24	0.28
0.27	0.25	0.18	0.24
0.28	0.20	0.26	0.34
0.32	0.27	0.28	0.20
0.24	0.19	0.24	0.18
0.31	0.24	0.25	0.24
0.26	0.22	0.20	0.28
0.21	0.29	0.21	0.22
0.28	0.16	0.19	0.21

The totals for the four samples are, respectively, 3.21, 2.72, 2.76, and 3.00, the grand total is 11.69, and the calculations required to obtain the necessary sums of squares are as follows:

$$C = \frac{(11.69)^2}{48} = 2.8470$$

$$SST = (0.25)^2 + (0.27)^2 + \ldots + (0.21)^2 - 2.8470 = 0.0809$$

$$SS(Tr) = \frac{(3.21)^2 + (2.72)^2 + (2.76)^2 + (3.00)^2}{12} - 2.8470 = 0.0130$$

$$SSE = 0.0809 - 0.0130 = 0.0679$$

Thus, we get the following *analysis of variance table*:

Source of variation	Degrees of freedom	Sum of squares	Mean square	F
Laboratories	3	0.0130	0.0043	2.87
Error	44	0.0679	0.0015	
Total	47	0.0809		

Since the value obtained for F exceeds 2.82, the value of $F_{.05}$ with 3 and 44 degrees of freedom, the null hypothesis can be rejected at the 0.05 level of significance; we conclude that the laboratories are *not* obtaining consistent results.

To estimate the parameters μ, α_1, α_2, α_3, and α_4 (or μ_1, μ_2, μ_3, and μ_4), we can use the method of least squares, minimizing

$$\sum_{i=1}^{k} \sum_{j=1}^{n} (y_{ij} - \mu - \alpha_i)^2$$

with respect to μ and the α_i, subject to the restriction that $\sum_{i=1}^{k} \alpha_i = 0$. This may be done by eliminating one of the α's or, better, by using the method of Lagrange multipliers, which is treated in most texts on advanced calculus. In either case we obtain the "intuitively obvious" estimates $\hat{\mu} = \bar{y}.$ and $\hat{\alpha}_i = \bar{y}_i - \bar{y}.$ for $i = 1, 2, \ldots, k$, and the corresponding estimates for the μ_i are given by $\hat{\mu}_i = \bar{y}_i$.

Continuation of Example. Thus, for the data from the four laboratories we get $\hat{\mu} = \frac{11.69}{48} = 0.244$, $\hat{\alpha}_1 = \frac{3.21}{12} - 0.244 = 0.024$, $\hat{\alpha}_2 =$

$\dfrac{2.72}{12} - 0.244 = -0.017, \quad \hat{\alpha}_3 = \dfrac{2.76}{12} - 0.244 = -0.014, \quad$ and $\hat{\alpha}_4 =$
$\dfrac{3.00}{12} - 0.244 = 0.006.$

The analysis of variance described in this section applies to one-way classifications where each sample has the same number of observations. If this is not the case and the sample sizes are $n_1, n_2, \ldots,$ and n_k, we have only to substitute $N = \displaystyle\sum_{i=1}^{k} n_i$ for nk throughout and write the computing formulas for SST and $SS(Tr)$ as

$$SST = \sum_{i=1}^{k} \sum_{j=1}^{n_i} y_{ij}^2 - C$$

$$SS(Tr) = \sum_{i=1}^{k} \frac{T_i^2}{n_i} - C$$

Otherwise, the procedure is the same as before. (See also Exercise 13 on page 345.)

EXERCISES

1. An experiment is performed to compare the cleansing action of two detergents, Detergent A and Detergent B. Twenty swatches of cloth are soiled with dirt and grease, each is washed with one of the detergents in an agitator-type machine, and then measured for "whiteness." Criticize the following aspects of the experiment:
   (a) The entire experiment is performed with soft water.
   (b) Fifteen of the swatches are washed with Detergent A and five with Detergent B.
   (c) To accelerate the testing procedure, very hot water and 30-second washing times are used in the experiment.
   (d) The "whiteness" readings of all swatches washed with Detergent A are taken first.

2. A certain *bon vivant*, wishing to ascertain the cause of his frequent hangovers, conducted the following experiment. On the first night, he drank nothing but whiskey and water; on the second night he drank vodka and water; on the third night he drank gin and water; and on the fourth night he drank rum and water. On each of the following mornings he had a hangover, and he concluded that it was the common factor, the water, that made him ill.

(a) This conclusion is obviously unwarranted, but can you state what principles of sound experimental design are violated?

(b) Give a less obvious example of an experiment having the same shortcomings.

(c) Suppose our friend had modified his experiment so each of the four alcoholic beverages was used both with and without water, so that the experiment lasted eight nights. Could the results of this enlarged experiment serve to support or refute the hypothesis that water was the cause of the hangovers? Explain.

3. To compare the effectiveness of three methods of teaching the programming of a certain computer—Method A which is straight teaching-machine instruction, Method B which involves the personal attention of an instructor and some direct experience working with the computer, and Method C which involves the personal attention of an instructor but no work with the computer—random samples of size four are taken from large groups of persons taught by the three methods and the following are the scores which they obtained in an appropriate achievement test:

Method A	Method B	Method C
73	91	72
77	81	77
67	87	76
71	85	79

(a) Without using the short-cut formulas, calculate $\sum\limits_{i=1}^{k} \sum\limits_{j=1}^{n} (y_{ij} - \bar{y}_{..})^2$, $\sum\limits_{i=1}^{k} \sum\limits_{j=1}^{n} (y_{ij} - \bar{y}_{i.})^2$, and $n \cdot \sum\limits_{i=1}^{k} (\bar{y}_{i.} - \bar{y}_{..})^2$, and verify the identity of Theorem 12.1.

(b) Verify the results obtained for the three sums of squares by using the short-cut formulas on pages 339 and 340.

4. Using the sums of squares obtained in Exercise 3, test at the level of significance $\alpha = 0.05$ whether the differences among the means obtained for the three samples are significant.

5. The following are the number of mistakes made in five successive days by four technicians working for a photographic laboratory:

Technician I	Technician II	Technician III	Technician IV
6	14	10	9
14	9	12	12
10	12	7	8
8	10	15	10
11	14	11	11

Test at the level of significance $\alpha = 0.01$ whether the differences among the four sample means can be attributed to chance.

6. The following are the weight losses of certain machine parts (in milligrams) due to friction, when three different lubricants were used under controlled conditions:

> *Lubricant A:* 12.2, 11.8, 13.1, 11.0, 3.9, 4.1, 10.3, 8.4
> *Lubricant B:* 10.9, 5.7, 13.5, 9.4, 11.4, 15.7, 10.8, 14.0
> *Lubricant C:* 12.7, 19.9, 13.6, 11.7, 18.3, 14.3, 22.8, 20.4

(a) Test at the 0.01 level of significance whether the differences among the sample means can be attributed to chance.

(b) Estimate the parameters of the model used in the analysis of this experiment.

7. To determine the effect on exit dust loading in a precipitator, the following measurements were made:

Total flow (cu ft/hr)	*Exist dust loading* (grains per cubic yard in flue gas)				
200	1.5	1.7	1.6	1.9	1.9
300	1.5	1.8	2.2	1.9	2.2
400	1.4	1.6	1.7	1.5	1.8
500	1.1	1.5	1.4	1.4	2.0

Use the level of significance $\alpha = 0.05$ to test whether the flow through the precipitator has an effect on the exit dust loading.

8. To study the performance of a newly-designed motorboat it was timed over a marked course under various wind and water conditions. Use the following data (in minutes) to test the null hypothesis that the boat's performance is not affected by the differences in wind and water conditions:

> *Calm conditions:* 20, 17, 14, 24
> *Moderate conditions:* 21, 23, 16, 25, 18, 23
> *Choppy conditions:* 26, 24, 23, 29, 21

Use the level of significance $\alpha = 0.05$.

9. To find the best arrangement of instruments on a control panel of an airplane, three different arrangements were tested by simulating an emergency condition and observing the reaction time required to correct the condition. The reaction times (in *tenths* of a second) of 28 pilots (randomly assigned to the different arrangements) were as follows:

> *Arrangement 1:* 14, 13, 9, 15, 11, 13, 14, 11
> *Arrangement 2:* 10, 12, 9, 7, 11, 8, 12, 9, 10, 13, 9, 10
> *Arrangement 3:* 11, 5, 9, 10, 6, 8, 8, 7

Test at the level of significance $\alpha = 0.01$ whether we can reject the null hypothesis that the differences among the arrangements have no effect.

10. Referring to the discussion on page 332, assume that the standard deviations of the tin-coating weights determined by any one of the three laboratories have the common value $\sigma = 0.012$, and that it is desired to be 95 percent confident of detecting a difference in means between any two of the laboratories in excess of 0.01 pound per base box. Show that these assumptions lead to the decision to send a sample of 12 discs to each laboratory.

11. Show that if $\mu_i = \mu + \alpha_i$, and μ is the mean of the μ_i, it follows that $\sum_{i=1}^{k} \alpha_i = 0$.

12. Verify the short-cut formulas for computing SST and $SS(Tr)$ given on page 339.

13. State and prove a theorem analogous to Theorem 12.1 for the case where the size of the ith sample is n_i; that is, where the sample sizes are not necessarily equal.

14. Samples of peanut butter produced by two different manufacturers are tested for aflatoxin content, with the following results:

Aflatoxin content (ppb)

Brand A	Brand B
0.5	4.7
0.0	6.2
3.2	0.0
1.4	10.5
0.0	2.1
1.0	0.8
8.6	
2.9	

(a) Use the method analysis of variance to test whether the two brands differ in aflatoxin content.

(b) Test the same hypothesis using a two-sample t-test.

(c) It can be shown that the t and F statistics were related by the formula

$$F(1, v) = t^2(v)$$

where v = degrees of freedom. Using this result, prove that the analysis of variance and two-sample t-test methods are equivalent in this case.

12.3   RANDOMIZED BLOCK DESIGNS

As we observed in Section 12.1, the estimate of chance variation (the experimental error) can often be reduced, that is, freed of variability due to extraneous causes, by dividing the observations in each classification into *blocks*. This is accomplished when known sources of variability (that is, extraneous variables) are held fixed in each block, but vary from block to block.

In this section we shall suppose that the experimenter has available measurements pertaining to a treatments distributed over b blocks. First, we shall consider the case where there is exactly one observation from each treatment in each block; with reference to the illustration on page 335, this case would arise if each laboratory tested one disc from each strip. Letting y_{ij} denote the observation pertaining to the ith treatment and the jth block $\bar{y}_{i.}$ the mean of the b observations for the ith treatment, $\bar{y}_{.j}$ the mean of the a observations in the jth block, and $\bar{y}_{..}$ the grand mean of all the ab observations, we shall use the following layout for this kind of *two-way classification*:

Blocks

	B_1	B_2	...	B_j	...	B_b	*Means*
Treatment 1:	$y_{11},$	$y_{12},$	$\ldots,$	$y_{1j},$	$\ldots,$	y_{1b}	$\bar{y}_{1.}$
Treatment 2:	$y_{21},$	$y_{22},$	$\ldots,$	$y_{2j},$	$\ldots,$	y_{2b}	$\bar{y}_{2.}$
Treatment i:	$y_{i1},$	$y_{i2},$	$\ldots,$	$y_{ij},$	$\ldots,$	y_{ib}	$\bar{y}_{i.}$
Treatment a:	$y_{a1},$	$y_{a2},$	$\ldots,$	$y_{aj},$	$\ldots,$	y_{ab}	$\bar{y}_{a.}$
Means	$\bar{y}_{.1}$	$\bar{y}_{.2}$	$\ldots$	$\bar{y}_{.j}$	$\ldots$	$\bar{y}_{.b}$	$\bar{y}_{..}$

This kind of arrangement is also called a *randomized block design*, since the treatments are presumably allocated at random *within* each block. Note that when a dot is used in place of a subscript, this means that the mean is obtained by summing over that subscript.

The underlying model which we shall assume for the analysis of this kind of experiment with one observation per "cell" (that is, there is one observation corresponding to each treatment within each block) is given by

$$y_{ij} = \mu + \alpha_i + \beta_j + \epsilon_{ij} \qquad \text{for } i = 1, 2, \ldots, a; j = 1, 2, \ldots, b$$

Here μ is the grand mean, α_i is the *effect* of the ith treatment, β_j is the *effect* of the jth block, and the ϵ_{ij} are values of *independent, normally distributed* random variables having *zero means* and the *common variance* σ^2. Analogous to the model for the one-way classification, we restrict the parameters by imposing the conditions that $\sum\limits_{i=1}^{a} \alpha_i = 0$ and $\sum\limits_{j=1}^{b} \beta_j = 0$ (see Exercise 10 on page 356).

In the analysis of a two-way classification where each treatment is represented once in each block, the major objective is to test for the significance of the difference among the $\bar{y}_{i.}$, that is, to test the null hypothesis

$$\alpha_1 = \alpha_2 = \ldots = \alpha_a = 0$$

In addition, it may also be desirable to test whether the blocking has been effective, that is, whether the null hypothesis

$$\beta_1 = \beta_2 = \ldots = \beta_b = 0$$

can be rejected. In either case, the alternative hypothesis is that at least one of the effects is different from zero.

As in the one-way analysis of variance, we shall base these significance tests on comparisons of estimates of σ^2—one based on the variation among treatments, one based on the variation among blocks, and one measuring the experimental error. Note that only the latter is an estimate of σ^2 when either (or both) of the null hypotheses do not hold. The required sums of squares are given by the three components into which the total sum of squares is "broken up" by means of the following theorem:

THEOREM 12.2

$$\sum_{i=1}^{a} \sum_{j=1}^{b} (y_{ij} - \bar{y}_{..})^2 = \sum_{i=1}^{a} \sum_{j=1}^{b} (y_{ij} - \bar{y}_{i.} - \bar{y}_{.j} + \bar{y}_{..})^2$$

$$+ b \sum_{i=1}^{a} (\bar{y}_{i.} - \bar{y}_{..})^2 + a \sum_{j=1}^{b} (\bar{y}_{.j} - \bar{y}_{..})^2$$

The left-hand side of this identity represents the total sum of squares, SST, and the terms on the right-hand side are, respectively, the error sum of squares, SSE, the treatment sum of squares, $SS(Tr)$, and the block sum of squares $SS(Bl)$. To prove this theorem, we make use of the identity

$$y_{ij} - \bar{y}_{..} = (y_{ij} - \bar{y}_{i.} - \bar{y}_{.j} + \bar{y}_{..}) + (\bar{y}_{i.} - \bar{y}_{..}) + (\bar{y}_{.j} - \bar{y}_{..})$$

and follow essentially the same argument as in the proof of Theorem 12.1.

In practice, we calculate the necessary sums of squares by means of short-cut formulas analogous to those on page 339, rather than by using the expressions which define these sums of squares in Theorem 12.2. We first calculate SST, $SS(Tr)$, and $SS(Bl)$ by means of the formulas

$$SST = \sum_{i=1}^{a} \sum_{j=1}^{b} y_{ij}^2 - C$$

$$SS(Tr) = \frac{\sum_{i=1}^{a} T_{i.}^2}{b} - C$$

$$SS(Bl) = \frac{\sum_{j=1}^{b} T_{.j}^2}{a} - C$$

where C, the correction term, *is given by*

$$C = \frac{T_{..}^2}{ab}$$

In these formulas $T_{i.}$ is the sum of the b observations for the ith treatment, $T_{.j}$ is the sum of the a observations in the jth block, and $T_{..}$ is the grand total of all the observations. Note that the divisors for $SS(Tr)$ and $SS(Bl)$ are the number of observations in the respective totals, $T_{i.}$ and $T_{.j}$. The error sum of squares is then obtained by subtraction; according to Theorem 12.2 we can write

$$SSE = SST - SS(Tr) - SS(Bl)$$

In Exercise 11 on page 356, the reader will be asked to verify that all these computing formulas are, indeed, equivalent to the corresponding terms of the identity of Theorem 12.2.

Using these sums of squares, we can reject the null hypothesis that the α_i are all equal to zero at the level of significance α if

$$F_{Tr} = \frac{MS(Tr)}{MSE} = \frac{SS(Tr)/(a-1)}{SSE/(a-1)(b-1)}$$

exceeds F_α with $a - 1$ and $(a - 1)(b - 1)$ degrees of freedom. The null hypothesis that the β_j are all equal to zero can be rejected at the level of significance α if

$$F_{Bl} = \frac{MS(Bl)}{MSE} = \frac{SS(Bl)/(b-1)}{SSE/(a-1)(b-1)}$$

exceeds F_α with $b-1$ and $(a-1)(b-1)$ degrees of freedom. Note that the mean squares, $MS(Tr)$, $MS(Bl)$, and MSE, are again defined as the corresponding sums of squares divided by their degrees of freedom.

The results obtained in this analysis are summarized in the following *analysis of variance table*:

Source of variation	Degrees of freedom	Sum of squares	Mean square	F
Treatments	$a-1$	$SS(Tr)$	$MS(Tr) = \frac{SS(Tr)}{(a-1)}$	$F_{Tr} = \frac{MS(Tr)}{MSE}$
Blocks	$b-1$	$SS(Bl)$	$MS(Bl) = \frac{SS(Bl)}{(b-1)}$	$F_{Bl} = \frac{MS(Bl)}{MSE}$
Error	$(a-1)(b-1)$	SSE	$MSE = \frac{SSE}{(a-1)(b-1)}$	
Total	$ab-1$	SST		

Example. Let us illustrate the analysis of a two-way classification with one observation for each treatment in each block, by considering an experiment designed to study the performance of four different detergents. The following "whiteness" readings were obtained with specially designed equipment for 12 loads of washing distributed over three different models of washing machines:

	Machine 1	Machine 2	Mchine 3	Totals
Detergent A	45	43	51	139
Detergent B	47	46	52	145
Detergent C	48	50	55	153
Detergent D	42	37	49	128
Totals	182	176	207	565

Looking upon the detergents as treatments and the machines as blocks, we obtain the necessary sums of squares as follows:

$$C = \frac{(565)^2}{12} = 26{,}602$$

$$SST = 45^2 + 43^2 + \ldots + 49^2 = 26{,}867 - 26{,}602 = 265$$

$$SS(Tr) = \frac{139^2 + 145^2 + 153^2 + 128^2}{3} - 26{,}602 = 111$$

$$SS(Bl) = \frac{182^2 + 176^2 + 207^2}{4} - 26{,}602 = 135$$

$$SSE = 265 - 111 - 135 = 19$$

Then dividing the sums of squares by their respective degrees of freedom to obtain the appropriate mean squares, we get the results shown in the following analysis of variance table:

Source of variation	Degrees of freedom	Sums of squares	Mean square	F
Detergents	3	111	37.0	11.6
Machines	2	135	67.5	21.1
Error	6	19	3.2	
Total	11	265		

Since $F_{Tr} = 11.6$ exceeds 9.78, the value of $F_{0.01}$ with 3 and 6 degrees of freedom, we conclude that there are differences in the effectiveness of the four detergents. Also, since $F_{Bl} = 21.1$ exceeds 10.9, the value of $F_{0.01}$ with 2 and 6 degrees of freedom, we conclude that the differences among the results obtained for the three washing machines are significant, namely, that the blocking has been effective. To make the effect of this blocking even more evident, the reader will be asked to verify in Exercise 6 on page 355 that the test for differences among the detergents would *not* yield significant results if we looked upon the data as a one-way classification.

The effect of the ith detergent can be estimated by means of the formula $\hat{\alpha}_i = \bar{y}_{i.} - \bar{y}_{..}$, which may be obtained by the method of least squares. The resulting estimates are

$$\hat{\alpha}_1 = 46.3 - 47.1 = -0.8, \qquad \hat{\alpha}_2 = 48.3 - 47.1 = 1.2,$$
$$\hat{\alpha}_3 = 51.0 - 47.1 = 3.9, \qquad \hat{\alpha}_4 = 42.7 - 47.1 = -4.4$$

It should be observed that a two-way classification automatically allows for repetitions of the experimental conditions; for example, in the preceding experiment each detergent was tested three times. Further repetitions may be handled in several ways, and care must be taken that the model used appropriately describes the situation. One way to provide further repetition in a two-way classification is to include additional blocks—for example, to test each detergent using several additional washing machines, randomizing the order of testing for each machine. Note that the model remains essentially the same as before, the only change being an increase in b, and a corresponding increase in the degrees of freedom for blocks and for error. The latter is important, because an increase in the degrees of freedom for error makes the test of the null hypothesis $\alpha_i = 0$ for all i *more sensitive* to small differences among the treatment means. In fact, the real purpose of this kind of repetition is to increase the degrees of freedom for error, thereby increasing the sensitivity of the F tests (see Exercise 9 on page 356).

A second method is to repeat the entire experiment, using a new pattern of randomization to obtain $a \cdot b$ additional observations. This is possible only if the blocks are *identifiable*, that is, if the conditions defining each block can be repeated. For example, in the tin-coating weight experiment described in Section 12.1, the blocks are strips across the rolling direction of a sheet of tin plate, and, given a new sheet, it is possible to identify which is strip 1, which is strip 2, and so forth. In the example of this section, this kind of repetition (usually called *replication*) would require that the settings of the machines be exactly duplicated. This kind of repetition will be used in connection with Latin-square designs in Section 12.5; see also Exercises 7 and 8 on pages 355 and 356.

A third method of repetition is to include n observations for each treatment in each block. When an experiment is designed in this way, the n observations in each "cell" are regarded as *duplicates*, and it is to be expected that their variability will be somewhat less than experimental error. To illustrate this point, suppose that the tin-coating weights of three discs from adjacent positions in a strip are measured in sequence by one of the laboratories, using the same chemical solutions. The variability of these measurements will probably be considerably less than that of three discs from the same strip measured in that laboratory at different times, using different chemical solutions, and perhaps different technicians. The analysis of variance appropriate for this kind of repetition reduces essentially to a two-way analysis of variance applied to the *means* of the n duplicates in the $a \cdot b$ cells; thus, *there would be no gain in degrees of freedom for error*, and, consequently, *no gain in sensitivity of the F tests*. It can be expected, however, that there will be some reduction in the error mean square, since it now measures the residual variance of the *means* of several observations.

12.4 MULTIPLE COMPARISONS

The F tests used so far in this chapter showed whether differences among several means are significant, but they did not tell us whether a given mean (or group of means) differs significantly from another given mean (or group of means). In actual practice, the latter is the kind of information an investigator really wants; for instance, having determined on page 341 that the means of the tin-coating weights obtained by the four laboratories differ significantly, it may be important to find out which laboratory (or laboratories) differs from which others.

If an experimenter is confronted with k means, it may seem reasonable at first to test for significant differences between all possible pairs, that is, to perform $\binom{k}{2} = \dfrac{k(k-1)}{2}$ two-sample t tests as described on page 218. Aside from the fact that this would require a large number of tests even if k is relatively small, these tests would not be independent, and it would be virtually impossible to assign an overall level of significance to this procedure.

Several *multiple-comparisons tests* have been proposed to overcome these difficulties, among them the Duncan *multiple-range test*, which we shall study in this section. (References to other multiple-comparisons tests are mentioned in the book by W. T. Federer listed in the Bibliography.) The assumptions underlying the Duncan multiple-range test are, essentially, those of the one-way analysis of variance for which the sample sizes are equal. The test compares the *range* of any set of p means with an appropriate *least significant range*, R_p, given by

$$R_p = s_{\bar{x}} \cdot r_p$$

Here $s_{\bar{x}}$ is an estimate of $\sigma_{\bar{x}} = \sigma/\sqrt{n}$, and it is computed by means of the formula

$$s_{\bar{x}} = \sqrt{\frac{MSE}{n}}$$

where MSE is the error mean square in the analysis of variance. The value of r_p depends on the desired level of significance and the number of degrees of freedom corresponding to MSE, and it may be obtained from Tables 12(a) and (b) for $\alpha = 0.05$ and 0.01, for $p = 2, 3, \ldots, 10$, and for various degrees of freedom from 1 to 120.

Example. To illustrate this multiple-comparisons procedure, let us refer to the tin-coating data on page 340 and arrange the four sample means as follows in an increasing order of magnitude:

Laboratory	B	C	D	A
Mean	0.227	0.230	0.250	0.268

Next, we compute $s_{\bar{x}}$, using the error mean square of 0.0015 in the analysis of variance on page 341, and we obtain

$$s_{\bar{x}} = \sqrt{\frac{0.0015}{12}} = 0.011$$

Then, we get (by linear interpolation) from Table 12(a) the following values of r_p for $\alpha = 0.05$ and 44 degrees of freedom:

p	2	3	4
r_p	2.85	3.00	3.09

Multiplying each value of r_p by $s_{\bar{x}} = 0.011$, we finally obtain

p	2	3	4
R_p	0.031	0.033	0.034

The range of *all four means* is $0.268 - 0.227 = 0.041$, which exceeds $R_4 = 0.034$, the least significant range. This result should have been expected, since the F test on page 341 showed that the differences among all four means are significant at $\alpha = 0.05$. To test for significant differences among *three adjacent means*, we obtain ranges of 0.038 and 0.023, respectively, for 0.230, 0.250, 0.268 and 0.227, 0.230, 0.250. Since the first of these values exceeds $R_3 = 0.033$, the differences observed in the first set are significant; since the second value does not exceed 0.033, the corresponding differences are not significant. Finally, for *adjacent pairs* of means we find that no adjacent pair has a range greater than the least significant range $R_2 = 0.031$. All these results can be summarized by writing

<u>0.227   0.230   0.250</u>   0.268

where *a line is drawn under any set of adjacent means for which the range is less than the appropriate value of R_p, that is, under any set of adjacent means for which differences are not significant.* We thus conclude in our example that Laboratory *A* averages higher tin-coating weights than laboratories B and C.

Example. If we apply this same method to the example of Section 12.3, where we compared the four detergents, we obtain (see also Exercise 12 on page 356)

Detergent

D	A	B	C
42.7	46.3	48.3	51.0

In other words, among triplets of adjacent means both sets of differences are significant. So far as pairs of adjacent means are concerned, we find that only the difference between 42.7 and 46.3 is significant. Interpreting these results, we conclude that detergent *D* is significantly inferior to any of the others.

EXERCISES

1. A laboratory technician measures the breaking strength of each of five kinds of linen threads by means of four different instruments, and obtains the following results (in ounces):

Measuring instruments

	I_1	I_2	I_3	I_4
Thread 1	20.6	20.7	20.0	21.4
Thread 2	24.7	26.5	27.1	24.3
Thread 3	25.2	23.4	21.6	23.9
Thread 4	24.5	21.5	23.6	25.2
Thread 5	19.3	21.5	22.2	20.6

Looking upon the threads as treatments and the instruments as blocks, perform an analysis of variance at the level of significance $\alpha = 0.01$.

2. Looking at the days (rows) as blocks, rework Exercise 5 on page 343 by the method of Section 12.3.

3. Four different, though supposedly equivalent, forms of a standardized reading achievement test were given to each of five students, and the following are the scores which they obtained:

	Student 1	Student 2	Student 3	Student 4	Student 5
Form A	75	73	59	69	84
Form B	83	72	56	70	92
Form C	86	61	53	72	88
Form D	73	67	62	79	95

Perform a two-way analysis of variance to test at the level of significance $\alpha = 0.01$ whether it is reasonable to treat the four forms as equivalent.

4. An experiment was performed to judge the effect of four different fuels and two different types of launchers on the range of a certain rocket. Test, on the basis of the following data (in nautical miles) whether there are significant differences (a) among the means obtained for the fuels, and (b) between the means obtained for the launchers:

	Fuel I	Fuel II	Fuel III	Fuel IV
Launcher X	62.5	49.3	33.8	43.6
Launcher Y	40.4	39.7	47.4	59.8

Use the level of significance $\alpha = 0.05$.

5. Referring to Exercise 7 on page 344, suppose the data in each of the five columns were obtained from a different precipitator. Repeat the analysis of variance, treating the experiment as a two-way classification, and observe what change results in the error mean square.

6. To emphasize the importance of "blocking," reanalyze the "whiteness" data on page 349 as a one-way classification with the four detergents being the different treatments.

7. If, in a two-way classification, the entire experiment is repeated r times, the model becomes

$$y_{ijk} = \mu + \alpha_i + \beta_j + \rho_k + \epsilon_{ijk}$$

for $i = 1, 2, \ldots, a, j = 1, 2, \ldots, b$, and $k = 1, 2, \ldots, r$, where the sum of the α's, the sum of the β's, and the sum of the ρ's are equal to zero. The ϵ_{ijk} are again values of independent normally distributed random variables with zero means and the common variance σ^2.

   (a) Write down (but do not prove) an identity analogous to the one of Theorem 12.2, which subdivides the total sum of squares into components attributable to treatments, blocks, replicates, and error.

   (b) Generalize the computing formulas on page 348, so that they apply to a replicated randomized block design. Note that the divisor in each case equals the number of observations in the respective totals.

   (c) If the number of degrees of freedom for the replicate sum of squares is $r - 1$, how many degrees of freedom are there for the error sum of squares?

8. The following are the number of defectives produced by four workmen operating, in turn, three different machines; in each case, the first figure represents the number of defectives produced on a Friday and the second figure represents the number of defectives produced on the following Monday:

Workman

		B_1	B_2	B_3	B_4
	A_1	37, 43	38, 44	38, 40	32, 36
Machine	A_2	31, 36	40, 44	43, 41	31, 38
	A_3	36, 40	33, 37	41, 39	38, 45

   Use the theory developed in Exercise 7 to analyze the combined figures for the two days as a two-way classification with replication. Use the level of significance $\alpha = 0.05$.

9. As was pointed out on page 351, two ways of increasing the size of a two-way classification experiment are (a) to double the number of blocks, and (b) to replicate the entire experiment. Discuss and compare the gain in degrees of freedom for the error sum of squares by the two methods.

10. Show that if $\mu_{ij} = \mu + \alpha_i + \beta_j$, the mean of the μ_{ij} (summed on j) is equal to $\mu + \alpha_i$, and the mean of the μ_{ij} summed on i and j is equal to μ, it follows that $\sum\limits_{i=1}^{a} \alpha_i = \sum\limits_{j=1}^{b} \beta_j = 0$.

11. Verify that the computing formulas for SST, $SS(Tr)$, $SS(Bl)$, and SSE, given on page 348, are equivalent to the corresponding terms of the identity of Theorem 12.2.

12. Verify the results of Duncan's test for the comparison of the four detergents, given on page 354.

13. Use the Duncan test with $\alpha = 0.05$ to compare the effectiveness of the three methods of teaching the programming of the computer in Exercise 3 on page 343.

14. Use the Duncan test with $\alpha = 0.05$ to compare the effectiveness of the three lubricants in Exercise 6 on page 344.

15. Use the Duncan test with $\alpha = 0.01$ to compare the strength of the five linen threads in Exercise 1 on page 354.

16. Does it make sense to use the Duncan test with $\alpha = 0.05$ to compare the results obtained with the four fuels in Exercise 4 on page 355? Why?

12.5 SOME FURTHER EXPERIMENTAL DESIGNS

The randomized-block design of Section 12.4 is appropriate when one extraneous source of variability is to be eliminated in comparing a set of sample means. An important feature of this kind of design is its *balance*, achieved by assigning the same number of observations for each treatment to each block. (In this connection, see also the comment on page 335, where we pointed out that differences due to blocks will not affect the means obtained for the different treatments.) The same kind of balance can be attained in more complicated kinds of designs, where it is desired to eliminate the effect of several extraneous sources of variability. In this section we shall introduce two further balanced designs, the Latin-square design and the Graeco-Latin-square design, which are used to eliminate the effects of two and three extraneous sources of variability, respectively.

Example. To introduce the Latin-square design, suppose it is desired to compare three treatments, A, B, and C, in the presence of two other sources of variability. For example, the three treatments may be three methods for soldering copper electrical leads, and the two extraneous sources of variability may be (1) different operators doing the soldering, and (2) the use of different solder fluxes. If three operators and three fluxes are to be considered, the experiment might be arranged in the following pattern:

	Flux 1	Flux 2	Flux 3
Operator 1	A	B	C
Operator 2	C	A	B
Operator 3	B	C	A

Here each soldering method is applied once by each operator in conjunction with each flux, and if there are systematic effects due to differences among operators or differences among fluxes, these effects are present equally for each treatment, that is, for each method of soldering.

An experimental arrangement such as the one just described is called a *Latin square*. An $n \times n$ Latin square is a square array of n distinct letters, with each letter appearing once and only once in each row and in each column. Examples of Latin squares with $n = 4$ and $n = 5$ are shown in Figure 12.2, and larger ones may be obtained from the book by W. G. Cochran and G. M. Cox mentioned in the Bibliography. Note that in a Latin-square experiment involving n treatments, it is necessary to include n^2 observations, n for each treatment.

4 x 4

A	B	C	D
B	C	D	A
C	D	A	B
D	A	B	C

5 x 5

A	B	C	D	E
B	A	E	C	D
C	D	A	E	B
D	E	B	A	C
E	C	D	B	A

Figure 12.2. Latin squares.

As we shall see on page 360, a Latin-square experiment without replication provides only $(n - 1)(n - 2)$ degrees of freedom for estimating the experimental error. Thus, such experiments are rarely run without *replication* if n is small, that is, without repeating the entire Latin-square pattern several times. If there is a total of r replicates, the analysis of the data presumes the following model, where $y_{ij(k)l}$ is the observation in the ith row and the jth column of the lth replicate, and the subscript k, in parentheses, indicates that it pertains to the kth treatment:

$$y_{ij(k)l} = \mu + \alpha_i + \beta_j + \gamma_k + \rho_l + \epsilon_{ij(k)l}$$

for $i, j, k = 1, 2, \ldots, n$, and $l = 1, 2, \ldots, r$, subject to the restrictions that $\sum_{i=1}^{n} \alpha_i = 0$, $\sum_{j=1}^{n} \beta_j = 0$, $\sum_{k=1}^{n} \gamma_k = 0$, and $\sum_{l=1}^{r} \rho_l = 0$. Here μ is the grand mean, α_i is the ith *row effect*, β_j is the jth *column effect*, γ_k is the kth *treatment effect*, ρ_l is the *effect* of the lth *replicate*, and the $\epsilon_{ij(k)l}$ are values of independent, normally distributed random variables with zero means and the common variance σ^2. Note that by "row effects" and "column effects" we mean the effects of the two extraneous variables, and that we are including replicate effects, since, as we shall see, replication can introduce a third extraneous variable. Note also that the subscript k is in parentheses in $y_{ij(k)l}$, because, for a given Latin-square design, k is automatically determined when i and j are known.

The main hypothesis we shall want to test is the null hypothesis $\gamma_k = 0$ for all k, namely, the null hypothesis that there is no difference in the effectiveness of the n treatments. However, we can also test whether the "cross blocking" of the Latin-square design has been effective; that is, we can test the two null hypotheses $\alpha_i = 0$ for all i and $\beta_j = 0$ for all j (against suitable alternatives) to see whether the two extraneous variables actually have an effect on the phenomenon under consideration. Furthermore, we can test the null hypothesis $\rho_l = 0$ for all l against the alternative that the ρ_l are not all equal to zero, and this test for *replicate effects* may be important if the parts of the experiment representing the individual Latin squares were performed on different days, by different technicians, at different temperatures, and so on.

$$C = \frac{(T...)^2}{r \cdot n^2}$$

$$SS(Tr) = \frac{1}{r \cdot n} \sum_{k=1}^{n} T_{(k)}^2 - C$$

$$SSR = \frac{1}{r \cdot n} \sum_{i=1}^{n} T_{i..}^2 - C \qquad \text{(for row effects)}$$

$$SSC = \frac{1}{r \cdot n} \sum_{j=1}^{n} T_{.j.}^2 - C \qquad \text{(for column effects)}$$

$$SS(Rep) = \frac{1}{n^2} \sum_{l=1}^{r} T_{..l}^2 - C \qquad \text{(for replicates)}$$

$$SST = \sum_{i=1}^{n} \sum_{j=1}^{n} \sum_{l=1}^{r} y_{ij(k)l}^2 - C$$

$$SSE = SST - SS(Tr) - SSR - SSC - SS(Rep)$$

The sums of squares required to perform these tests are usually obtained by means of the above short-cut formulas, where $T_{i..}$ is the total of the $r \cdot n$ observations in all of the ith rows, $T_{.j.}$ is the total of the $r \cdot n$ observations in all of the jth columns, $T_{..l}$ is the total of the n^2 observations in the lth replicate, $T_{(k)}$ is the total of all the $r \cdot n$ observations pertaining to the kth treatment, and $T_{...}$ is the grand total of all the $r \cdot n^2$ observations:

Note that again each divisor equals the number of observations in the corresponding squared totals. Finally, the results of the analysis are as shown in the following analysis of variance table:

Source of variation	Degrees of freedom	Sum of squares	Mean square	F
Treatments	$n - 1$	$SS(Tr)$	$MS(Tr)$ $= \dfrac{SS(Tr)}{n-1}$	$\dfrac{MS(Tr)}{MSE}$
Rows	$n - 1$	SSR	MSR $= \dfrac{SSR}{n-1}$	$\dfrac{MSR}{MSE}$
Columns	$n - 1$	SSC	MSC $= \dfrac{SSC}{n-1}$	$\dfrac{MSC}{MSE}$
Replicates	$r - 1$	$SS(Rep)$	$MS(Rep)$ $= \dfrac{SS(Rep)}{r-1}$	$\dfrac{MS(Rep)}{MSE}$
Error	$(n - 1)(rn + r - 3)$	SSE	MSE $= \dfrac{SSE}{(n-1)(rn+r-3)}$	
Total	$rn^2 - 1$	SST		

As before, the degrees of freedom for the total sum of squares equals the *sum* of the degrees of freedom for the individual components; thus, the degrees of freedom for error are usually found last, by subtraction.

Example. To illustrate the analysis of a replicated Latin-square experiment, let us suppose that two replicates of the soldering experiment were run, using the following arrangement:

	Replicate I flux			Replicate II flux		
	1	2	3	1	2	3
Operator 1	A	B	C	C	B	A
Operator 2	C	A	B	A	C	B
Operator 3	B	C	A	B	A	C

The results, showing the number of pounds tensile force required to separate the soldered leads, were as follows:

Replicate I			Replicate II		
14.0	16.5	11.0	10.0	16.5	13.0
9.5	17.0	15.0	12.0	12.0	14.0
11.0	12.0	13.5	13.5	18.0	11.5

The total for the first treatment (Method A) is

$$14.0 + 17.0 + 13.5 + 13.0 + 12.0 + 18.0 = 87.5$$

while the totals for the other two treatments (Methods B and C) are

$$16.5 + 15.0 + 11.0 + 16.5 + 14.0 + 13.5 = 86.5$$

and

$$11.0 + 9.5 + 12.0 + 10.0 + 12.0 + 11.5 = 66.0$$

respectively. Also, the totals for the three rows are 81.0, 79.5, and 79.5, those for the three columns are 70.0, 92.0, and 78.0, the totals for the two replicates are 119.5 and 120.5, and the grand total is 240. We thus obtain

$$C = \frac{(240)^2}{18} = 3200.0$$

$$SS(Tr) = \tfrac{1}{6}[(87.5)^2 + (86.5)^2 + (66.0)^2] - 3200.0 = 49.1$$

$$SSR = \tfrac{1}{6}[(81.0)^2 + (79.5)^2 + (79.5)^2] - 3200.0 = 0.2$$

$$SSC = \tfrac{1}{6}[(70.0)^2 + (92.0)^2 + (78.0)^2] - 3200.0 = 41.3$$
$$SS(Rep) = \tfrac{1}{9}[(119.5)^2 + (120.5)^2] - 3200.0 = 0.1$$
$$SST = (14.0)^2 + (16.5)^2 + \ldots + (11.5)^2 - 3200.0 = 104.5$$
$$SSE = 104.5 - 49.1 - 41.3 - 0.2 - 0.1 = 13.8$$

and the results are as shown in the following analysis of variance table:

Source of variation	Degrees of freedom	Sum of squares	Mean square	F
Methods	2	49.1	24.6	17.6
Operators	2	0.2	0.1	0.1
Fluxes	2	41.3	20.6	14.7
Replicates	1	0.1	0.1	0.1
Error	10	13.8	1.4	
Total	17	104.5		

Since the F ratios for methods and fluxes both exceed 7.56, the value of $F_{.01}$ for 2 and 10 degrees of freedom, the differences due to methods as well as fluxes are significant. As can be seen by inspection, the differences due to the other two sources of variation (operators and replicates) are not significant. To go one step further, the Duncan multiple-range test of Section 12.4 gives the following *decision pattern* at the 0.01 level of significance:

	Method C	Method B	Method A
Mean	11.0	14.4	14.6

Thus, we conclude that Method C definitely yields weaker solder bonds than Methods A or B.

The elimination of *three* extraneous sources of variability can be accomplished by means of a design called a Graeco-Latin square. This design is a square array of n Latin letters and n Greek letters, with the Latin and Greek letters each forming a Latin square; furthermore, each Latin letter appears once and only once in conjunction with each Greek letter. The following is an example of a 4 × 4 Graeco-Latin square:

$A\alpha$	$B\beta$	$C\gamma$	$D\delta$
$B\delta$	$A\gamma$	$D\beta$	$C\alpha$
$C\beta$	$D\alpha$	$A\delta$	$B\gamma$
$D\gamma$	$C\delta$	$B\alpha$	$A\beta$

The construction of Graeco-Latin squares, also called *orthogonal* Latin squares, poses interesting mathematical problems, some of which are mentioned in the book by H. B. Mann listed in the Bibliography.

Example. To give an example where the use of a Graeco-Latin square might be appropriate, suppose that in the soldering example an additional source of variability is the temperature of the solder. If three solder temperatures, denoted by α, β, and γ, are to be used together with the three methods, three operators (rows), and three solder fluxes (columns), a replicate of a suitable Graeco-Latin-square experiment could be laid out as follows:

	Flux 1	*Flux 2*	*Flux 3*
Operator 1	$A\alpha$	$B\gamma$	$C\beta$
Operator 2	$C\gamma$	$A\beta$	$B\alpha$
Operator 3	$B\beta$	$C\alpha$	$A\gamma$

Thus, Method A would be used by operator 1 using solder flux 1 and temperature α, by operator 2 using solder flux 2 and temperature β, and by operator 3 using solder flux 3 and temperature γ. Similarly, Method B would be used by operator 1 using solder flux 2 and temperature γ, and so forth.

In a Graeco-Latin square each variable (represented by rows, columns, Latin letters, or Greek letters) is "distributed evenly" over the other variables. Thus, in comparing the means obtained for one variable, the effects of the other variables are all averaged out. The analysis of a Graeco-Latin square is similar to that of a Latin square, with the addition of an extra source of variability corresponding to the Greek letters.

There exists a wide variety of experimental designs, other than the ones discussed in this chapter, that are suitable for many diverse purposes. Among the more widely used designs are the *incomplete block designs*, which are characterized by the feature that each treatment is not represented in each block. If the number of treatments under investigation in an experiment is large, it often happens that it is impossible to find homogeneous blocks such that all of the treatments can be accommodated in each block.

> **Example.** For instance, if n paints are to be compared by applying each paint to a sheet of metal and then baking the sheets in an oven, it may be impossible to put all of the sheets in the oven at one time. Consequently, it would be necessary to use an experimental design in which $k < n$ treatments (paints) are included in each block (oven run). One way of doing this is to assign the treatments to each oven run in such a way that each treatment occurs together with each other treatment in the same number of blocks. For example, for $n = 4$ and $k = 2$ we might use the following scheme:

Oven run	Paints
1	1 and 2
2	3 and 4
3	1 and 3
4	2 and 4
5	1 and 4
6	2 and 3

This kind of design is called a *balanced incomplete-block design*, and it has the important feature that *comparisons between any two treatments can be made with equal precision.*

Since balanced incomplete-block designs may require too many blocks, many other schmes have been developed. Most of these experimental designs arose to meet the specific needs of experimenters, notably in the field of agriculture. As we have pointed out earlier, much of the language of experimental design, including such terms as "treatments," "blocks," "plots," etc., has been borrowed from agriculture. Only in recent years have the more sophisticated designs been applied to industrial and engineering experimentation, and, with more widespread application, it is to be expected that many new designs will be developed to meet the requirements in these fields.

EXERCISES

1. Referring to the problem in which samples of tin plate are to be distributed among four laboratories (Section 12.1) suppose that we are concerned with systematic differences in tin-coating weight along the direction of rolling as

well as across the rolling direction. To eliminate these two sources of variability, each of two sheets of tin plate is divided into 16 parts, representing four positions across and four positions along the rolling direction. Then, four samples from each sheet are sent to each of the Laboratories A, B, C, and D, as shown, and the resulting tin-coating weights are determined.

<table>
<tr><td colspan="4" align="center">*Replicate I*</td><td></td><td colspan="4" align="center">*Replicate II*</td></tr>
<tr>
<td>C
0.20</td><td>A
0.24</td><td>D
0.20</td><td>B
0.27</td>
<td rowspan="4">Rolling
direction</td>
<td>B
0.29</td><td>A
0.25</td><td>C
0.18</td><td>D
0.28</td>
</tr>
<tr>
<td>B
0.28</td><td>C
0.19</td><td>A
0.22</td><td>D
0.28</td>
<td>D
0.28</td><td>B
0.18</td><td>A
0.21</td><td>C
0.25</td>
</tr>
<tr>
<td>D
0.34</td><td>B
0.23</td><td>C
0.21</td><td>A
0.28</td>
<td>C
0.28</td><td>D
0.23</td><td>B
0.20</td><td>A
0.28</td>
</tr>
<tr>
<td>A
0.32</td><td>D
0.22</td><td>B
0.16</td><td>C
0.27</td>
<td>A
0.30</td><td>C
0.19</td><td>D
0.24</td><td>B
0.25</td>
</tr>
</table>

Determine from these data whether the laboratories were obtaining consistent results. Also determine whether there are actual tin-coating weight differences across and along the rolling direction. (Use the 0.05 level of significance.)

2. A Latin-square design was used to compare the bond strengths of gold semiconductor lead wires bonded to the lead terminal by five different methods, A, B, C, D, and E. The bonds were made by five different operators, and the devices were encapsulated using five different plastics, with the following results, expressed as pounds force required to break the bond.

Operators

		O_1	O_2	O_3	O_4	O_5
	P_1	A 3.0	B 2.4	C 1.9	D 2.2	E 1.7
	P_2	B 2.1	C 2.7	D 2.3	E 2.5	A 3.1
Plastics	P_3	C 2.1	D 2.6	E 2.5	A 2.9	B 2.1
	P_4	D 2.0	E 2.5	A 3.2	B 2.5	C 2.2
	P_5	E 2.1	A 3.6	B 2.4	C 2.4	D 2.1

Analyze these results and apply the Duncan multiple-range test with $\alpha = 0.01$ to the mean breaking strengths of the five bonding methods.

3. A test was made to determine which of three different golf-ball designs, A, B, and C will give the greatest distance. Golf balls were driven by three different golf pros, P_1, P_2, and P_3, using three different drivers, D_1, D_2, and D_3. The experiment was performed on three different fairways, and the distances from the tee to the point where the ball came to rest were measured in yards, as follows:

First fairway

	D_1	D_2	D_3
P_1	B 265	A 311	C 249
P_2	A 350	C 284	B 330
P_3	C 258	B 305	A 351

Second fairway

	D_1	D_2	D_3
P_1	B 220	A 276	C 189
P_2	A 319	C 232	B 264
P_3	C 175	B 254	A 307

Third fairway

	D_1	D_2	D_3
P_1	B 159	A 205	C 142
P_2	A 262	C 168	B 246
P_3	C 198	B 237	A 283

Is any of the golf-ball designs superior to the others with respect to distance?

4. To study the effectiveness of five different kinds of front-seat passenger restraint systems in automobiles, A, B, C, D, and E, the following Graeco-Latin-square experiment was performed. The rows represent different automotive size classes (from sub-compact to full-size), the columns represent different barrier impact speeds, and the Greek letters α, β, γ, δ, and ϵ represent different impact angles. The experimental results are given in terms of an index of forces at critical points on the test dummy that relates to the probability of fatal injury.

$A\alpha$ 0.50	$B\beta$ 0.21	$C\gamma$ 0.43	$D\delta$ 0.35	$E\epsilon$ 0.46
$B\gamma$ 0.51	$C\delta$ 0.20	$D\epsilon$ 0.40	$E\alpha$ 0.25	$A\beta$ 0.39
$C\epsilon$ 0.45	$D\alpha$ 0.07	$E\beta$ 0.29	$A\gamma$ 0.20	$B\delta$ 0.31
$D\beta$ 0.39	$E\gamma$ 0.10	$A\delta$ 0.31	$B\epsilon$ 0.24	$C\alpha$ 0.27
$E\delta$ 0.43	$A\epsilon$ 0.17	$B\alpha$ 0.31	$C\beta$ 0.22	$D\gamma$ 0.32

Analyze this experiment.

5. A clothing manufacturer wishes to determine which of four different needle designs is best for his sewing machines. The sources of variability which must be eliminated to make this comparison are the actual sewing machine used, the operator, and the type of thread. Using the design shown (the rows represent the operators, the columns represent the machines, the Latin letters represent the needles, and the Greek letters stand for the types of thread), the manufacturer recorded the number of rejected garments at the end of each of two weeks, with the following results:

First week

$D\gamma$ 47	$B\alpha$ 40	$A\beta$ 23	$C\delta$ 72
$C\beta$ 74	$A\delta$ 37	$B\gamma$ 28	$D\alpha$ 75
$A\alpha$ 52	$C\gamma$ 95	$D\delta$ 57	$B\beta$ 15
$B\delta$ 10	$D\beta$ 45	$C\alpha$ 93	$A\gamma$ 52

Second week

$C\alpha$ 105	$A\gamma$ 38	$B\delta$ 15	$D\beta$ 60
$D\delta$ 70	$B\beta$ 20	$A\alpha$ 60	$C\gamma$ 85
$B\gamma$ 13	$D\alpha$ 82	$C\beta$ 90	$A\delta$ 28
$A\beta$ 33	$C\delta$ 75	$D\gamma$ 53	$B\alpha$ 31

Using the 0.05 level of significance, determine whether there is a difference among the needles. Also, determine whether there are significant differences in the operators, the machines, and the types of thread.

6. Give an example of a balanced incomplete block design for which
(a) $n = 3$ and $k = 2$;   (b) $n = 4$ and $k = 3$;
(c) $n = 6$ and $k = 4$.
What is the minimum number of times each treatment must be replicated in each of these designs?

12.6 ANALYSIS OF COVARIANCE

The purpose of the methods of Sections 12.3 and 12.5 was to free the experimental error from variability due to identifiable and controllable extraneous causes. In this section, we shall introduce a method, called the *analysis of covariance*, which applies when such extraneous, or *concomitant*, variables cannot be held fixed, but can be measured, nevertheless. This would be the case, for example, if we wanted to compare the effectvieness of several industrial training programs and the results depended on the trainees' I.Q.'s; if we wanted to compare the durability of several kinds of leather soles and the results depended on the weight of the persons wearing the shoes; or if we wanted to compare the merits of several cleaning agents and the results depended on the original condition of the surfaces cleaned.

The method by which we analyze data of this kind is a combination of the linear regression method of Section 11.1 and the analysis of variance of Section 12.2. The underlying model is given by

$$y_{ij} = \mu + \alpha_i + \delta x_{ij} + \epsilon_{ij}$$

for $i = 1, 2, \ldots, k; j = 1, 2, \ldots, n$. As in the model on page 337, μ is the grand mean, α_i is the effect of the ith treatment, and the ϵ_{ij} are values of independent, normally distributed variables with zero means and the common variance σ^2; as in the model on page 295, δ is the slope of the linear regression equation.

In the analysis of such data, the values of the concomitant variable, the x_{ij}, are eliminated by regression methods, namely, by estimating δ by the method of least squares, and then an analysis of variance is performed on the adjusted y's, namely, on the quantities $y_{ij}' = y_{ij} - \hat{\delta} x_{ij}$. This procedure is referred to as an *analysis of covariance*, as it involves a partitioning of the *total sum of products*

$$SPT = \sum_{i=1}^{k} \sum_{j=1}^{n} (y_{ij} - \bar{y}_.)(x_{ij} - \bar{x}_.)$$

in the same way as an ordinary analysis of variance involves the partitioning of the total sum of squares. In actual practice, the calculations are performed as follows:

1. The total, treatment, and error sums of squares are calculated for the x's by means of the formulas for a one-way classification on pages 339 and 340; they will be denoted SST_x, $SS(Tr)_x$, and SSE_x.
2. The total, treatment, and error sums of squares are calculated for the y's by means of the formulas for a one-way classification on pages 339 and 340; they will be denoted SST_y, $SS(Tr)_y$, and SSE_y.
3. The total, treatment, and error sums of products are calculated by means of the formulas

$$SPT = \sum_{i=1}^{k} \sum_{j=1}^{n} x_{ij} \cdot y_{ij} - C$$

$$SP(Tr) = \frac{\sum_{i=1}^{k} T_{x_i} \cdot T_{y_i}}{n} - C$$

$$SPE = SPT - SP(Tr)$$

where the correction term, C, is given by

$$C = \frac{T_x \cdot T_y}{k \cdot n}$$

and where T_{x_i} is the total of the x's for the ith treatment, T_{y_i} is the total of the y's for the ith treatment, T_x is the total of all the x's, and T_y is the total of all the y's.

4. The total, error, and treatment sums of squares are calculated for the adjusted y's by means of the formulas

$$SST_{y'} = SST_y - \frac{(SPT)^2}{SST_x}$$

$$SSE_{y'} = SSE_y - \frac{(SPE)^2}{SSE_x}$$

$$SS(Tr)_{y'} = SST_{y'} - SSE_{y'}$$

The results obtained in these calculations are conveniently summarized by means of the following kind of *analysis-of-covariance table*

Source of variation	Sum of squares for x	Sum of squares for y	Sum of products	Sum of squares for y'	Degrees of freedom	Mean squares
Treatments	$SS(Tr)_x$	$SS(Tr)_y$	$SP(Tr)$	$SS(Tr)_{y'}$	$k-1$	$MS(Tr)_{y'}$ $= \dfrac{SS(Tr)_{y'}}{k-1}$
Error	SSE_x	SSE_y	SPE	$SSE_{y'}$	$nk-k-1$	$MSE_{y'}$ $= \dfrac{SSE_{y'}}{nk-k-1}$
Total	SST_x	SST_y	SPT	$SST_{y'}$	$nk-2$	

Note that each *mean square* is obtained by dividing the corresponding sum of squares by its degrees of freedom.

Finally, the null hypothesis $\alpha_1 = \alpha_2 = \ldots = \alpha_{k_0} = 0$ is tested against the alternative hypothesis that the α_i are not all equal to zero on the basis of the statistic

$$F = \frac{MS(Tr)_{y'}}{MSE_{y'}}$$

It is rejected at the level of significance α if the value obtained for F exceeds F_α with $k-1$ and $nk-k-1$ degrees of freedom.

Example. To illustrate this kind of analysis, suppose a research worker has three different cleaning agents, A_1, A_2, and A_3, and he wishes to select the most efficient agent for cleaning a metallic surface. The cleanliness of a surface is measured by its reflectivity, expressed in arbitrary units as the ratio of the reflectivity observed to that of a standard mirror surface. Analysis of covariance must be used because the effect of a cleaning agent on reflectivity will depend on the original cleanliness, namely, the original reflectivity, of the surface. The research worker obtained the following results:

A_1	Original reflectivity, x	0.50	0.55	0.60	0.35
	Final reflectivity, y	1.00	1.20	0.80	1.40
A_2	Original reflectivity, x	0.75	1.65	1.00	1.10
	Final reflectivity, y	0.75	0.60	0.55	0.50
A_3	Original reflectivity, x	0.60	0.90	0.80	0.70
	Final reflectivity, y	1.00	0.70	0.80	0.90

and the totals are $T_{x_1} = 2.00$, $T_{x_2} = 4.50$, $T_{x_3} = 3.00$, $T_x = 9.50$, $T_{y_1} = 4.40$, $T_{y_2} = 2.40$, $T_{y_3} = 3.40$, and $T_y = 10.20$.

For the x's, the correction term is $\dfrac{(9.50)^2}{3 \cdot 4} = 7.52$, and the sums of squares are

$$SST_x = 0.50^2 + 0.55^2 + \ldots + 0.70^2 - 7.52 = 1.31$$

$$SS(Tr)_x = \frac{2.00^2 + 4.50^2 + 3.00^2}{4} - 7.52 = 0.79$$

$$SSE_x = 1.31 - 0.79 = 0.52$$

For the y's, the correction term is $\dfrac{(10.20)^2}{3 \cdot 4} = 8.67$, and the sums of squares are

$$SST_y = 1.00^2 + 1.20^2 + \ldots + 0.90^2 - 8.67 = 0.79$$

$$SS(Tr)_y = \frac{4.40^2 + 2.40^2 + 3.40^2}{4} - 8.67 = 0.50$$

$$SSE_y = 0.79 - 0.50 = 0.29$$

For the sums of products, the correction term is $\dfrac{(9.50)(10.20)}{3 \cdot 4} = 8.08$, and we get

$$SPT = (0.50)(1.00) + (0.55)(1.20) + \ldots + (0.70)(0.90) - 8.08$$
$$= -0.80$$

$$SP(Tr) = \frac{(2.00)(4.40) + (4.50)(2.40) + (3.00)(3.40)}{4} - 8.08$$
$$= -0.63$$

$$SPE = -0.80 - (-0.63) = -0.17$$

Finally, for the adjusted y's we get

$$SST_{y'} = 0.79 - \frac{(-0.80)^2}{1.31} = 0.30$$

$$SSE_{y'} = 0.29 - \frac{(-0.17)^2}{0.52} = 0.23$$

$$SS(Tr)_{y'} = 0.30 - 0.23 = 0.07$$

All these results are summarized in the following analysis-of-covariance table:

Source of variation	Sum of squares for x	Sum of squares for y	Sum of products	Sum of squares for y'	Degrees of freedom	Mean squares
Treatments	0.79	0.50	−0.63	0.07	2	0.035
Error	0.52	0.29	−0.17	0.23	8	0.026
Total	1.31	0.79	−0.80	0.30	10	

Thus, $F = \dfrac{0.035}{0.026} = 1.34$, and since this does *not* exceed $F_{0.05} = 4.46$ for 2 and 8 degrees of freedom, the null hypothesis cannot be rejected. In other words, one cannot conclude that any one of the cleaning agents is more effective than the others.

Analysis of covariance methods have not been widely used until recent years, due mainly to the rather extensive calculations that are required. Of course, with the widespread availability of computers and appropriate programs, this is no longer a problem. There are several ways in which the analysis of covariance method presented here can be generalized. First, there can be more than one concomitant variable; then, the method can be applied to more complicated kinds of designs, say, to a randomized block design, where the regression coefficient could even assume a different value for each block.

EXERCISES

1. To compare the life expectancy of a transistor under three storage conditions and account at the same time for a leakage current (collector to base), a laboratory technician obtained the following results, where the leakage current, x, is in microamperes, and the life times, y, are in hours:

Storage condition 1		Storage condition 2		Storage condition 3	
x	y	x	y	x	y
4.8	9,912	6.4	9,952	8.8	9,596
7.2	9,383	8.7	9,482	6.2	9,697
5.5	9,734	7.1	9,435	7.5	9,700
6.0	9,551	5.3	9,915	4.9	9,610
8.3	8,959	4.6	9,492	5.4	10,145
7.6	9,474	6.0	9,565	5.8	10,191
5.9	9,179	7.2	9,704	7.3	9,855
8.0	9,359	8.8	9,636	8.6	9,682
4.3	9,580	5.4	9,608	8.8	10,160
5.1	9,245	7.8	9,548	6.0	9,982

Perform an analysis of covariance, using the level of significance $\alpha = 0.05$. Also, estimate the value of the regression coefficient.

2. Four different railroad-track cross-section configurations were tested to determine which is most resistant to breakage under use conditions. Ten miles of each kind of track were laid in each of six locations, and the number of cracks and other fracture-related conditions (y) was measured over a two-year usage period. To compare these track designs adequately, however, it is necessary to correct for extent of usage (x), measured in terms of the average number of trains per day that ran over each section of track. Use the following experimental results to test (0.01 level of significance) whether the track designs were equally resistant to breakage and to estimate the effect of usage on breakage resistance.

Track design A		Track design B		Track design C		Track design D	
x	y	x	y	x	y	x	y
10.4	3	16.9	8	17.8	5	19.6	9
19.3	7	23.6	11	24.4	9	25.4	8
13.7	4	14.4	7	13.5	5	35.5	16
7.2	0	17.2	10	20.1	6	16.8	7
16.3	5	9.1	4	11.0	4	31.2	11

3. Three different instrument-panel configurations were tested by placing airline pilots in flight simulators and testing their reaction time to simulated flight emergencies. Each pilot was faced with ten emergency conditions in a randomized sequence, and the total time required to take corrective action for all ten conditions was measured, with the following results:

Instrument panel 1		Instrument panel 2		Instrument panel 3	
x	y	x	y	x	y
8.1	6.55	12.1	5.74	15.2	6.37
19.4	6.40	2.1	5.93	8.7	6.97
11.6	5.93	3.9	6.16	7.2	7.38
24.9	6.79	5.2	5.68	6.1	6.43
6.2	7.16	4.6	5.41	11.8	7.59
3.8	5.64	14.4	6.29	12.1	7.16
18.4	5.87	16.1	5.55	9.5	7.02
9.4	6.32	8.5	4.82	2.6	6.85

In this table, x is the number of years of experience of the pilot, and y is the total reaction time in seconds. Perform an analysis of covariance to test whether the instrument-panel configurations yield significantly different results ($\alpha = 0.05$). Also, perform a one-way analysis of variance (ignoring the covariate, x) and determine in that way what effect experience has on the results.

13

Factorial
Experimentation

13.1 TWO-FACTOR EXPERIMENTS

In Chapter 12 we were interested mainly in the effects of one variable, whose values we referred to as "treatments." Extraneous variables were accommodated by means of blocks, replicates, or the rows and columns of Latin squares and more complicated kinds of designs. In this chapter we shall be concerned with the individual and joint effects of several variables, and combinations of the values, or *levels*, of these variables will now play the roles of the different treatments. Extraneous variables, if any, will be handled as before.

Example. To consider a simple *two-factor* (two-variable) experiment, suppose it is desired to determine the effects of flue temperature and oven width on the time required to make coke. The experimental con-

375

ditions used are

Oven width (inches)	Flue temperature (degrees F)
4	1600
4	1900
8	1600
8	1900
12	1600
12	1900

and if several blocks (or replicates) were run, each consisting of these six "treatments," it would be possible to analyze the data as a two-way classification and test for significant differences among the six treatment means. In this instance, however, the experimenter is interested in knowing far more than that—he wishes to know whether variations in oven width or in flue temperature affect the coking time, and perhaps also whether any changes in coking time attributable to variations in oven width are the same at different temperatures.

It is possible to answer questions of this kind if the experimental conditions, the treatments, consist of appropriate combinations of the *levels* (or values) of the various *factors*. The factors in the preceding example are oven width and flue temperature; oven width has the *three levels* 4, 8, and 12 inches, while flue temperature has the *two levels* 1,600 and 1,900 degrees Fahrenheit. Note that the six treatments were chosen in such a way that each level of oven width is used once in conjunction with each level of flue temperature. In general, if two factors A and B are to be investigated at a levels and b levels, respectively, and if there are $a \cdot b$ experimental conditions (treatments) corresponding to all possible combinations of the levels of the two factors, the resulting experiment is referred to as a *complete $a \times b$ factorial experiment*. Note that if one or more of the $a \cdot b$ experimental conditions is omitted, the experiment can still be analyzed as a two-way classification, but it cannot readily be analyzed as a factorial experiment. It is customary to omit the word "complete," so that an $a \times b$ *factorial* experiment is understood to contain experimental conditions corresponding to all possible combinations of the levels of the two factors.

In order to obtain an estimate of the experimental error in a two-factor experiment it is necessary to replicate, that is, to repeat the entire set of $a \cdot b$ experimental conditions, say, a total of r times, randomizing the order of applying the conditions in each replicate. If y_{ijk} is the observation in the kth replicate, taken at the ith level of factor A and the jth level of factor B, the model assumed for the analysis of this kind of experiment is usually written as

$$y_{ijk} = \mu + \alpha_i + \beta_j + (\alpha\beta)_{ij} + \rho_k + \epsilon_{ijk}$$

for $i = 1, 2, \ldots, a$, $j = 1, 2, \ldots, b$, and $k = 1, 2, \ldots, r$. Here μ is the grand mean, α_i is the effect of the ith level of factor A, β_j is the effect of the jth level of factor B, $(\alpha\beta)_{ij}$ is the *interaction*, or joint effect, of the ith level of factor A and the jth level of factor B, and ρ_k is the effect of the kth replicate. As in the models used in Chapter 12 we shall assume that the ϵ_{ijk} are values of independent random variables having normal distributions with zero means and the common variance σ^2. Also, analogous to the restrictions imposed on the models on pages 337 and 346, we shall assume that

$$\sum_{i=1}^{a} \alpha_i = \sum_{j=1}^{b} \beta_j = \sum_{i=1}^{a} (\alpha\beta)_{ij} = \sum_{j=1}^{b} (\alpha\beta)_{ij} = \sum_{k=1}^{r} \rho_k = 0$$

It can be shown that these restrictions will assure unique estimates for the parameters μ, α_i, β_j, $(\alpha\beta)_{ij}$, and ρ_k.

Example. To illustrate the model underlying a two-factor experiment, let us consider an experiment with two replicates in which factor A occurs at two levels, factor B occurs at two levels, and the replication effects are zero, that is, $\rho_1 = \rho_2 = 0$. In view of the restrictions on the parameters we also have

$$\alpha_2 = -\alpha_1, \beta_2 = -\beta_1 \quad \text{and} \quad (\alpha\beta)_{21} = (\alpha\beta)_{12} = -(\alpha\beta)_{22} = -(\alpha\beta)_{11}$$

and the population means corresponding to the four experimental conditions defined by the two levels of factor A and the two levels of factor B can be written as

$$\mu_{111} = \mu_{112} = \mu + \alpha_1 + \beta_1 + (\alpha\beta)_{11}$$
$$\mu_{121} = \mu_{122} = \mu + \alpha_1 - \beta_1 - (\alpha\beta)_{11}$$
$$\mu_{211} = \mu_{212} = \mu - \alpha_1 + \beta_1 - (\alpha\beta)_{11}$$
$$\mu_{221} = \mu_{222} = \mu - \alpha_1 - \beta_1 + (\alpha\beta)_{11}$$

Substituting for $\mu_{ij1} = \mu_{ij2}$ the mean of all observations obtained for the ith level of factor A and the jth level of factor B, we get four simultaneous linear equations which can be solved for the parameters μ, α_1, β_1, and $(\alpha\beta)_{11}$ (see Exercise 8 on page 393).

To continue our illustration, let us now suppose that $\mu = 10$. If all of the other effects equalled zero, each of the μ_{ijk} would equal 10, and the response surface would be the horizontal plane shown in Figure 13.1(a). If we now add an effect of factor A, with $\alpha_1 = -4$, the response surface becomes the tilted plane shown in Figure 13.1(b), and if we add to this an effect of factor B, with $\beta_1 = 5$, we get the plane shown in Figure 13.1(c). Note that, so far, the effects of factors A and B are

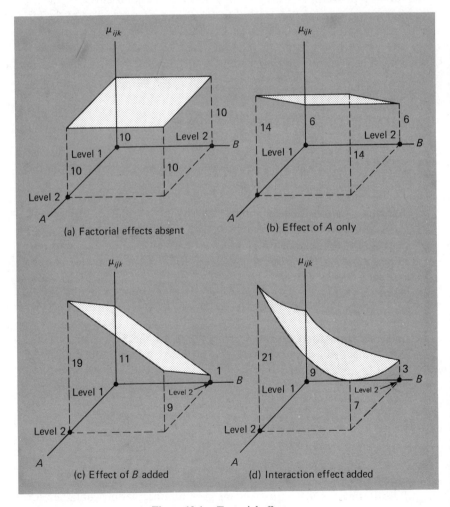

Figure 13.1. Factorial effects.

additive, that is, the change in the mean for either factor in going from level 1 to level 2 does not depend on the level of the other factor, and the response surface is a plane. If we now include an *interaction*, with $(\alpha\beta)_{11} = -2$, the plane becomes twisted as shown in Figure 13.1(d), the effects are no longer additive, and the response surface is no longer a plane. Note, also, that if the replication effects had not equalled zero, we would have obtained a different surface for each replicate; for each replicate the surface of Figure 13.1(d) would have been shifted an appropriate number of units up or down.

The analysis of an $a \times b$ *factorial experiment* is based on the following breakdown of the total sum of squares. First, we subdivide SST into components attributed to treatments, replicates (or blocks), and error, by means of the identity

$$\sum_{i=1}^{a} \sum_{j=1}^{b} \sum_{k=1}^{r} (y_{ijk} - \bar{y}_{...})^2 = r \sum_{i=1}^{a} \sum_{j=1}^{b} (\bar{y}_{ij.} - \bar{y}_{...})^2$$

$$+ ab \sum_{k=1}^{r} (\bar{y}_{..k} - \bar{y}_{...})^2$$

$$+ \sum_{i=1}^{a} \sum_{j=1}^{b} \sum_{k=1}^{r} (y_{ijk} - \bar{y}_{ij.} - \bar{y}_{..k} + \bar{y}_{...})^2$$

Except for notation, this identity is equivalent to that of Theorem 12.2. The total sum of squares, on the left-hand side of the identity, has $abr - 1$ degrees of freedom. The terms on the right are, respectively, the treatment sum of squares having $ab - 1$ degrees of freedom, the replicate (block) sum of squares having $r - 1$ degrees of freedom, and the error sum of squares having $(ab - 1)(r - 1)$ degrees of freedom. (Note that the various degrees of freedom are the same as those of the analysis of variance table on page 349 if we substitute ab for a and r for b.)

So far there is nothing new about the analysis of the data; it is the analysis of a two-way classification, *but the distinguishing feature of a factorial experiment is that the treatment sum of squares can be further subdivided into components corresponding to the various factorial effects*. Thus, for a two-factor experiment we have the following subdivision, or breakdown, of the treatment sum of squares:

$$r \sum_{i=1}^{a} \sum_{j=1}^{b} (\bar{y}_{ij.} - \bar{y}_{...})^2 = rb \sum_{i=1}^{a} (\bar{y}_{i..} - \bar{y}_{...})^2 + ra \sum_{j=1}^{b} (\bar{y}_{.j.} - \bar{y}_{...})^2$$

$$+ r \sum_{i=1}^{a} \sum_{j=1}^{b} (\bar{y}_{ij.} - \bar{y}_{i..} - \bar{y}_{.j.} + \bar{y}_{...})^2$$

The first term on the right measures the variability of the means corresponding to the different levels of factor A, and we refer to it as the *factor*

A sum of squares, SSA. Similarly, the second term is the sum of squares for factor *B*, *SSB*, and the third term is the *interaction sum of squares* *SS(AB)*, which measures the variability of the means $\bar{y}_{ij.}$ that is not attributable to the individual (or separate) effects of factors *A* and *B*. The *ab* − 1 degrees of freedom for treatments are, accordingly, subdivided into *a* − 1 degrees of freedom for the effect of factor *A*, *b* − 1 for the effect of factor *B*, and

$$ab - 1 - (a - 1) - (b - 1) = (a - 1)(b - 1)$$

degrees of freedom for interaction.

Example. To illustrate the analysis of a two-factor experiment, let us refer again to the coking experiment described on page 375, and let us suppose that three replicates yielded the following coking times (in hours):

Factor A Oven width	Factor B Flue temp.	Rep. 1	Rep. 2	Rep. 3	Total
4	1600	3.5	3.0	2.7	9.2
4	1900	2.2	2.3	2.4	6.9
8	1600	7.1	6.9	7.5	21.5
8	1900	5.2	4.6	6.8	16.6
12	1600	10.8	10.6	11.0	32.4
12	1900	7.6	7.1	7.3	22.0
	Total	36.4	34.5	37.7	108.6

Following the procedure used in analyzing a two-way classification, we first compute the correction term

$$C = \frac{(108.6)^2}{18} = 655.22$$

Then, the total sum of squares is given by

$$SST = (3.5)^2 + (2.2)^2 + \ldots + (7.3)^2 - 655.22 = 149.38$$

and the treatment and replicate (instead of blocks) sums of squares are given by

$$SS(Tr) = \tfrac{1}{3}[(9.2)^2 + (6.9)^2 + \ldots + (22.0)^2] - 655.22 = 146.05$$

$$SSR = \tfrac{1}{6}[(36.4)^2 + (34.5)^2 + (37.7)^2] - 655.22 = 0.86$$

Finally, by subtraction, we obtain

$$SSE = 149.38 - 146.05 - 0.86 = 2.47$$

Subdivision of the treatment sum of squares into components for factors A and B, and for interaction, can be facilitated by constructing the following kind of two-way table, where the entries are the totals in the right-hand column of the table giving the original data:

| | | Factor B Flue temperature | | |
		1600	1900	
	4	9.2	6.9	16.1
Factor A Oven width	8	21.5	16.6	38.1
	12	32.4	22.0	54.4
		63.1	45.5	108.6

Using formulas analogous to the ones with which we computed the sums of squares for various effects in Chapter 12, we now have for the two *main effects*

$$SSA = \frac{1}{b \cdot r} \sum_{i=1}^{a} T_{i..}^2 - C$$
$$= \tfrac{1}{6}[(16.1)^2 + (38.1)^2 + (54.4)^2] - 655.22 = 123.14$$

$$SSB = \frac{1}{a \cdot r} \sum_{j=1}^{b} T_{.j.}^2 - C$$
$$= \tfrac{1}{9}[(63.1)^2 + (45.5)^2] - 655.22 = 17.21$$

and for the *interaction*

$$SS(AB) = SS(Tr) - SSA - SSB$$
$$= 146.05 - 123.14 - 17.21 = 5.70$$

Finally, dividing the various sums of squares by their degrees of freedom, and dividing the appropriate mean squares by the error mean square, we obtain the results shown in the following analysis of variance table:

Source of variation	Degrees of freedom	Sum of squares	Mean square	F
Replication	2	0.86	0.43	1.72
Main effects:				
A	2	123.14	61.57	246
B	1	17.21	17.21	68.8
Interaction	2	5.70	2.85	11.4
Error	10	2.47	0.25	
Total	17	149.38		

The F test for replications is not significant at the 0.05 or 0.01 levels of significance, but the other three F tests are significant at the 0.01 level of significance. Hence, we reject the null hypothesis that the α_i are all equal to zero, the null hypothesis that the β_j are all equal to zero, and the null hypothesis that the $(\alpha\beta)_{ij}$ are all equal to zero. These results are illustrated in Figure 13.2, showing the trend of mean coking times for changing oven width at each of the flue temperatures. It is apparent from this figure that the increase in coking time for changing oven width is *greater* at the lower flue temperature. In view of this interaction, great care must be exercised in stating the results of this experiment. For instance, it would be very misleading to state merely that the effect of

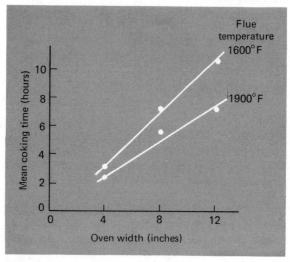

increasing flue temperature from 1,600 to 1,900 degrees Fahrenheit is to lower the coking time by $\frac{63.1}{9} - \frac{45.5}{9} = 1.96$ hours. In fact, the coking time is lowered on the average by as little as 0.77 hours when the oven width is 4 inches, and by as much as 3.47 hours when the oven width is 12 inches.

13.2 MULTIFACTOR EXPERIMENTS

Much industrial research and experimentation is conducted to discover the individual and joint effects of several factors on variables which are most relevant to phenomena under investigation. The experimental designs most often used are of the simple randomized-block or two-way classification type, but the distinguishing feature of most of them is the factorial arrangement of the treatments, or experimental conditions. As we observed in the preceding section, r sets of data pertaining to $a \cdot b$ experimental conditions can be analyzed as a factorial experiment in r replicates if the experimental conditions represent all possible combinations of the levels of two factors A and B. In this section, we shall extend the discussion to factorial experiments involving more than two factors, that is, to experiments where the experimental conditions represent all possible combinations of the levels of three or more factors.

Example. To illustrate the analysis of a multifactor experiment, let us consider the following situation. A warm sulfuric pickling bath is used to remove oxides from the surface of a metal prior to plating, and it is desired to determine what factors in addition to the concentration of the sulfuric acid might affect the electrical conductivity of the bath. As it is felt that the salt concentration as well as the bath temperature might also affect the electrical conductivity, an experiment is planned to determine the individual and joint effects of these three variables on the electrical conductivity of the bath. In order to cover the ranges of concentrations and temperatures normally encountered, it is decided to use the following levels of the three factors:

Factor	Level 1	Level 2	Level 3	Level 4
A. Acid concentration (percent)	0	6	12	18
B. Salt concentration (percent)	0	10	20	
C. Bath temperature (degrees F)	80	100		

The resulting factorial experiment requires $4 \cdot 3 \cdot 2 = 24$ experimental conditions in each replicate, where each experimental condition is a pickling bath made up according to specifications. The order in which these pickling baths are made up should be random. Let us suppose that two replicates of the experiment have actually been completed, that is, the electrical conductivities of the various pickling baths have been measured, and that the results are as shown in the following table:

RESULTS OF ACID-BATH EXPERIMENT

Level of factor			Conductivity ($mhos/cm^3$)		
A	B	C	Rep. 1	Rep. 2	Total
1	1	1	0.99	0.93	1.92
1	1	2	1.15	0.99	2.14
1	2	1	0.97	0.91	1.88
1	2	2	0.87	0.86	1.73
1	3	1	0.95	0.86	1.81
1	3	2	0.91	0.85	1.76
2	1	1	1.00	1.17	2.17
2	1	2	1.12	1.13	2.25
2	2	1	0.99	1.04	2.03
2	2	2	0.96	0.98	1.94
2	3	1	0.97	0.95	1.92
2	3	2	0.94	0.99	1.93
3	1	1	1.24	1.22	2.46
3	1	2	1.12	1.15	2.27
3	2	1	1.15	0.95	2.10
3	2	2	1.11	0.95	2.06
3	3	1	1.03	1.01	2.04
3	3	2	1.12	0.96	2.08
4	1	1	1.24	1.20	2.44
4	1	2	1.32	1.24	2.56
4	2	1	1.14	1.10	2.24
4	2	2	1.20	1.19	2.39
4	3	1	1.02	1.01	2.03
4	3	2	1.02	1.00	2.02
		Total	25.53	24.64	50.17

The model we shall assume for the analysis of this experiment (or any similar three-factor experiment) is an immediate extension of the one used in Section 13.1. If y_{ijkl} is the conductivity measurement obtained at the ith level of acid concentration, the jth level of salt concentration, the kth level of bath temperature, in the lth replicate, we write

$$y_{ijkl} = \mu + \alpha_i + \beta_j + \gamma_k + (\alpha\beta)_{ij} + (\alpha\gamma)_{ik} + (\beta\gamma)_{jk}$$
$$+ (\alpha\beta\gamma)_{ijk} + \rho_l + \epsilon_{ijkl}$$

for $i = 1, 2, \ldots, a, j = 1, 2, \ldots, b, k = 1, 2, \ldots, c$, and $l = 1, 2, \ldots, r$. We also assume that the sums of the main effects (α's, β's, and γ's) as well as the sum of the replication effects are equal to zero, that the sums of the two-way interaction effects summed on either subscript equal zero for any value of the other subscript, and that the sum of the three-way interaction effects summed on any one of the subscripts is zero for any values of the other two subscripts. As before, the ϵ_{ijkl} are assumed to be values of independent random variables having zero means and the common variance σ^2.

Continuation of example. We begin the analysis of the data by treating the experiment as a two-way classification with $a \cdot b \cdot c$ treatments and r replicates (blocks), and using the short-cut formulas on page 348, we obtain

$$C = \frac{(50.17)^2}{48} = 52.4381$$

$$SST = (0.99)^2 + (1.15)^2 + \ldots + (1.00)^2 - 52.4381 = 0.6624$$

$$SS(Tr) = \tfrac{1}{2}[(1.92)^2 + (2.14)^2 + \ldots + (2.02)^2] - 52.4381 = 0.5712$$

$$SSR = \tfrac{1}{24}[(25.53)^2 + (24.64)^2] - 52.4381 = 0.0165$$

$$SSE = 0.6624 - 0.5712 - 0.0165 = 0.0747$$

The degrees of freedom for these sums of squares are, respectively, 47, 23, 1, and 23.

Next, we shall want to subdivide the treatment sum of squares into the three *main effect sums of squares SSA, SSB, SSC*, the three *two-way interaction sums of squares, SS(AB), SS(AC)*, and *SS(BC)*, and the *three-way interaction sum of squares SS(ABC)*. To facilitate the calculation of these sums of squares we first construct the following three tables analogous to the one on page 381:

		B						**C**		
		1	*2*	*3*				*1*	*2*	
	1	4.06	3.61	3.57	11.24		*1*	5.61	5.63	11.24
	2	4.42	3.97	3.85	12.24		*2*	6.12	6.12	12.24
A	*3*	4.73	4.16	4.12	13.01	*A*	*3*	6.60	6.41	13.01
	4	5.00	4.63	4.05	13.68		*4*	6.71	6.97	13.68
		18.21	16.37	15.59	50.17			25.04	25.13	50.17

$$B$$

		1	2	3	
	1	8.99	8.25	7.80	25.04
C					
	2	9.22	8.12	7.79	25.13
		18.21	16.37	15.59	50.17

The entries of these tables are the totals of all measurements obtained at the respective levels of the two variables. Note the self-checking feature of these tables; the same marginal totals appear several times, thus providing a rapid and effective check on the calculations.

To calculate SSA, SSB, and $SS(AB)$ we refer to the first of the above tables and an identity analogous to the one on page 379. As a matter of fact, the calculations parallel those for calculating SSA, SSB, and $SS(AB)$ in the two-factor experiment. To take the place of the treatment sum of squares we first calculate

$$rc \sum_{i=1}^{a} \sum_{j=1}^{b} (\bar{y}_{ij..} - \bar{y}_{....})^2 = \frac{1}{r \cdot c} \sum_{i=1}^{a} \sum_{j=1}^{b} T_{ij..}^2 - C$$
$$= \tfrac{1}{4}[(4.06)^2 + (4.42)^2 + \ldots + (4.05)^2]$$
$$- 52.4381$$
$$= 0.5301$$

and we then obtain

$$SSA = \frac{1}{bcr} \sum_{i=1}^{a} T_{i...}^2 - C$$
$$= \tfrac{1}{12}[(11.24)^2 + \ldots + (13.68)^2 - 52.4381$$
$$= 0.2750$$
$$SSB = \frac{1}{acr} \sum_{j=1}^{b} T_{.j..}^2 - C$$
$$= \tfrac{1}{16}[(18.21)^2 + (16.37)^2 + (15.59)^2] - 52.4381$$
$$= 0.2262$$

and

$$SS(AB) = 0.5301 - 0.2750 - 0.2262$$
$$= 0.0289$$

Performing the same calculations for the second of the tables on page 385 we obtain, similarly,

$$SSC = 0.0002 \quad \text{and} \quad SS(AC) = 0.0085$$

and the analysis of the third table yields

$$SS(BC) = 0.0042$$

For the three-way interaction sum of squares we finally obtain by subtraction

$$
\begin{aligned}
SS(ABC) &= SS(Tr) - SSA - SSB - SSC - SS(AB) - SS(AC) \\
&\quad - SS(BC) \\
&= 0.5712 - 0.2750 - 0.2262 - 0.0002 - 0.0289 - 0.0085 \\
&\quad - 0.0042 \\
&= 0.0282
\end{aligned}
$$

Note that the degrees of freedom for each main effect is one less than the number of levels of the corresponding factor. The degrees of freedom for each interaction is the *product* of the degrees of freedom for those factors appearing in the interaction. Thus, the degrees of freedom for the three main effects are 3, 2, and 1 in this example, while the degrees of freedom for the two-way interactions are 6, 3, and 2, and the degrees of freedom for the three-way interaction is 6.

The following table shows the complete analysis of variance for the acid-bath experiment:

Source of variation	Degrees of freedom	Sum of squares	Mean square	F
Replicates	1	0.0165	0.0165	5.16
Main effects:				
A	3	0.2750	0.0917	28.66
B	2	0.2262	0.1131	35.34
C	1	0.0002	0.0002	< 1
Two-factor interactions:				
AB	6	0.0289	0.0048	1.50
AC	3	0.0085	0.0028	< 1
BC	2	0.0042	0.0021	< 1
Three-factor interaction:				
ABC	6	0.0282	0.0047	1.47
Error	23	0.0747	0.0032	
Total	47	0.6624		

Obtaining the appropriate values of $F_{.05}$ and $F_{.01}$ from Table 6, we find that the test for replicates is significant at the 0.05 level (perhaps the two replicates were performed under different atmospheric conditions or the thermometer used to measure bath temperatures went out of calibration, etc.), the tests for the factor A and factor B main effects are significant at the 0.01 level, while none of the other F's are significant at either level. We conclude from this analysis that variations in acid concentration and salt concentration affect the electrical conductivity, variations in bath temperature do not, and that there are no interactions. To go one step further, we might investigate the *magnitudes* of the effects by studying graphs of means like those shown in Figures 13.3 and 13.4. Here we find that the conductivity increases as

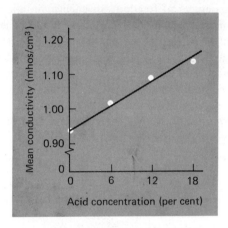

Figure 13.3. Effect of acid concentration.

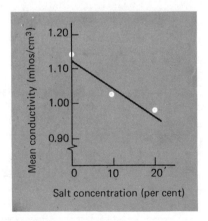

Figure 13.4. Effect of salt concentration.

acid is added and decreases as salt is added; using the methods of Chapter 11 we might even fit lines, curves, or surfaces to describe the response surface relating conductivity to the variables under consideration.

The general computational procedure for a multifactor experiment is similar to the method illustrated here for a $4 \times 3 \times 2$ factorial experiment. We first analyze the data as a two-way classification (or whatever other design is being used) and then we analyze the treatment sum of squares into the components attributed to the various main effects and interactions. In general, the sum of squares for each main effect is found by adding the squares of the totals corresponding to the different levels of that factor, dividing by the number of observations comprising each of these totals, and then subtracting the correction term. The sum of squares for any interaction is found by adding the squares of all totals obtained by summing over those subscripts pertaining to factors not involved in the interaction, dividing by the number of observations comprising each of these totals, and then subtracting the correction term *and* all sums of squares corresponding to main effects and fewer-factor interactions involving the factors contained in that interaction.

EXERCISES

1. To determine optimum conditions for a plating bath, the effects of sulfone concentration and bath temperature on the reflectivity of the plated metal are studied in a 2×5 factorial experiment. The results of three replicates are as follows:

Concentration (grams/liter)	Temperature (degrees F)	Rep. 1	Reflectivity Rep. 2	Rep. 3
5	75	35	39	36
5	100	31	37	36
5	125	30	31	33
5	150	28	20	23
5	175	19	18	22
10	75	38	46	41
10	100	36	44	39
10	125	39	32	38
10	150	35	47	40
10	175	30	38	31

Analyze these results and determine the bath condition or conditions that produce the highest reflectivity. Also construct a 0.95 confidence interval for the reflectivity of the plating bath corresponding to these optimum conditions.

2. A spoilage retarding ingredient is added in brewing beer. To determine the extent to which the taste of the beer is affected by the amount of this ingredient added to each batch, and how such taste changes might depend on the age of the beer, a 3 × 4 factorial experiment in two replications was designed. The taste of the beer was rated on a scale of 0, 1, 2, or 3 (3 being the most desirable) by a panel of trained experts, who reported the following mean ratings:

Amount of ingredient (grams per batch)	Aging period (weeks)	Mean ratings Rep. 1	Rep. 2
2	2	2.1	1.6
2	4	2.6	1.9
2	6	2.9	2.4
3	2	1.4	1.7
3	4	1.9	2.2
3	6	2.3	2.7
4	2	0.5	0.9
4	4	1.2	0.8
4	6	1.7	1.4
5	2	1.0	1.6
5	4	2.2	1.3
5	6	2.3	2.1

(a) Analyze the data first as a two-way classification with 12 treatments and 2 blocks (replicates).

(b) Calculate the sums of squares corresponding to the main effects and interaction, and present the results in an analysis of variance table.

(c) Interpret the results of the experiment.

3. Suppose that in the experiment described on page 349 it is desired to determine also whether there is an interaction between the detergents and the washing machines, that is, whether one detergent might perform better in Machine 1, another might perform better in Machine 2, and so forth. Combining the data on page 349 with the replicate of the experiment given below, test for a significant interaction and discuss the results.

	Machine 1	Machine 2	Machine 3
Detergent A	39	42	58
Detergent B	44	46	48
Detergent C	34	47	45
Detergent D	47	45	57

4. Referring to Exercise 4 on page 355, it can be argued that the rocket-launcher experiment was poorly designed, as it was not replicated, and, thus, it is impossible to test whether there is an interaction between launchers and rocket fuels. Suppose a second replicate were performed, with the following results:

	Fuel I	Fuel II	Fuel III	Fuel IV
Launcher X	66.8	51.1	40.1	49.2
Launcher Y	47.7	38.2	50.6	64.3

Combining the results of both replicates, perform an appropriate analysis of variance, and test for the presence of an interaction.

5. The following tables give the weights (in grams) of food ingested by two different strains of rats after having been deprived of food for the stated number of hours and then having been given the stated dosage of a certain drug:

		Replicate 1		Replicate 2	
		Strain A	Strain B	Strain A	Strain B
Dosage 0.2 mg/kg	1 hour	8.77	7.59	9.07	6.02
	4 hours	11.82	9.21	9.16	7.05
	8 hours	14.65	15.35	16.08	12.01
Dosage 0.4 mg/kg	1 hour	8.76	6.13	5.63	5.87
	4 hours	11.53	8.30	11.57	9.56
	8 hours	14.46	9.26	10.30	10.13
Dosage 0.8 mg/kg	1 hour	3.01	3.81	4.42	4.35
	4 hours	9.21	10.10	5.22	8.01
	8 hours	6.10	11.16	7.27	8.17

Perform an appropriate analysis of variance and interpret the results.

6. A study was conducted to measure the effect of three different meat tenderizers on the weight loss of steaks having the same initial (precooked) weights. The effects of cooking temperatures and cooking times also were measured by performing a 3 × 2 × 2 factorial experiment in 3 replicates. The results are as follows:

Tenderizer	Cooking time (minutes)	Cooking temperature (°F)	Weight loss (ounces) Rep. 1	Rep. 2	Rep. 3
A	20	350	1.5	1.3	1.4
A	20	400	1.6	1.4	1.5
A	30	350	1.7	1.8	1.7
A	30	400	1.8	1.9	2.0
B	20	350	1.9	2.1	2.0
B	20	400	2.2	2.4	2.5
B	30	350	2.6	2.3	2.4
B	30	400	2.6	2.7	2.5
C	20	350	0.9	0.8	0.8
C	20	400	1.1	1.0	0.9
C	30	350	0.8	0.9	1.0
C	30	400	1.2	1.0	1.1

(a) Analyze this experiment first as a two-way classification with 12 treatments and 3 replicates (blocks).

(b) Complete the analysis by computing the sums of squares corresponding to the various main effects and interactions.

(c) Present the results in an analysis of variance table and interpret the experiment.

7. To study the effects of ingot location (A), slab position (B), specimen preparation (C), and twisting temperature (D) on the number of turns required to break a steel specimen by twisting, the following observations were recorded:

A	B	C	D	No. of turns Rep. 1	Rep. 2
top	1	turn	2,100°F	24	22
top	1	turn	2,200	25	28
top	1	turn	2,300	41	39
top	1	grind	2,100	18	18
top	1	grind	2,200	33	27
top	1	grind	2,300	35	41
top	2	turn	2,100	22	19
top	2	turn	2,200	26	31
top	2	turn	2,300	37	43
top	2	grind	2,100	23	7
top	2	grind	2,200	30	26
top	2	grind	2,300	34	30
mid	1	turn	2,100	26	19
mid	1	turn	2,200	30	31
mid	1	turn	2,300	39	42
mid	1	grind	2,100	19	19
mid	1	grind	2,200	31	31
mid	1	grind	2,300	26	35
mid	2	turn	2,100	30	26
mid	2	turn	2,200	31	34

A	B	C	D	No. of turns Rep. 1	Rep. 2
mid	2	turn	2,300	39	42
mid	2	grind	2,100	22	20
mid	2	grind	2,200	32	26
mid	2	grind	2,300	38	22
bot	1	turn	2,100	18	21
bot	1	turn	2,200	35	32
bot	1	turn	2,300	34	37
bot	1	grind	2,100	21	19
bot	1	grind	2,200	20	29
bot	1	grind	2,300	44	31
bot	2	turn	2,100	23	22
bot	2	turn	2,200	31	26
bot	2	turn	2,300	38	41
bot	2	grind	2,100	18	19
bot	2	grind	2,200	31	24
bot	2	grind	2,300	35	41

Analyze this experiment.

8. Solve the four equations on page 377 for μ, α_1, β_1, and $(\alpha\beta)_{11}$ in terms of the population means μ_{ij1} corresponding to the four experimental conditions in the first replicate. Note that these equations serve as a guide for estimating the parameters in terms of the *sample means* corresponding to the various experimental conditions.

13.3 2^n FACTORIAL EXPERIMENTS

There are several reasons why factorial experiments are often performed with each factor taken at only two levels. Primarily, the number of experimental conditions in a factorial experiment increases multiplicatively with the number of levels of each factor; thus, if many factors are to be investigated simultaneously, it may be economically impossible to include more than two levels of each factor. Another important reason for treating 2^n factorial experiments separately is that there exist computational short-cuts which apply only to this case. In fact, the remainder of this section will be devoted to these computational short-cuts, while other advantages, such as the ease of confounding higher-order interactions and the adaptability of 2^n factorials to experiments involving fractional replication, will be discussed in succeeding sections.

Before we introduce some of the special notation used in connection with 2^n factorial experiments, let us point out that such experiments do have some drawbacks. Since each factor is measured only at two levels, it is impossible to judge whether the effects produced by variations in a factor are linear or, perhaps, parabolic or exponential. For this reason 2^n factorial experiments are often used in "screening experiments," which are followed up by experiments involving fewer factors (ordinarily those found to be "significant" individually or jointly in the screening experiment) taken at more than two levels.

In the analysis of a 2^n factorial experiment it is convenient to denote the two levels of each factor by 0 and 1 (instead of 1 and 2). Thus, the models used for the analysis of this kind of experiment differ from those of Section 13.2 only inasmuch as we now have $i = 0, 1$ instead of $i = 1, 2, \ldots, a$, $j = 0, 1$ instead of $j = 1, 2, \ldots, b$, and so forth. For instance, for a 2^3 factorial experiment the model on page 384 becomes

$$y_{ijkl} = \mu + \alpha_i + \beta_j + \gamma_k + (\alpha\beta)_{ij} + (\alpha\gamma)_{ik} + (\beta\gamma)_{jk}$$
$$+ (\alpha\beta\gamma)_{ijk} + \rho_l + \epsilon_{ijkl}$$

for $i = 0, 1, j = 0, 1, k = 0, 1$, and $l = 1, 2, \ldots, r$. The ϵ_{ijkl} are defined as before, and the parameters are now subject to the restrictions $\alpha_1 = -\alpha_0$, $\beta_1 = -\beta_0, \gamma_1 = -\gamma_0, (\alpha\beta)_{10} = (\alpha\beta)_{01} = -(\alpha\beta)_{11} = -(\alpha\beta)_{00}, \ldots$, and $\sum_{l=1}^{r} \rho_l = 0$. Note that besides the parameters for replicates we need only *one parameter of each kind:* that is, besides the parameters for replicates, we can express the entire model in terms of the parameters $\mu, \alpha_0, \beta_0, \gamma_0, (\alpha\beta)_{00}$, $(\alpha\gamma)_{00}, (\beta\gamma)_{00}$, and $(\alpha\beta\gamma)_{000}$.

A 2^n factorial experiment requires 2^n experimental conditions; since their number can be fairly large, it will be convenient to represent the experimental conditions by means of a special notation and list them in a so-called standard order. The notation consists of representing each experimental condition by the product of lower-case letters corresponding to the factors which are taken at level 1, called the "higher level." If a lower-case letter corresponding to a factor is missing, this means that the factor is taken at level 0, called the "lower level." Thus, in a three-factor experiment, ac represents the experimental condition where factors A and C are taken at the higher level and factor B is taken at the lower level, c represents the experimental condition where factor C is taken at the higher level and factors A and B are taken at the lower level, and so forth. The symbol "1" is used to denote the experimental condition in which all factors are taken at the lower level.

Although the experimental conditions are applied in a random order during the experiment itself, for the purpose of analyzing the results it is convenient to arrange them in a so-called *standard order*. For $n = 2$, this order is 1, a, b, ab, and for $n = 3$ it is the order shown in the following table:

Experimental condition	Level of factor A	B	C
1	0	0	0
a	1	0	0
b	0	1	0
ab	1	1	0
c	0	0	1
ac	1	0	1
bc	0	1	1
abc	1	1	1

Note that the symbols for the first four experimental conditions are like those for a two-factor experiment, and that the second four are obtained by multiplying each of the first four symbols by c. Similarly, the arrangement for $n = 4$ on page 398 is obtained by first listing the eight symbols for a three-factor experiment and then repeating the set with each symbol multiplied by d.

Throughout this and the preceding chapter we referred to the total of all observations corresponding to a given experimental condition as a treatment total, and we represented these totals by means of symbols such as $T_{i.}$, $T_{ij..}$, and so forth. Having introduced a special notation for the experimental conditions in a 2^n factorial experiment, we extend this notation by letting (1), (a), (b), (ab), (c), ..., be the treatment totals corresponding to experimental conditions 1, $a, b, ab, c, \ldots$. Thus, in a three-factor experiment

$$(1) = \sum_{l=1}^{r} y_{000l} \qquad (a) = \sum_{l=1}^{r} y_{100l}$$

$$\cdots\cdots\cdots\cdots\cdots\cdots\cdots\cdots\cdots\cdots\cdots$$

$$(bc) = \sum_{l=1}^{r} y_{011l} \qquad (abc) = \sum_{l=1}^{r} y_{111l}$$

The computational short-cuts referred to on page 393 consist, essentially, of expressing estimates of the various main effects and interactions as well as the corresponding sums of squares, in terms of *linear combinations* of the treatment totals. To illustrate what we mean, let us consider the quantity

$$-(1) + (a) - (b) + (ab) - (c) + (ac) - (bc) + (abc)$$

which is a linear combination, with coefficients $+1$ and -1, of the treatment totals corresponding to the eight experimental conditions. Referring to the model equation on page 394 and making use of the relationships among the parameters (but leaving all details to the reader in Exercises 6, 7, and 8 on pages 404 and 405) it can be shown that

$$-(1) + (a) - (b) + (ab) - (c) + (ac) - (bc) + (abc) = -8r\alpha_0 + \epsilon_A$$

where ϵ_A is a corresponding linear combination of sums of the ϵ_{ijkl}. From Theorem 7.1 on page 215 it follows that ϵ_A is a value of a random variable whose distribution has zero mean; as a matter of fact, it can be shown that ϵ_A is a value of a random variable having a normal distribution with zero mean and the variance $8r\sigma^2$. Referring to the above linear combination as the *effect total* $[A]$ for factor A, we find that $-[A]/8r$ provides an estimate fo α_0, the main effect for factor A, and it can be shown that $[A]^2/8r$ actually equals SSA, the sum of squares for the main effect of factor A.

Similarly analyzing the linear combination

$$(1) + (a) - (b) - (ab) - (c) - (ac) + (bc) + (abc)$$

the reader will be asked to show in Exercise 8 on page 405 that it equals $8r(\beta\gamma)_{00} + \epsilon_{BC}$, where ϵ_{BC} is a corresponding linear combination of sums of the ϵ_{ijkl}. Referring to this linear combination of treatment totals as the *effect total* $[BC]$ for the two-way interaction of factors B and C, we find that $[BC]/8r$ provides an estimate of $(\beta\gamma)_{00}$, the effect of the BC interaction, and it can also be shown that $[BC]^2/8r$ equals $SS(BC)$, the sum of squares for the BC interaction. Proceeding in this fashion, we can present linear combinations of the treatment totals which yield estimates of the various other main effects and interactions, and whose squares, divided by $8r$, yield the corresponding sums of squares. These linear combinations, or *effect totals*, can easily be obtained with the use of the following *table of signs:*

(I)	(a)	(b)	(ab)	(c)	(ac)	(bc)	(abc)	Effect totals
1	1	1	1	1	1	1	1	$[I]$
−1	1	−1	1	−1	1	−1	1	$[A]$
−1	−1	1	1	−1	−1	1	1	$[B]$
1	−1	−1	1	1	−1	−1	1	$[AB]$
−1	−1	−1	−1	1	1	1	1	$[C]$
1	−1	1	−1	−1	1	−1	1	$[AC]$
1	1	−1	−1	−1	−1	1	1	$[BC]$
−1	1	1	−1	1	−1	−1	1	$[ABC]$

The entries of this table are the coefficients of the linear combinations of the treatment totals for the various main effects and interactions. As an aid for constructing similar tables for $n = 4$, $n = 5$, etc., note that for each main effect there is a "$+1$" when the factor is at the higher level and a "-1" when the factor is at the lower level. The signs for an interaction effect are obtained by multiplying the corresponding coefficients of *all* factors contained in the interaction. Thus, for $[AB]$ we multiply each sign for $[A]$ by the corresponding sign for $[B]$, getting

$$(-1)(-1) \quad (1)(-1) \quad (-1)(1) \quad (1)(1) \quad (-1)(-1) \quad (1)(-1) \quad (-1)(1) \quad (1)(1)$$

or

$$1 \quad -1 \quad -1 \quad 1 \quad 1 \quad -1 \quad -1 \quad 1$$

Note also that in the above table $[I]$ stands for the grand total of all the observations, so that $[I]^2/8r$ gives the correction term for calculating SST, SSE, SSR, and $SS(Tr)$.

Although we have illustrated the above short-cut method for obtaining the various main-effect and interaction sums of squares with reference to a 2^3 factorial experiment, the only difference in a 2^n factorial experiment with $n > 3$ is that we require a more extensive table of signs and that the respective sums of squares are obtained by dividing the squares of the effect totals by $r \cdot 2^n$.

Example. To illustrate this technique and introduce a further simplification, let us consider the following 2^4 factorial experiment, designed to determine the effects of certain variables on the reliability of a rotary stepping switch. The factors studied were as follows:

Factor	Low level	High level
A. Lubrication	dry	lubricated
B. Dust protection	unprotected	enclosed in dust cover
C. Spark suppression	no	yes
D. Current	0	0.5 amp

Each switch was operated continuously until a malfunction occurred, and the number of hours of operation was recorded. The whole experiment was performed twice, with the following results:

Experimental condition	Hours of operation Rep. 1	Rep. 2	Total
1	828	797	1,625
a	997	948	1,945
b	735	776	1,511
ab	807	1,003	1,810
c	994	949	1,943
ac	1,069	1,094	2,163
bc	989	1,215	2,204
abc	889	1,010	1,899
d	593	813	1,406
ad	773	1,026	1,799
bd	740	922	1,662
abd	936	1,138	2,074
cd	748	970	1,718
acd	1,202	1,182	2,384
bcd	1,103	966	2,069
abcd	985	1,154	2,139
Totals	14,388	15,963	30,351

Analyzing these data first as a two-way classification with 16 treatments and 2 replications (blocks), we obtain

$$C = \frac{(30,351)^2}{32} = 28,786,975$$

$$SST = (828)^2 + (997)^2 + \ldots + (1,154)^2 - 28,786,975$$
$$= 744,876$$

$$SS(Tr) = \tfrac{1}{2}[(1,625)^2 + (1,945)^2 + \ldots + (2,139)^2] - 28,786,975$$
$$= 547,288$$

$$SSR = \tfrac{1}{16}[(14,388)^2 + (15,963)^2] - 28,786,975$$
$$= 77,520$$

$$SSE = 744,876 - 547,288 - 77,520 = 120,068$$

In order to subdivide the treatment sum of squares into SSA, SSB, . . . , and $SS(ABCD)$, we could construct a table of signs like the one on page 396, calculate the effect totals, and then divide the squares of the effect totals by $r \cdot 2^n = 2 \cdot 2^4 = 32$. For the A factor main effect we would thus obtain

$$[A] = -1,625 + 1,945 - 1,511 + 1,810 - 1,943 + 2,163 - 2,204$$
$$+ 1,899 - 1,406 + 1,799 - 1,662 + 2,074 - 1,718 + 2,384$$
$$- 2,069 + 2,139$$
$$= 2,075$$

and

$$SSA = \frac{(2,075)^2}{32} = 134,551$$

These calculations are quite tedious, but they can be simplified considerably by using a further short-cut, called the *method of Yates*. This method of calculating the effect totals is illustrated in the table shown below. The experimental conditions and the corresponding totals are listed in *standard order*. In the column marked (1), the upper half is obtained by adding successive pairs of treatment totals, and the lower half is obtained by subtracting successive pairs. Thus, in column (1) we obtained

$$1,625 + 1,945 = 3,570$$
$$1,511 + 1,810 = 3,321$$
$$\cdots\cdots\cdots\cdots\cdots$$
$$2,069 + 2,139 = 4,208$$
$$\overline{1,945 - 1,625 = 320}$$
$$1,810 - 1,511 = 299$$
$$\cdots\cdots\cdots\cdots\cdots$$
$$2,139 - 2,069 = 70$$

Experimental condition	Treatment totals	(1)	(2)	(3)	(4)	Identification	Sum of squares
1	1,625	3,570	6,891	15,100	30,351	[I]	28,786,975
a	1,945	3,321	8,209	15,251	2,075	[A]	134,551
b	1,511	4,106	6,941	534	385	[B]	4,632
ab	1,810	4,103	8,310	1,541	−1,123	[AB]	39,410
c	1,943	3,205	619	−252	2,687	[C]	225,624
ac	2,163	3,736	−85	637	−773	[AC]	18,673
bc	2,204	4,102	805	−546	−179	[BC]	1,001
abc	1,899	4,208	736	−577	−1,119	[ABC]	39,130
d	1,406	320	−249	1,318	151	[D]	713
ad	1,799	299	−3	1,369	1,007	[AD]	31,689
bd	1,662	220	531	−704	889	[BD]	24,698
abd	2,074	−305	106	−69	−31	[ABD]	30
cd	1,718	393	−21	246	51	[CD]	81
acd	2,384	412	−525	−425	635	[ACD]	12,601
bcd	2,069	666	19	−504	−671	[BCD]	14,070
abcd	2,139	70	−596	−615	−111	[ABCD]	385

Note that the first total in each pair is subtracted from the second.

Column (2) is then obtained by performing the identical operations on the entries of column (1), and columns (3) and (4) are obtained in the same manner from the entries in columns (2) and (3), respectively. Column (4), and in general column (n), gives the effect totals in standard order, as shown. Each sum of squares is then obtained as before, by squaring the corresponding effect total and then dividing the result by $r \cdot 2^n = 2 \cdot 2^4 = 32$.

Dividing the sums of squares by their degrees of freedom to obtain the mean squares, and dividing the various mean squares by the error mean square, we get the following analysis of variance table for the 2^4 factorial experiment:

Source of variation	Degrees of freedom	Sum of squares	Mean square	F
Replicates	1	77,520	77,520	9.68
Main effects:				
A	1	134,551	134,551	16.81
B	1	4,632	4,632	< 1
C	1	225,624	225,624	28.19
D	1	713	713	< 1
Two-factor interactions:				
AB	1	39,410	39,410	4.92
AC	1	18,673	18,673	2.33
AD	1	31,689	31,689	3.96
BC	1	1,001	1,001	< 1
BD	1	24,698	24,698	3.09
CD	1	81	81	< 1
Three-factor interactions:				
ABC	1	39,130	39,130	4.89
ABD	1	30	30	< 1
ACD	1	12,601	12,601	1.57
BCD	1	14,070	14,070	1.76
Four-factor interactions:				
ABCD	1	385	385	< 1
Error	15	120,068	8,005	
Total	31	744,876		

Since $F_{.05} = 4.54$ and $F_{.01} = 8.68$ for 1 and 15 degrees of freedom, we find that the replication effects as well as the effects of lubrication and spark suppression are significant at the 0.01 level, and that there are significant interactions at the 0.05 level between lubrication, dust protection, and spark suppression. The reader will be asked to interpret these results and estimate the magnitude of some of the effects in Exercise 9 on page 405.

EXERCISES

1. To determine the effect on taste of three different factors in manufacturing soft-drink cans, an experiment was performed where the taste of one soft drink was rated by a judge on a scale from 1 to 10. The results are as follows:

A Lubricant	*B* Heat	*C* Resin	Ratings Rep. 1	Ratings Rep. 2
fresh	unheated	*A*	6	8
fresh	unheated	*B*	8	7
fresh	heated	*A*	9	9
fresh	heated	*B*	1	2
aged	unheated	*A*	6	7
aged	unheated	*B*	6	8
aged	heated	*A*	9	8
aged	heated	*B*	2	3

(a) Analyze the results first as a two-way classification with 7 degrees of freedom for treatments and 1 degree of freedom for blocks (replicates).

(b) Use an appropriate table of signs to calculate the effect totals [*A*], [*B*], [*C*], [*AB*], [*AC*], [*BC*], [*ABC*].

(c) Using the results obtained in (b) find the sums of squares corresponding to the main effects and interactions, and check their total against the treatment sum of squares obtained in (a).

(d) Arrange the data with treatment combinations in standard order, and use the Yates method to find the effect totals. Compare with the results obtained in (b).

(e) Construct an analysis of variance table and analyze the experiment.

2. An experiment was conducted to determine the effects of certain alloying elements on the ductility of a metal, and the following results were obtained:

Nickel	Carbon	Manganese	Breaking strength (ft-lb) Rep. 1	Rep. 2	Rep. 3
0.0%	0.3%	0.5%	36.7	39.6	38.2
0.0	0.3	1.0	47.5	43.5	45.9
0.0	0.6	0.5	40.6	36.8	36.0
0.0	0.6	1.0	41.1	45.8	46.4
4.0	0.3	0.5	37.8	32.7	31.6
4.0	0.3	1.0	34.2	37.2	36.5
4.0	0.6	0.5	39.5	41.7	39.1
4.0	0.6	1.0	46.4	43.7	49.4

Perform an appropriate analysis of variance and interpret the results.

3. A screening experiment was conducted to determine what factors are influential in controlling the final phosphorus content of steel produced in a converter. The levels of the factors studied and the experimental results are contained in the following table:

E Pouring temp. (°F)	D Lime ratio	C Oxygen	B Original phosphorus	A Original manganese	Final phosphorus Rep. 1	Rep. 2
2,400	3	5%	0.15%	1%	0.003%	0.001%
2,400	3	5	0.15	3	0.004	0.009
2,400	3	5	0.30	1	0.002	0.008
2,400	3	5	0.30	3	0.015	0.007
2,400	3	15	0.15	1	0.002	0.005
2,400	3	15	0.15	3	0.011	0.006
2,400	3	15	0.30	1	0.004	0.001
2,400	3	15	0.30	3	0.002	0.004
2,400	4	5	0.15	1	0.000	0.003
2,400	4	5	0.15	3	0.008	0.002
2,400	4	5	0.30	1	0.003	0.007
2,400	4	5	0.30	3	0.005	0.012
2,400	4	15	0.15	1	0.010	0.006
2,400	4	15	0.15	3	0.006	0.001
2,400	4	15	0.30	1	0.006	0.014
2,400	4	15	0.30	3	0.011	0.015
2,600	3	5	0.15	1	0.003	0.007
2,600	3	5	0.15	3	0.007	0.004
2,600	3	5	0.30	1	0.011	0.005
2,600	3	5	0.30	3	0.010	0.017
2,600	3	15	0.15	1	0.004	0.008
2,600	3	15	0.15	3	0.019	0.013
2,600	3	15	0.30	1	0.004	0.008
2,600	3	15	0.30	3	0.017	0.023
2,600	4	5	0.15	1	0.007	0.004
2,600	4	5	0.15	3	0.015	0.009
2,600	4	5	0.30	1	0.004	0.011

E Pouring temp. (°F)	D Lime ratio	C Oxygen	B Original phosphorus	A Original manganese	Final phosphorus Rep. 1	Rep. 2
2,600	4	5	0.30	3	0.010	0.006
2,600	4	15	0.15	1	0.017	0.011
2,600	4	15	0.15	3	0.005	0.010
2,600	4	15	0.30	1	0.014	0.009
2,600	4	15	0.30	3	0.016	0.011

Analyze the results of this experiment.

4. An experiment was conducted to determine the effects of the following factors on the gain of a semiconductor device:

Factor	Level 0	Level 1
A. Location of assembly	Laboratory	Production line
B. Partial pressure of controlling material	10^{-15}	10^{-4}
C. Relative humidity	1%	30%
D. Aging time	72 hours	144 hours

The results were as follows:

Experimental condition	Gain Rep. 1	Rep. 2
1	39.0	43.2
a	31.8	43.7
b	47.0	51.4
ab	40.9	40.3
c	43.8	40.5
ac	29.3	52.9
bc	34.8	48.2
abc	45.6	58.2
d	40.1	41.9
ad	42.0	40.5
bd	54.9	53.0
abd	39.9	40.2
cd	43.1	40.2
acd	30.1	39.9
bcd	35.6	53.7
abcd	41.4	49.5

Perform an appropriate analysis of variance and interpret the results.

5. A study was performed to examine the effect of five different factors on the time required for odorants in natural gas to reach the surface once a leak

has occurred. The results of the experiment were reported as follows:

A Odorant	B Odorant concentration mg/ml	C Soil moisture	D Flow rate ml/min	E Tempera- ture	Time (hours) Rep. 1	Rep. 2
A	4	8%	20	50°F	20	20
A	4	8	20	70	17	18
A	4	8	80	50	20	19
A	4	8	80	70	18	19
A	4	16	20	50	12	13
A	4	16	20	70	14	16
A	4	16	80	50	15	14
A	4	16	80	70	17	16
A	16	8	20	50	10	9
A	16	8	20	70	9	8
A	16	8	80	50	8	9
A	16	8	80	70	7	8
A	16	16	20	50	7	6
A	16	16	20	70	6	4
A	16	16	80	50	5	6
A	16	16	80	70	4	5
B	4	8	20	50	28	29
B	4	8	20	70	27	26
B	4	8	80	50	25	27
B	4	8	80	70	28	28
B	4	16	20	50	24	25
B	4	16	20	70	26	25
B	4	16	80	50	23	24
B	4	16	80	70	24	25
B	16	8	20	50	26	24
B	16	8	20	70	23	25
B	16	8	80	50	27	26
B	16	8	80	70	25	26
B	16	16	20	50	19	18
B	16	16	20	70	16	17
B	16	16	80	50	20	19
B	16	16	80	70	17	15

Analyze the results of this experiment.

6. Writing the treatment total (a) as the sum of the corresponding observations y_{100l} and substituting for these observations the expressions given by the model equation on page 394, it can be shown that

$$(a) = r[\mu + \alpha_1 + \beta_0 + \gamma_0 + (\alpha\beta)_{10} + (\alpha\gamma)_{10} + (\beta\gamma)_{00} + (\alpha\beta\gamma)_{100}]$$
$$+ \sum_{l=1}^{r} \epsilon_{100l}$$

Making use of the restrictions imposed on the parameters, rewrite this expression for (a) in terms of the parameters μ, α_0, β_0, γ_0, $(\alpha\beta)_{00}$, $(\alpha\gamma)_{00}$, $(\beta\gamma)_{00}$, and $(\alpha\beta\gamma)_{000}$.

7. Duplicating the work of Exercise 6, express (1), (b), (ab), (c), (ac), (bc), and (abc) in terms of the parameters μ, α_0, β_0, γ_0, $(\alpha\beta)_{00}$, $(\alpha\gamma)_{00}$, $(\beta\gamma)_{00}$, and $(\alpha\beta\gamma)_{000}$.

8. Using the results of Exercises 6 and 7, verify the expressions for $[A]$ and $[BC]$ obtained on page 396. Also express ϵ_A in terms of the quantities ϵ_{ijkl}.

9. Interpret the results of the analysis of variance given by the table on page 400, and estimate the magnitude of the significant effects.

10. A computational check on the sums of squares obtained for the various main effects and interactions is that their *sum* must equal the treatment sum of squares obtained by analyzing the data first as a two-way classification. Perform this check on the sums of squares given in the table on page 400.

11. If it is desired to find an expression for an effect total without constructing a complete table of signs, one can use the following method, illustrated by finding $[ABC]$ for a 2^4 factorial experiment. We take the expression $(a \pm 1)$ $(b \pm 1)(c \pm 1)(d \pm 1)$ with a "$+$" if the corresponding letter does *not* appear in the symbol for the main effect or interaction for which we want to calculate an effect total, and a "$-$" if the corresponding letter does appear. Thus, for finding $[ABC]$ we write

$$(a - 1)(b - 1)(c - 1)(d + 1) = abcd + abc - abd - acd - bcd - ab$$
$$- ac + ad - bc + bd + cd + a + b$$
$$+ c - d - 1$$

and after arranging the terms in standard order and adding parentheses we finally obtain

$$[ABC] = -(1) + (a) + (b) - (ab) + (c) - (ac) - (bc) + (abc) - (d)$$
$$+ (ad) + (bd) - (abd) + (cd) - (acd) - (bcd) + (abcd)$$

(a) Use this method to express $[B]$, $[AC]$, and $[ABC]$ in terms of the treatment totals in a 2^3 factorial experiment.

(b) Use this method to express $[AC]$ and $[BCD]$ in terms of the treatment totals in a 2^4 factorial experiment.

13.4 CONFOUNDING IN A 2^n FACTORIAL EXPERIMENT

In some experiments it is impossible to run all the required experimental conditions in one block. For example, if a 2^3 factorial experiment involves eight combinations of paint pigments that are to be applied to a surface and baked in an oven which can accommodate only four specimens, it becomes necessary to divide the eight treatments into two blocks (oven runs) in each replicate. As we have pointed out earlier, if the block size is too small to accommodate all treatments, this requires special, so-called *incomplete block* designs.

When the experimental conditions are distributed over several blocks, one or more of the effects may become confounded (inseparable) with possible block effects, that is, between-block differences. For example, if in the 2^3 factorial experiment referred to in the preceding paragraph, experimental conditions *a*, *ab*, *ac*, and *abc* are included in one oven run (Block 1) and experimental conditions 1, *b*, *c*, and *bc* are included in a second oven run (Block 2), then the "block effect," the difference between the two block totals, is given by

$$[(a) + (ab) + (ac) + (abc)] - [(1) + (b) + (c) + (bc)]$$

Referring to the table of signs on page 396, we observe that this quantity is, in fact, the effect total $[A]$, so that the estimate of the main effect of factor *A* is *confounded with blocks*. Note that all other factorial effects remain unconfounded; for each other effect total there are two $+1$ coefficients and two -1 coefficients in each block, so that the block effects cancel out. This kind of argument can also be used to decide what experimental conditions to put into each block to confound a given main effect or interaction. For instance, had we wanted to confound the *ABC* interaction with blocks in the above example, we could have put experimental conditions *a*, *b*, *c*, and *abc*, whose totals have $+1$ coefficients in $[ABC]$, into one block, and experimental conditions 1, *ab*, *ac*, and *bc*, whose totals have -1 coefficients, into another.

In general, confounding in a 2^n factorial experiment can be much more complicated than in the example just given. To avoid serious difficulties, we shall require that the number of blocks used is a power of 2, say 2^p. It turns out that the price paid for running a 2^n factorial experiment in 2^p blocks is that a total of $2^p - 1$ effects are confounded with blocks. To make it clear just which effects are confounded, and to indicate a method that can be used to confound only certain effects and no others, it is helpful to define the term "generalized interaction" as follows: the *generalized interaction* of two effects is the "product" of these effects, with like letters cancelled. Thus, the generalized interaction of *AB* and *CD* is *ABCD*, and the generalized interaction of *ABC* and *BCD* is *AB̶C̶B̶C̶D*, or *AD*. To confound a 2^n factorial experiment in 2^p blocks, the following method can be used: one selects any *p* effects for confounding, making sure that none is the generalized interaction of any of the others selected. Then, it can be shown that a further $2^p - (p + 1)$ effects are automatically confounded with blocks; together with the p effects originally chosen, this gives a total of $2^p - 1$ confounded effects in the experiment. The additional confounded effects are, in fact, the generalized interactions of the p effects originally chosen.

Example. To illustrate the construction of a confounded design, let us divide a 2^4 factorial experiment into four blocks so that desired effects are confounded with blocks. In actual practice one ordinarily confounds only the higher-order interactions (in the hope that they are nonexistent anyhow). Since we have decided upon four blocks, we have $2^p = 4$ and $p = 2$, and we shall arbitrarily select two higher-order interactions for confounding. If we were to select $ABCD$ and BCD, then their generalized interaction, A, would also be confounded. Thus, to avoid confounding any main effects, and to confound as few two-factor interactions as possible, we shall select ABD and ACD, noting that the BC interaction is also confounded. (Observe that it is impossible to avoid confounding at least one main effect or two-factor interaction in this experiment.)

In order to assign the 16 experimental conditions to four blocks, we first distribute them into two blocks, so that the ABD interaction is confounded with blocks. Referring to an appropriate table of signs, we put all treatments whose totals have a "+1" in the row for [ABD] into one block, all those whose totals have a "−1" into a second block, and we get the following blocks:

First block: a b ac bc d abd cd $abcd$
Second block: 1 ab c abc ad bd acd bcd

Note that each experimental condition in the first block has an *odd number* of letters in common with ABD, while each experimental condition in the second block has an *even number* of letters in common with ABD. This odd-even rule provides an alternate way of distributing experimental conditions among two blocks to confound a given effect, and it has the advantage that it does not require the construction of a complete table of signs.

So far, we have confounded the ABD interaction by dividing the 16 experimental conditions into two blocks; now we shall confound the ACD interaction by dividing each of these blocks into two blocks of four conditions each. Using the odd-even rule just described (or a table of signs), we obtain the following four blocks:

$$
\begin{array}{lllll}
\textit{Block 1:} & a & bc & d & abcd \\
\textit{Block 2:} & b & ac & abd & cd \\
\textit{Block 3:} & ab & c & bd & acd \\
\textit{Block 4:} & 1 & abc & ad & bcd
\end{array}
$$

By comparing these blocks with a table of signs, or, equivalently, by applying the odd-even rule, the reader will be asked to verify in Exercise 7 on page 416 that the BC interaction is also confounded with blocks, while all other effects are left unconfounded.

The analysis of a confounded 2^n factorial experiment is similar to that of an unconfounded experiment, with the exception that the sums of squares for the confounded effects are not computed, and we compute a block sum of squares as if the experiment consisted of br blocks rather than b blocks in each of r replicates.

Continuation of example. Referring to our example of a 2^4 factorial experiment with the ABD, ACD, and BC interactions confounded, and using two replicates, we have the following *dummy* analysis of variance table.

Source of variation	Degrees of freedom
Blocks	7
Main effects	4
Unconfounded two-factor interactions	5
Unconfounded three-factor interactions	2
Four-factor interaction	1
Intrablock error	12
Total	31

The sum of squares for blocks is obtained, as usual, by adding the squares of the eight block totals, dividing the result by 4 (the number of observations in each block), and subtracting the correction term. The total sum of squares and the sums of squares for the unconfounded factorial effects are obtained in the usual way, and the sum of squares for the *intrablock error*, a measure of the variability *within blocks*, is obtained by subtraction.

Example. To illustrate the analysis of a confounded 2^n factorial experiment, let us suppose that each replicate of the stepping-switch experiment described in the preceding section was actually run in four blocks, because only four mountings were available for the 16 switches. (The order of running the blocks is assumed to have been randomized within each replicate, and the assignment of switches is assumed to have been randomized within each block.) Assuming also that the *ABD*, *ACD*, and *BC* interactions were confounded with the blocks as shown on pages 407 and 408, we obtain the following block totals from the data on page 398:

	Block 1	Block 2	Block 3	Block 4
Replicate 1	3,564	3,488	3,743	3,593
Replicate 2	4,130	3,978	4,056	3,799

Thus, the sum of squares for blocks is given by

$$SS(Bl) = \frac{(3,564)^2 + (3,488)^2 + \ldots + (3,799)^2}{4} - 28,786,975$$

$$= 101,240$$

where the correction factor is the same as in the analysis on page 398.

Copying the total sum of squares and the sums of squares for the various unconfounded effects from the table on page 400, we obtain the analysis of variance table for the confounded factorial experiment shown on page 410. In this analysis, the *A* and *C* main effects are again significant at the 0.01 level, but none of the other main effects or interactions is significant.

Source of variation	Degrees of freedom	Sum of squares	Mean square	F
Blocks	7	101,240	14,463	1.58
Main effects:				
A	1	134,551	134,551	14.68
B	1	4,632	4,632	< 1
C	1	225,624	225,624	24.62
D	1	713	713	< 1
Unconfounded two-factor interactions:				
AB	1	39,410	39,410	4.30
AC	1	18,673	18,673	2.04
AD	1	31,689	31,689	3.46
BD	1	24,698	24,698	2.69
CD	1	81	81	< 1
Unconfounded three-factor interactions:				
ABC	1	39,130	39,130	4.27
BCD	1	14,070	14,070	1.54
Four-factor interaction:				
ABCD	1	385	385	< 1
Intrablock error	12	109,980	9,165	
Total	31	744,876		

If there is replication, some of the lost information about the con-founded effects can be recovered by a further breakdown of the above blocks sum of squares. This analysis consists of dividing the sum of squares for blocks into a component for each of the confounded effects, a component for replications, and a residual component called the "interblock error," which is a measure of the variability *between blocks*. Copying the sum of squares for replicates as well as those for the *BC*, *ABD*, and *ACD* interactions from the analysis of variance table on page 400, and copying the blocks sum of squares from the *intrablock* analysis of variance table above we obtain the following *interblock* analysis of variance:

Source of variation	Degrees of freedom	Sum of squares	Mean square	F
Replicates	1	77,520	77,520	23.05
Confounded effects:				
BC	1	1,001	1,001	< 1
ABD	1	30	30	< 1
ACD	1	12,601	12,601	3.75
Interblock error	3	10,088	3,363	
Total (blocks)	7	101,240		

Note that the interblock error is obtained by subtraction and that the F ratios are obtained by dividing the mean squares for the confounded effects and the mean square for replicates by the mean square for the interblock error. Only the F test for replicates is significant (at the 0.05 level). The small number of degrees of freedom for the interblock error implies that the sensitivity of these significance tests is *very poor;* in fact, it rarely pays to make this kind of interblock analysis unless the number of replications is relatively large.

13.5 FRACTIONAL REPLICATION

In studies involving complex production lines, chemical processes such as may be encountered in the petroleum, plastics, or metals industries, physical-chemical processes such as may be encountered in the electronics or space-technology industries, and in many other engineering studies, the experimenter is often faced with a large and bewildering array of inter-related variables. The principles of factorial experimentation treated so far in this chapter help him to "sort out" these variables, to discover which of them have the greatest influence on the process under consideration, and what important interrelationships may exist.

There are, however, some serious limitations to the simultaneous study of a large number of factors. Even if each factor is assigned only two levels, one replicate of a six-factor experiment requires 64 observations; there are 128 observations in one replicate of a seven-factor experiment, and 1,024 observations in one replicate of a ten-factor experiment. The economic and practical limitations of these large numbers make it necessary to seek out ways in which the size of factorial experiments can be kept within manageable bounds. Of course, it must be emphasized that there is no substitute for careful preliminary planning which, coupled with engineering insight, can result in the elimination of many needless factors.

In spite of the most careful preliminary planning, however, it is often difficult to avoid having to include as many as six or ten (or more) factors in a single experiment. One way to reduce the size of such an experiment would be to break it up into several parts, each part involving the deliberate variation of one factor while all others are held fixed. This would have the undesirable consequence that we could not study any of the interactions. Even if we were to include half the factors in one part and the other half in another, such as replacing a ten-factor experiment (requiring 1,024 observations) by two five-factor experiments (each requiring 32 observations), any interaction between factors in the first part and factors in the second part would be irretrievably lost. It is possible to overcome some of these difficulties by observing that most of the time we are not interested in *all* the interactions. For instance, it is possible to perform only a fraction of a 2^n factorial and yet obtain most of the desired information, say, about the main effects and two-factor interactions (but not the higher interactions).

The principles involved in *fractional replication*, that is, in performing only a fraction of a complete 2^n factorial experiment, are similar to those used in confounding. To obtain a *half-replicate* one selects only one of the two blocks into which the experimental conditions have been divided by confounding one effect; to obtain a *quarter-replicate* one selects only one of the four blocks into which the experimental conditions have been divided by confounding two effects, and so forth. In contrast to a confounded experiment as discussed in Section 13.4, we find that in a fractional replicate *the effects are confounded, not with blocks, but with each other.*

Example. To illustrate, suppose that only the experimental conditions *a*, *b*, *c*, and *abc* are included in a half-replicate of a 2^3 factorial experiment. (This is one block of a 2^3 factorial experiment with the *ABC* interaction confounded.) Considering the table of signs on page 396 with all columns except those corresponding to *a*, *b*, *c*, and *abc* crossed out, we find that the effect total for factor *A* is now given by

$$[A] = (a) - (b) - (c) + (abc)$$

If we write $(a) = \sum_{l=1}^{r} y_{100l}$, $(b) = \sum_{l=1}^{r} y_{010l}, \ldots$, and substitute for the y_{ijkl}, the expressions given by the model equation on page 394, we obtain

$$[A] = -4r[\alpha_0 - (\beta\gamma)_{00}] + \epsilon$$

where ϵ is the value of a random variable having zero mean (see Exercise 16 on page 418). We thus find that $[A]$ measures the main effect of factor A as well as the BC interaction, so that these two effects have become inseparable, or confounded. Note also that in the reduced table of signs (having columns only corresponding to a, b, c, and abc) the signs for $[A]$ and $[BC]$ are identical, and hence $[A] = [BC]$. In Exercise 17 on page 418, the reader will be asked to show that for the given fractional replicate the main effect for factor B is, similarly, confounded, or *aliased*, with the AC interaction, while the main effect for factor C is confounded, or aliased, with the AB interaction. The ABC interaction cannot be estimated.

With careful design, it is generally possible to confound all main effects and two-factor interactions *only* with interactions of higher order. This is illustrated in the following example, where we shall construct a half-replicate of a 2^5 factorial.

Example. First, we select an effect (usually a higher-order interaction) to split the experiment into two blocks, as in confounding. The effect chosen is called the *defining contrast*, and it cannot be estimated at all by the fractional replicate. Every other effect is aliased with another effect, namely, its generalized interaction (see page 406) with the defining contrast. Thus, if the defining contrast is $ABCDE$, the main effect for factor A has the four-factor interaction $BCDE$ as its alias, BC and ADE are an alias pair, and so forth. As we have seen, only the combined effect (the sum or difference of the aliased effects) can be estimated in the experiment. However, if it can be assumed that there are no higher-order interactions, one can attribute the effect of the alias pair BC and ADE entirely to the two-factor interaction BC, one can attribute the effect of the alias pair A and $BCDE$ entirely to the main effect of factor A, and so forth. A complete listing of the aliases in a half-replicate of a 2^5 factorial experiment having the defining contrast $ABCDE$ is as follows:

A and $BCDE$, B and $ACDE$, C and $ABDE$, D and $ABCE$
E and $ABCD$, AB and CDE, AC and BDE, AD and BCE
AE and BCD, BC and ADE, BD and ACE, BE and ACD
CD and ABE, CE and ABD, DE and ABC

Note that no main effect or two-factor interaction is aliased with another main effect or two-factor interaction.

The sixteen experimental conditions to be included in the half-replicate are given by those in either of the two blocks obtained by confounding the defining contrast. Choosing "evens" in the odd-even rule, namely, those conditions which have an even number of letters in common with the defining contrast *ABCDE*, we obtain the following half-replicate:

1	*ad*	*ae*	*de*
ab	*bd*	*be*	*abde*
ac	*cd*	*ce*	*acde*
bc	*abcd*	*abce*	*bcde*

To go one step further, let us illustrate how to construct a quarter-replicate of the given 2^5 factorial. We shall do this by dividing the above half-replicate in half, confounding the three-factor interaction *ABC*. Again using "evens," we get the following eight experimental conditions:

1	*de*	*ab*	*abde*
ac	*acde*	*bc*	*bcde*

Since *DE*, the generalized interaction of the two confounded effects, is also confounded, we now have the three defining contrasts *ABCDE*, *ABC*, and *DE*. None of these effects can be estimated in the quarter-replicate, and each other effect is aliased with its three generalized interactions with the three defining contrasts. The complete aliasing is as follows:

Alias sets

A,	*BCDE*,	*BC*,	*ADE*
B,	*ACDE*,	*AC*,	*BDE*
C,	*ABDE*,	*AB*,	*CDE*
D,	*ABCE*,	*ABCD*,	*E*
AD,	*BCE*,	*BCD*,	*AE*
BD,	*ACE*,	*ACD*,	*BE*
CD,	*ABE*,	*ABD*,	*CE*

We have given this quarter-replicate merely as an illustration; it would hardly seem useful in actual practice because of the hopeless aliasing of main effects and two-factor interactions. Nevertheless, quarter-replicates of six- and seven-factor experiments (and even eighth-replicates of seven- and eight-factor experiments) can often provide much useful information.

The analysis of a fractional factorial is practically the same as that of a fully replicated factorial experiment. Given the fraction $1/2^p$ of a 2^n factorial, there are 2^{n-p} experimental conditions, and the method of Yates can be used as if the experiment were a 2^{n-p} factorial. Some care must be exercised to arrange the experimental conditions in a modified "standard order" as indicated in Exercise 14 on page 417.

When dealing with fractional factorials, there is the problem of obtaining an estimate of the experimental error. For example, in the half-replicate of a 2^5 factorial described on page 413, the breakdown of the total sum of squares having 15 degrees of freedom yields sums of squares for main effects (5 degrees of freedom), sums of squares for two-factor interactions (10 degrees of freedom), but no component (0 degrees of freedom) for the experimental error. In a situation like this, and in all other cases where the number of degrees of freedom for error is small, it is best to include a limited amount of replication. This may be accomplished by randomly selecting several experimental conditions, and making additional observations corresponding to these conditions. Furthermore, if it can be assumed that there are no higher-order interactions, the total of the sums of squares corresponding to higher-order interactions *which are not aliased with main effects or lower-order interactions* can be attributed to "error," and used in the denominator of the F test (see Exercise 15 below). In this way, use can be made of the "hidden replication" inherent in most large factorial experiments.

To summarize, fractional replication is useful whenever the number of factors to be included in an experiment is large and it is not economically feasible to include all possible experimental conditions. The reduction in size (and, therefore, in cost) of a fractionally replicated experiment is partially offset by the loss of information caused by aliasing, and by the difficulties inherent in estimating the experimental error. A more detailed discussion of fractional replication, including fractional replicates of 3^n experiments and a variety of other designs, can be found in the book by W. G. Cochran and G. M. Cox listed in the Bibliography.

EXERCISES

1. A 2^5 factorial experiment, having the factors A, B, C, D, and E, is to be run in several blocks. Show which treatments are assigned to each block if
   (a) there are to be two blocks, with the $ACDE$ interaction confounded;
   (b) there are to be four blocks, with the BDE and $ABCE$ interactions confounded. What other factorial effect or effects are also confounded?

2. List the experimental conditions included in each block if a 2^4 factorial experiment is to be confounded
   (a) in 2 blocks on $ABCD$;
   (b) in 4 blocks on ABC and BCD.

3. What is the largest number of blocks in which one can perform a 2^6 factorial experiment without confounding any main effects?

4. List the experimental conditions included in each block if a 2^6 factorial experiment is to be confounded in 8 blocks on $ABDE$, $BCDF$, and ABC. What other factorial effects are confounded?

5. Suppose that in Exercise 1 on page 401 the judge rates the soft drinks in sets of four, with a rest period in between, and that the experiment was actually performed so that each replicate consisted of two blocks with ABC confounded. Perform an intrablock analysis of the data.

6. Four new drugs are to be investigated to determine their effectiveness as tranquilizers, individually and in combination with each other. Each patient was given regular doses of one of the sixteen tranquilizers formed from these drugs (including a placebo corresponding to the 0 level for each drug) and, after a two-week period, the effect of these tranquilizers on the emotional stability of each patient was judged (on a scale from 1 to 5) by five psychiatrists. To keep the staff work load within reasonable bounds, two hospitals were used in this experiment, and eight patients were selected by the staff of each hospital for each trial (replicate); thus, the experiment involved two replicates of two blocks each, with the $ABCD$ interaction confounded. The results of the two 2-week trials were as follows:

FIRST HOSPITAL			SECOND HOSPITAL		
Treatment combination	Mean rating Trial 1	Trial 2	Treatment combination	Mean rating Trial 1	Trial 2
1	2.0	2.6	a	2.8	2.6
ab	3.8	3.4	b	3.6	2.0
ac	4.2	4.8	c	2.4	1.8
bc	4.8	4.0	abc	4.0	3.8
ad	1.8	2.4	d	1.8	2.2
bd	3.4	3.8	abd	1.6	2.0
cd	4.6	2.8	acd	3.6	2.4
abcd	4.2	4.6	bcd	3.4	3.8

If a high rating indicates satisfactory progress, which drug combination (or combinations) seems to be the most promising? Perform an intrablock analysis of variance to test for significant effects.

7. Verify that in the example on page 407 all treatment totals having $+1$ as a coefficient in $[BC]$ are in Blocks 1 and 4, that all treatment totals having -1 as a coefficient in $[BC]$ are in Blocks 2 and 3, and, hence, that the BC interaction is confounded with blocks.

8. Suppose that in Exercise 3 on page 402 only 8 specimens could be tested in any one shift, and that the experiment was actually performed so that each replicate consists of four blocks with ABC, ADE, and $BCDE$ confounded.

(a) If for each factor levels 0 and 1 are, respectively, the lower and higher values, construct a table showing which experimental conditions go into each of the four blocks.

(b) Perform an intrablock analysis of the experiment.

9. Show that the "odd-even" rule for assigning treatments to blocks, given on page 407, is equivalent to the method described in Exercise 11 on page 405.

10. Referring to Exercise 1, suppose that in each case the block containing treatment combination 1 was chosen as a fractional replicate of a 2^5 factorial.

(a) Show the alias pairs in the resulting half-replicate.

(b) Show the alias sets in the resulting quarter-replicate.

(In practice, random selection should be used to choose the block to be included in a fractional replicate.)

11. Referring to Exercise 2, list the alias sets if an experiment consists of

(a) a half-replicate of a 2^4 factorial experiment with $ABCD$ confounded,

(b) a quarter-replicate of a 2^4 factorial experiment with ABC and BCD confounded.

12. Construct a half-replicate of a 2^6 factorial having the defining contrast $ABCDEF$. List the 32 treatments in the block containing experimental condition 1, and show the alias pairs.

13. Design an experiment consisting of a quarter-replicate of a 2^7 factorial with $ABCDE$, $ABCFG$, and $DEFG$ confounded, if the experimental condition with each factor at the 0 level is to be included. Also exhibit all alias sets.

14. Referring to Exercise 12, we can define a *modified standard order* for the 32 treatment combinations in the half-replicate as follows. First, we list the 32 treatment combinations corresponding to the five factors A, B, C, D, and E in standard order. Then we append the letter f to 16 of these treatment combinations, so that the list contains the same ones as the block chosen for the half-replicate.

(a) Use this method to list the treatment combinations obtained in Exercise 12 in modified standard order.

(b) Generalize the above rule for arranging the treatment combinations in modified standard order so that it applies to a half-replicate of a 2^n factorial experiment. (For a $1/2^p$ fractional replicate of a 2^n factorial, a modified standard order can be obtained by noting that any block chosen contains a subset of $n - p$ letters which form a complete replicate of a 2^{n-p} factorial. The modified standard order is then obtained by using these letters only and later appending the remaining p letters to get the required treatment combinations.)

15. The following factors are to be studied in a half-replicate of a 2^6 factorial experiment (defining contrast $ABCDEF$), designed to evaluate several chemicals as insecticides.

Factor	Level 0	Level 1
A. BMC	0%	5%
B. Malathion	3%	6%
C. Tedion	1%	2%
D. Chlordane	2%	5%
E. Lindane	1%	4%
F. Pyrethrum	2%	4%

Each experimental unit consists of 10 insects, and the average lifetimes (in seconds) after application of the respective insecticides are as follows, in the random order in which they are obtained:

ce	181	acdf	162	bd	135	abdf	131
ae	172	1	182	df	171	ab	136
abef	140	bf	171	acef	159	bcde	105
bcdf	165	cf	176	bc	179	abcdef	109
acde	139	be	187	ac	165	af	176
ef	186	abce	131	bcef	181	ad	150
de	164	abcf	125	cdef	163	abde	115
abcd	112	adef	158	bdef	128	cd	166

(a) Write down the alias pairs.

(b) Arrange the results in modified standard order (see Exercise 14) and use the method of Yates to find the effect totals.

(c) Identify the effect totals as follows. In the last column of the Yates table, write down the 32 combinations $[I]$, $[A]$, $[B]$, $[AB]$, ..., $[ABCDE]$ in standard order. Each of these is aliased with another one; identify the alias pairs by using the main effect or lower-order interaction. (For example, $ABCDE$ is aliased with F—identify the effect total $[ABCDE]$ as that of the main effect F.) Note that ten of the alias pairs are three-factor interactions aliased with three-factor interactions—label these "error."

(d) Obtain the mean squares and complete the analysis of variance. Note that the error mean square, having 10 degrees of freedom, is the average of the mean squares of the ten effects labeled "error."

16. Verify the expression obtained for $[A]$ on page 412.

17. Duplicating the steps indicated on page 412, show that in the given design the main effect for factor B is confounded with the AC interaction. [*Hint:* Express $[B]$ and $[AC]$ in terms of the parameters of the model.]

14

Applications to Quality Assurance

14.1 QUALITY ASSURANCE

Although there is the tendency to think of the subject of quality assurance as a recent development, there is nothing new about the basic idea of making a quality product characterized by a high degree of uniformity. For centuries skilled artisans have striven to make products distinctive through superior quality, and once a standard of quality was achieved, to eliminate insofar as possible all variability between products that were nominally alike.

The idea that statistics might be instrumental in assuring the quality of manufactured products goes back no farther than the advent of mass production, and the widespread use of statistical methods in problems of quality assurance is even more recent than that. Many problems in the manufacture of a product are amenable to statistical treatment, and we

419

have already studied some of them in earlier parts of this book.When we speak of (statistical) quality assurance, we are referring essentially to the three special techniques to which we shall devote the remainder of this chapter: *quality control, acceptance sampling,* and *the establishment of tolerance limits.* Note that the word "quality," when used technically as in this discussion, refers to some *measurable* or *countable* property of a product, such as the outside diameter of a ball bearing, the breaking strength of a yarn, the number of imperfections in a piece of cloth, the potency of a drug, and so forth.

14.2 QUALITY CONTROL

It may surprise some persons to learn that two apparently identical parts made under carefully controlled conditions, from the same batch of raw material, and only seconds apart by the same machine, can nevertheless be different in many respects. Indeed, any manufacturing process, however good, is characterized by a certain amount of variability which is of a random nature, and which cannot be completely eliminated.

When the variability present in a production process is confined to *chance variation,* the process is said to be in a state of *statistical control.* Such a state is usually attained by finding and eliminating trouble of the sort causing another kind of variation, called *assignable variation,* which may be due to poorly trained operators, poor-quality raw materials, faulty machine settings, worn parts, and the like. Since manufacturing processes are rarely free from trouble of this kind, it is important to have some

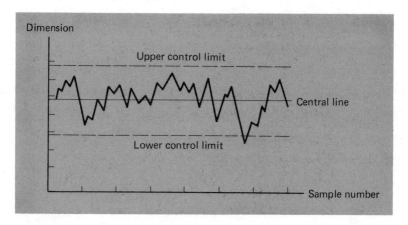

Figure 14.1. Control chart.

systematic method of detecting serious deviations from a state of statistical control when, or if possible *before*, they actually occur. It is to this end that *control charts* are principally used.

In what follows, we shall differentiate between *control charts for measurements* and *control charts for attributes*, depending on whether the observations with which we are concerned are measurements or count data (say, the number of defectives in a sample of a given size). In either case, a control chart consists of a *central line* (see Figure 14.1) corresponding to the *average* quality at which the process is to perform, and lines corresponding to the *upper and lower control limits*. These limits are chosen so that values falling between them can be attributed to chance, while values falling beyond them are interpreted as indicating a lack of control. By plotting the results obtained from samples taken periodically at frequent intervals, it is possible to check by means of such a chart whether the process is under control, or whether trouble of the sort indicated above has entered the process. When a sample point falls beyond the control limits, one looks for trouble, but even if the points fall between the control limits, a trend or some other systematic pattern may serve notice that action should be taken to avoid serious trouble.

The ability to "read" control charts and to determine from them just what corrective action should be taken is a matter of experience and highly developed judgment. A quality-control engineer must not only understand the statistical foundation of his subject, but he must also be thoroughly acquainted with the processes themselves. The engineering and managerial aspects of quality control (and quality assurance in general), which nowadays includes incoming raw materials, outgoing products, and in-process control, constitute an extensive subject in themselves. In this chapter we shall present only the statistical aspects of the subject; more complete discussions of other aspects can be found in the book by D. J. Cowden listed in the Bibliography.

14.3 CONTROL CHARTS FOR MEASUREMENTS

When dealing with measurements, it is customary to exercise control over the average quality of a process as well as its variability. The first goal is accomplished by plotting the means of periodic samples on a so-called *control chart for means*, or more succinctly, an $\bar{x}$ *chart*. Variability is controlled by plotting the sample ranges or standard deviations, respectively, on an *R chart* or a σ *chart*, depending on which statistic is used to estimate the population standard deviation.

If the process mean and standard deviation, μ and σ, are known, and it is reasonable to treat the measurements as samples from a normal population, we can assert with a probability of $1 - \alpha$ that the mean of a random sample of size n will fall between $\mu - z_{\alpha/2}\dfrac{\sigma}{\sqrt{n}}$ and $\mu + z_{\alpha/2}\dfrac{\sigma}{\sqrt{n}}$. These two limits on $\bar{x}$ provide upper and lower control limits, and, under the given assumptions, they enable the quality-control engineer to determine whether or not to make an adjustment in the process.

In actual practice, μ and σ are usually unknown and it is necessary to estimate their values from a large sample (or samples) taken while the process is "in control." For this reason and because there may be no assurance that the measurements can be treated as samples from a normal population, the confidence level $1 - \alpha$ associated with the control limits is only approximate, and such "probability limits" are seldom used in practice. Instead, it is common industrial practice to use "three-sigma limits" obtained by substituting 3 for $z_{\alpha/2}$. With three-sigma limits one usually is highly confident that the process will not be declared out of control when, in fact, it is actually in control.

If there exists a long history of a process in good control, μ and σ can be estimated from past data practically without error. Thus, the *central line* of an $\bar{x}$ chart is given by μ, and the *upper and lower three-sigma control limits* are given by $\mu \pm A\sigma$, where $A = 3/\sqrt{n}$ and n is the size of each sample.* For convenience, values of A for $n = 2, 3, \ldots$, and 15 are given in Table 13. The use of a constant sample size n simplifies the maintenance and interpretation of an $\bar{x}$ chart, but as the reader will observe in Exercise 4 on page 432, this restriction is not absolutely necessary.

In the more common case where the population parameters are *unknown*, it is necessary to estimate these parameters on the basis of preliminary samples. For this purpose, it is usually desirable to obtain the results of 20 or 25 consecutive samples taken when the process is in control. If k samples are used, each of size n, we shall denote the mean of the ith sample by $\bar{x}_i$, and the grand mean of the k sample means by $\bar{x}$, that is,

$$\bar{x} = \frac{1}{k}\sum_{i=1}^{k} \bar{x}_i$$

---

*Throughout this chapter we shall depart somewhat from the customary quality-control notation, in order to be consistent with the more widely accepted statistical notation used elsewhere in this book. (For instance, in quality control it is customary to denote the sample mean and standard deviation by $\bar{x}$ and σ, and the corresponding population parameters by $\bar{x}'$ and σ'.)

The process variability σ can be estimated either from the standard deviations or the ranges of the k samples. Since the sample size commonly used in connection with control charts for measurements is small, there is usually very little loss of efficiency in estimating σ from the sample ranges. (For an example where sample standard deviations are used in this connection, see Exercise 5 on page 432.) Denoting the range of the ith sample by R_i, we shall thus make use of the statistic

$$\bar{R} = \frac{1}{k} \sum_{i=1}^{k} R_i$$

Since $\bar{x}$ provides an unbiased estimate of the population mean μ, the central line for the $\bar{x}$ chart is given by $\bar{x}$. The statistic $\bar{R}$ does not provide an unbiased estimate of σ, but multiplying $\bar{R}$ by the constant A_2, we obtain an unbiased estimate of $3\sigma/\sqrt{n}$. The constant multiplier A_2, tabulated in Table 13 for various values of n, depends on the assumption that the measurements constitute a sample from a normal population. Thus, the central line and the upper and lower three-sigma control limits, *UCL* and *LCL*, for an $\bar{x}$ chart (*with μ and σ estimated from past data*) are given by

$$central\ line = \bar{x}$$
$$UCL = \bar{x} + A_2\bar{R}$$
$$LCL = \bar{x} - A_2\bar{R}$$

An example of this kind of control chart for the mean is shown in Figure 14.2.

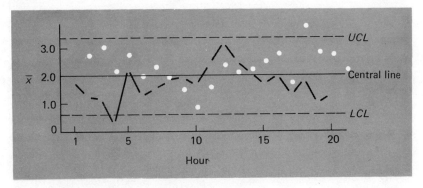

Figure 14.2. $\bar{x}$ chart.

In controlling a process, it may not be enough to monitor the population mean. Although an increase in process variability may become apparent from increased fluctuations of the $\bar{x}$'s, a more sensitive test of shifts in process variability is provided by a separate control chart, an R chart based on the sample ranges or a so-called σ chart based on the sample standard deviations. An example of the latter may be found in Exercise 5 on page 432.

The central line and control limits of an R chart are based on the distribution of the range of samples of size n from a normal population. As we observed on page 228, the mean and the standard deviation of this sampling distribution are given by $d_2\sigma$ and $d_3\sigma$, respectively, when σ is known. Thus, three-sigma control limits for the range are given by $d_2\sigma \pm 3d_3\sigma$, and the complete set of control-chart values for an R chart (*with σ known*) is given by

$$central\ line = d_2\sigma$$
$$UCL = D_2\sigma$$
$$LCL = D_1\sigma$$

Here $D_1 = d_2 - 3d_3$ and $D_2 = d_2 + 3d_3$, and values of these constants can be found in Table 13 for various values of n.

If σ is unknown, it is estimated from past data as previously described, and the control-chart values for an R chart (*with σ unknown*) are as follows:

$$central\ line = \bar{R}$$
$$UCL = D_4\bar{R}$$
$$LCL = D_3\bar{R}$$

Here $D_3 = D_1/d_2$ and $D_4 = D_2/d_2$, and values of these constants can also be found in Table 13 for various values of n.

Example. To illustrate the construction of an $\bar{x}$ chart and an R chart, let us consider the following example. A manufacturer of a certain bearing knows from a preliminary record of 20 hourly samples of size 4 that for the diameters of these bearings $\bar{x} = 0.9752$ and $\bar{R} = 0.0002$. Coding his data by means of the equation $\dfrac{x - 0.9750}{0.0001}$, that is, expressing each measurement as a deviation from 0.9750 in 0.0001 inch, he obtains

$\bar{x}$ chart (coded)			R chart (coded)	
central line	$\bar{\bar{x}} = 2.0$		central line	$\bar{R} = 2.0$
UCL	$\bar{\bar{x}} + A_2\bar{R} = 3.4$		UCL	$D_4\bar{R} = 4.6$
LCL	$\bar{\bar{x}} - A_2\bar{R} = 0.6$		LCL	$D_3\bar{R} = 0$

The values of $A_2 = 0.729$, $D_3 = 0$, and $D_4 = 2.282$ for samples of size 4 were obtained from Table 13. Graphically, these control charts are shown in Figures 14.2 and 14.3, where we have also indicated the results subsequently obtained in the following 20 samples:

Hour	Coded sample values				$\bar{x}$	R
1	1.7	2.2	1.9	1.2	1.75	1.0
2	0.8	1.5	2.1	0.9	1.32	1.3
3	1.0	1.4	1.0	1.3	1.18	0.4
4	0.4	−0.6	0.7	0.2	0.18	1.3
5	1.4	2.3	2.8	2.7	2.30	1.4
6	1.8	2.0	1.1	0.1	1.25	1.9
7	1.6	1.0	1.5	2.0	1.52	1.0
8	2.5	1.6	1.8	1.2	1.78	1.3
9	2.9	2.0	0.5	2.2	1.90	2.4
10	1.1	1.1	3.1	1.6	1.72	2.0
11	1.7	3.6	2.5	1.8	2.40	1.9
12	4.6	2.8	3.5	1.9	3.20	2.7
13	2.6	2.8	3.2	1.5	2.52	1.7
14	2.3	2.1	2.1	1.7	2.05	0.6
15	1.9	1.6	1.8	1.4	1.68	0.5
16	1.3	2.0	3.9	0.8	2.00	3.1
17	2.8	1.5	0.6	0.2	1.28	2.6
18	1.7	3.6	0.9	1.5	1.92	2.7
19	1.6	0.6	1.0	0.8	1.00	1.0
20	1.7	1.0	0.5	2.2	1.35	1.7

Inspection of Figure 14.2 shows that only one of the points falls outside of the control limits, but it also shows that there may nevertheless have been a downward shift in the process average. Figure 14.3 shows a definite downward shift in the process variability; note especially that most of the sample ranges fall below the central line of the R chart.

The reader may have observed the close connection between the use of control charts and the testing of hypotheses. A point on an $\bar{x}$ chart that is out of control corresponds to a sample for which the null hypothesis that $\mu = \mu_0$ is rejected. To be more precise, we should say that control-chart

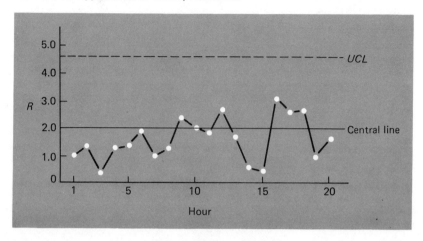

Figure 14.3. *R* chart.

techniques provide sequential, *temporally ordered* sets of tests. We are interested not only in the positions of individual points, but also in possible trends or other patterns exhibited by the points representing successive samples. We might, thus, use the sign test (see page 272) to check whether there has been a shift in the product average in the above example, or a run test (see page 284) to determine whether there is a significant trend.

14.4 CONTROL CHARTS FOR ATTRIBUTES

Although more complete information can usually be gained from measurements made on a finished product, it is often quicker and cheaper to check the product against specifications on an "attribute" or "go, no-go" basis. For example, in checking the diameter and eccentricity of a ball bearing it is far simpler to determine whether it will pass through circular holes cut in a template than to make several measurements of the diameter with a micrometer. In this section we shall discuss two fundamental kinds of control charts used in connection with attribute sampling, the *fraction-defective chart*, also called a "*p* chart," and the *number-of-defects chart*, also called a "*c* chart." To clarify the distinction between "number defective" and "number of defects," note that a unit tested can have several defects, while, on the other hand, it is either defective or it is not. In many applications, a unit is referred to as defective if it has at least one defect.

Control limits for a fraction-defective chart are based on the sampling theory for proportions introduced in Section 9.1, namely, on the normal curve approximation to the binomial distribution. Thus, if a standard is given, that is, if the fraction defective should take on some preassigned value p, the central line is p and three-sigma control limits for the fraction defective in random samples of size n are given by $p \pm 3\sqrt{\dfrac{p(1-p)}{n}}$. If no standard is given, which is more frequently the case in actual practice, p will have to be estimated from past data. If k samples are available, d_i is the number of defectives in the ith sample, and n_i is the number of observations in the ith sample, it is customary to estimate p as the proportion of defectives in the combined sample, namely, as

$$\bar{p} = \frac{d_1 + d_2 + \ldots + d_k}{n_1 + n_2 + \ldots + n_k}$$

Substituting $\bar{p}$ for p in the above formulas, we obtain the following control-chart values for a *fraction-defective chart based on the analysis of past data:*

$$\text{central line} = \bar{p}$$

$$UCL = \bar{p} + 3\sqrt{\frac{\bar{p}(1-\bar{p})}{n}}$$

$$LCL = \bar{p} - 3\sqrt{\frac{\bar{p}(1-\bar{p})}{n}}$$

Note that if p is small, as is often the case in practice, substitution in the formula for the lower control limit might yield a negative number. When this occurs, it is customary to regard the lower control limit as if it were zero and, in effect, to use only the upper control limit. Another complication that can arise if p is small is that the binomial distribution may not be adequately approximated by the normal distribution. Generally speaking, the use of the above control limits for p charts is unrealistic whenever n and p are such that the underlying binomial (or hypergeometric) distribution cannot be approximated by a normal curve (see page 108). In such cases it is best to use an upper control limit obtained directly from a table of binomial probabilities or, perhaps, use the Poisson approximation to the binomial distribution.

Equivalent to the p chart for the fraction defective is the control chart for the *number of defectives*. Instead of plotting the fraction defective in a sample of size n, one plots the number of defectives, and the control-chart values for this kind of chart are obtained by multiplying the above values for the central line and the control limits by n. Thus, if p is estimated by $\bar{p}$, the control-chart values for a *number-of-defectives* chart are as follows:

$$\text{central line} = n\bar{p}$$
$$UCL = n\bar{p} + 3\sqrt{n\bar{p}(1 - \bar{p})}$$
$$LCL = n\bar{p} - 3\sqrt{n\bar{p}(1 - \bar{p})}$$

Example. As an illustration of a p chart, suppose it is desired to control the output of a certain integrated circuit production line to maintain a "yield" of 60 percent, that is, a proportion defective of 40 percent. To this end, daily samples of 100 units are checked to electrical specifications, with the following results:

Date	Number of defectives	Date	Number of defectives	Date	Number of defectives
3-12	24	3-26	44	4- 9	23
3-13	38	3-27	52	4-10	31
3-16	62	3-30	45	4-13	26
3-17	34	3-31	30	4-14	32
3-18	26	4- 1	34	4-15	35
3-19	36	4- 2	33	4-16	15
3-20	38	4- 3	22	4-17	24
3-23	52	4- 6	34	4-20	38
3-24	33	4- 7	43	4-21	21
3-25	44	4- 8	28	4-22	16

Since the standard is given as $p = 0.40$, the control-chart values are

$$\text{central line} = 0.40$$

$$UCL = 0.40 + 3\sqrt{\frac{(0.40)(0.60)}{100}} = 0.55$$

$$LCL = 0.40 - 3\sqrt{\frac{(0.40)(0.60)}{100}} = 0.25$$

The corresponding control chart with points for the 30 sample fractions defective is shown in Figure 14.4, and it exhibits some interesting characteristics. Note that there is only one point out of control on the high side, but there are seven points out of control on the low side.

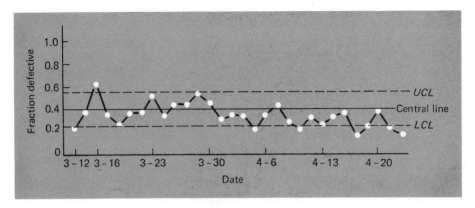

Figure 14.4. *p*-chart.

Most of these seven "low points" occurred after April 1, and there appears to be a general downward trend. In fact, there is an unbroken run of eleven points below the central line after April 7. It would appear from this chart that the yield is not yet stabilized, and that the process is potentially capable of maintaining a yield well above the nominal 60 percent value.

There are situations where it is necessary to control the number of *defects* in a unit of product, rather than the fraction defective. For example, in the production of carpeting it is important to control the number of defects per hundred yards; in the production of newsprint one may wish to control the number of defects per roll. These situations are similar to the one described in Section 3.7, which led to the Poisson distribution. Thus, if c is the number of defects per manufactured unit, c is taken to be a value of a random variable having the Poisson distribution.

It follows that the center line for a *number-of-defects chart* is the parameter λ of the corresponding Poisson distribution, and that three-sigma control limits can be based on the fact that the standard deviation of this distribution is $\sqrt{\lambda}$. If λ is unknown, that is, if no standard is given, its value is usually estimated from at least 20 values of c observed from past data. If k is the number of units of product available for estimating λ, and if c_i is the number of defects in the ith unit, then λ is estimated by

$$\bar{c} = \frac{1}{k} \sum_{i=1}^{k} c_i$$

and the control-chart values for the *c chart* are

$$central\ line = \bar{c}$$
$$UCL = \bar{c} + 3\sqrt{\bar{c}}$$
$$LCL = \bar{c} - 3\sqrt{\bar{c}}$$

Example. To illustrate this kind of control chart, suppose that it is known from past experience that on the average an aircraft assembly made by a certain company has $\bar{c} = 4$ missing rivets. The corresponding control chart for the number of missing rivets is shown in Figure 14.5, on which we have also plotted the results of inspection which revealed 4, 6, 5, 1, 2, 3, 5, 7, 1, 2, 2, 4, 6, 5, 3, 2, 4, 1, 8, 4, 5, 6, 3, 4, and 2 missing rivets in 25 assemblies.

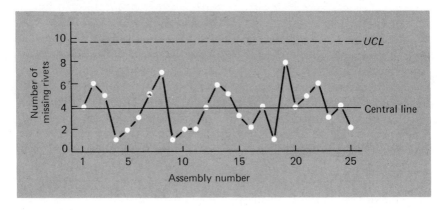

Figure 14.5. *c*-chart.

EXERCISES

1. A plastics manufacturer extrudes blanks for use in the manufacture of eyeglass temples. Specifications require that the thickness of these blanks have $\mu = 0.150$ inch and $\sigma = 0.002$ inch.
   (a) Use the specifications to calculate a central line and three-sigma control limits for an $\bar{x}$ chart with $n = 5$.
   (b) Use the specifications to calculate a central line and three-sigma control limits for an R chart with $n = 5$.

(c) Plot the following means and ranges, obtained in 20 successive random samples of size 5, on charts based on the control-chart constants obtained in parts (a) and (b), and discuss the process.

Sample	$\bar{x}$	R	Sample	$\bar{x}$	R
1	0.152	0.004	11	0.149	0.003
2	0.147	0.006	12	0.153	0.004
3	0.153	0.004	13	0.150	0.005
4	0.153	0.002	14	0.152	0.001
5	0.151	0.003	15	0.149	0.003
6	0.148	0.002	16	0.146	0.002
7	0.149	0.006	17	0.154	0.004
8	0.144	0.001	18	0.152	0.005
9	0.149	0.003	19	0.151	0.002
10	0.152	0.005	20	0.149	0.004

2. Calculate $\bar{x}$ and $\bar{R}$ for the data of part (c) of Exercise 1, and use these values to construct the central lines and three-sigma control limits for new $\bar{x}$ and R charts to be used in the control of the thickness of the extruded plastic blanks.

3. The following data give the means and ranges of 25 samples, each consisting of four compression test results on steel forgings, in thousands of pounds per square inch:

Sample	1	2	3	4	5	6	7	8
$\bar{x}$	45.4	48.1	46.2	45.7	41.9	49.4	52.6	54.5
R	2.7	3.1	5.0	1.6	2.2	5.7	6.5	3.6

Sample	9	10	11	12	13	14	15	16
$\bar{x}$	45.1	47.6	42.8	41.4	43.7	49.2	51.1	42.8
R	2.5	1.0	3.9	5.6	2.7	3.1	1.5	2.2

Sample	17	18	19	20	21	22	23	24	25
$\bar{x}$	51.1	52.4	47.9	48.6	53.3	49.7	48.2	51.6	52.3
R	1.4	4.3	2.2	2.7	3.0	1.1	2.1	1.6	2.4

(a) Use these data to find the central line and control limits for an $\bar{x}$ chart.
(b) Use these data to find the central line and control limits for an R chart.
(c) Plot the given data on $\bar{x}$ and R charts based on the control-chart constants computed in parts (a) and (b), and interpret the results.
(d) Using runs above and below the central line (similar to runs above and below the median discussed on page 284), test at a level of significance of 0.05 whether there is a trend in the $\bar{x}$-values.

(e) Would it be reasonable to use the control limits found in this exercise in connection with subsequent compression test measurements from the same process? Why?

4. Reverse-current readings (in nanoamperes) are made on a sample of ten transistors every half hour. Since some of the units may prove to be "shorts" or "opens," it is not always possible to obtain ten readings. The following table shows the number of readings made at the end of each half-hour interval during an eight-hour shift, and the mean reverse currents obtained.

Sample	1	2	3	4	5	6	7	8
n	10	6	9	8	8	10	7	9
$\bar{x}$	12.5	11.1	10.2	11.6	21.9	12.3	9.7	15.6

Sample	9	10	11	12	13	14	15	16
n	7	8	10	9	7	8	9	10
$\bar{x}$	16.7	9.8	11.6	17.2	10.1	9.5	13.1	14.2

(a) Find the central line for an $\bar{x}$ chart by taking the weighted mean of the sixteen $\bar{x}$'s, weighting each value with the size of the corresponding sample.
(b) Construct a table showing the central line found in part (a) and three-sigma control limits corresponding to $n = 6, 7, 8, 9,$ and 10. Use $R = 4.0$, a value based on prior data.
(c) Plot the data on a control chart like the one of Figure 14.6 and interpret the results.

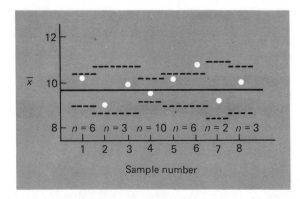

Figure 14.6. Exercise 4.

5. If the sample standard deviations instead of the sample ranges are used to estimate σ, the control limits for the resulting $\bar{x}$ chart are given by $\bar{x} \pm A_1\bar{s}$, where $\bar{s}$ is the mean of the sample standard deviations obtained from given

data, and A_1 can be found in Table 13. Note that in connection with problems of quality control the sample standard deviation is defined using the divisor n instead of $n - 1$. The corresponding R chart is replaced by a σ chart, having the central line $c_2 \bar{s}$ and the upper and lower control limits $B_3 \bar{s}$ and $B_4 \bar{s}$, where B_3 and B_4 can be obtained from Table 13.

(a) Construct an $\bar{x}$ chart and a σ chart for 20 samples of size 3 which had $\bar{x}$ equal to 21.2, 19.4, 20.4, 20.4, 20.4, 19.0, 20.3, 21.1, 21.6, 22.1, 24.4, 23.9, 24.9, 24.1, 21.8, 19.5, 20.3, 22.5, 23.4, 23.3, and s equal to 2.0, 0.8, 1.1, 0.9, 1.0, 0.3, 1.3, 2.0, 0.8, 1.0, 1.5, 1.0, 1.5, 0.8, 1.3, 2.9, 4.3, 1.2, 0.3, 3.1.

(b) Would it be reasonable to use these control limits for subsequent data? Why?

6. In order to establish control charts for a boring process, thirty samples of 5 measurements of the inside diameters are taken, and the results are $\bar{x} = 1.317$ inches and $\bar{s} = 0.002$ inch. Using the method of Exercise 5, construct an $\bar{x}$ chart for $n = 5$, and on it plot the following means obtained in 25 successive samples: 1.328, 1.330, 1.321, 1.325, 1.332, 1.340, 1.327, 1.321, 1.324, 1.325, 1.329, 1.326, 1.330, 1.324, 1.328, 1.322, 1.326, 1.327, 1.329, 1.325, 1.324, 1.329, 1.330, 1.321, and 1.329. Discuss the results.

7. Suppose that in the example of Exercise 6 it is desired to establish control also over the variability of the process. Using the method of Exercise 5 and the values of $\bar{x}$ and $\bar{s}$ given in Exercise 6, calculate the central line and control limits for a σ chart with $n = 5$.

8. Thirty-five successive samples of 100 castings each, taken from a production line, contained, respectively, 3, 3, 5, 3, 5, 0, 3, 2, 3, 5, 6, 5, 9, 1, 2, 4, 5, 2, 0, 10, 3, 6, 3, 2, 5, 6, 3, 3, 2, 5, 1, 0, 7, 4, and 3 defectives. If the fraction defective is to be maintained at 0.02, construct a p chart for these data and state whether or not this standard is being met.

9. The data of Exercise 8 may be looked upon as evidence that the standard of 2 percent defectives is being exceeded.

(a) Use the data of Exercise 8 to construct new control limits for the fraction defective.

(b) Using the control limits found in part (a), continue the control of the process by plotting the following data on the number of defectives obtained in 20 subsequent samples of size $n = 100$: 2, 4, 2, 4, 7, 5, 3, 2, 2, 3, 5, 6, 4, 5, 8, 0, 5, 5, 4, and 2.

10. The specifications for a certain mass-produced valve prescribe a testing procedure according to which each valve can be classified as satisfactory or unsatisfactory (defective). Past experience has shown that the process can perform so that $\bar{p} = 0.03$. Construct a three-sigma control chart for the number of defectives obtained in samples of size 100, and on it plot the following numbers of defectives obtained in such samples randomly selected from 30 successive half-day's production: 3, 4, 2, 1, 5, 2, 1, 2, 3, 1, 3, 2, 2, 2, 1, 1, 2, 0, 4, 3, 1, 0, 2, 4, 0, 1, 5, 7, 3, and 2.

11. The standard for a process producing tin plate in a continuous strip is 5 defects in the form of pinholes or visual blemishes per hundred feet. Based on the following set of 25 observations, giving the number of defects per hundred feet, can it be concluded that the process is in control to this standard?

Inspection number	1	2	3	4	5	6	7	8	9	10	11	12
Number of defects	3	2	2	4	4	4	6	4	1	7	5	5

Inspection number	13	14	15	16	17	18	19	20	21	22	23	24	25
Number of defects	4	6	6	9	5	2	6	5	11	6	6	8	2

12. A process for the manufacture of 4 feet by 8 feet woodgrained panels has performed in the past with an average of 2.7 imperfections per 100 panels. Construct a chart to be used in the inspection of the panels and discuss the control if 25 successive 100-panel lots contained, respectively, 4, 1, 0, 3, 5, 3, 5, 4, 1, 4, 0, 1, 4, 2, 3, 7, 4, 2, 1, 3, 0, 2, 6, 1, and 3 imperfections.

14.5   TOLERANCE LIMITS

Inherent in every phase of industrial quality control is the problem of comparing some quality characteristic or measurement of a finished product against given specifications. Sometimes the specifications, or *tolerance limits*, are so stated by the customer or by the design engineer that any appreciable departure will make the product unusable. There remains, however, the problem of producing the part so that an acceptably high proportion of units will fall within tolerance limits specified for the given quality characteristic. Also, if a product is made without prior specifications, or if modifications are made, it is desirable to know within what limits the process can hold a quality characteristic a reasonably high percentage of the time. We, thus, speak of "natural" tolerance limits; that is, we let the process establish its own limits which, according to experience, can be met in actual practice.

If reliable information is available about the distribution underlying the measurement in question, it is a relatively simple matter to find natural tolerance limits. For instance, if long experience with a product enables us to assume that a certain dimension is normally distributed with the mean μ and the standard deviation σ, it is easy to construct limits between which we can expect to find any given proportion P of the population. For $P = 0.90$ we have the tolerance limits $\mu \pm 1.645\sigma$, and for $P = 0.95$ we have $\mu \pm 1.96\sigma$, as can easily be verified from a table of normal curve areas.

In most practical situations the true values of μ and σ are not known, and tolerance limits must be based on the mean $\bar{x}$ and the standard deviation s of a random sample. Whereas $\mu \pm 1.96\sigma$ are limits including 95 percent of a normal population, the same cannot be said for the limits $\bar{x} \pm 1.96s$. These limits are values of random variables and they may or may not include a given proportion of the population. Nevertheless, it is possible to determine a constant K so that *one can assert with a degree of confidence* $1 - \alpha$ *that the proportion of the population contained between* $\bar{x} - Ks$ *and* $\bar{x} + Ks$ *is at least P.* Such values of K for random samples from normal populations are given in Table 14 for $P = 0.90$, 0.95, and 0.99, degrees of confidence of 0.95 and 0.99, and selected values of n from 2 to 1000.

Example. To illustrate this technique, suppose that a manufacturer takes a sample of size $n = 100$ from a very large lot of mass-produced compression springs and that he obtains $\bar{x} = 1.507$ and $s = 0.004$ inch for the free lengths of the springs. Choosing a level of confidence of 0.99 and a minimum proportion of $P = 0.95$, he obtains the tolerance limits $1.507 \pm (2.355)(0.004)$; in other words, the manufacturer can assert with a degree of confidence of 0.99 that at least 95 percent of the springs in the entire lot have free lengths from 1.497 to 1.517 inches. Note that in problems like these the minimum proportion P as well as the degree of confidence $1 - \alpha$ must be specified; also note that the lower tolerance limit is rounded *down* and the upper tolerance limit is rounded *up*.

To avoid confusion, let us also point out that there is an essential difference between confidence limits and tolerance limits. Whereas confidence limits are used to estimate a parameter of a population, tolerance limits are used to indicate between what limits one can find a certain proportion of a population. This distinction is emphasized by the fact that when n becomes large the length of a confidence interval approaches zero, while the tolerance limits will approach the corresponding values for the population. Thus, for large n, K approaches 1.96 in the columns for $P = 0.95$ in Table 14.

EXERCISES

1. To check the strength of carbon steel for use in chain links, the yield stress of a random sample of 25 pieces was measured, yielding a mean and a standard deviation of 52,800 psi and 4,600 psi, respectively. Establish tolerance limits

with $\alpha = 0.05$ and $P = 0.99$, and express *in words* what these tolerance limits mean.

2. In a study designed to determine the number of turns required for an artillery-shell fuse to arm, 75 fuses, rotated on a turntable, averaged 38.7 turns with a standard deviation of 4.3 turns. Establish tolerance limits for which one can assert with a degree of confidence 0.99 that *at least 95 percent* of the fuses will arm within these limits.

3. In a random sample of 40 piston rings chosen from a production line, the mean edge width was 0.1063 inch, and the standard deviation was 0.0004 inch.
   (a) Between what limits can it be said with 95 percent confidence that at least 90 percent of the edge widths of piston rings produced by this production line will lie?
   (b) Find 95 percent confidence limits for the true mean edge width, and explain the difference between these limits and the tolerance limits found in part (a).

4. *Nonparametric tolerance limits* can be based on the extreme values in a random sample of size n taken from any continuous population. The following equation relates the quantities n, P, and α, where P is the minimum proportion of the population contained between the smallest and largest observations with confidence $1 - \alpha$:

$$nP^{n-1} - (n - 1)P^n = \alpha$$

An approximate solution for n is given by

$$n \simeq \frac{1}{2} + \frac{1 + P}{1 - P} \cdot \frac{\chi_\alpha^2}{4}$$

where χ_α^2 is the value of chi square for 4 degrees of freedom that corresponds to a right-hand tail of area α.
   (a) How large a sample is required to be 95 percent certain that at least 90 percent of the population will be included between the extreme values of the sample?
   (b) At least what proportion of the population can be expected to be included between the extreme values of a sample of size 100 with 95 percent confidence?

14.6 ACCEPTANCE SAMPLING

Manufactured goods are shipped to the purchaser in lots ranging in size from only a few to many thousands of individual items. Ideally, each lot should not contain any defectives, but practically speaking it is rarely possible to meet this goal. Recognizing the fact that some defective goods are bound to be delivered, even if each lot were to be inspected 100 percent, most consumers require that evidence, based on careful inspection, be given that the proportion of defectives in each lot is not excessive.

A frequently used and highly effective method for providing such evidence is that of sampling inspection, where items are selected from each lot prior to shipment (or prior to acceptance by the consumer), and a decision is made on the basis of this sample whether to accept or reject the lot. Acceptance of a lot ordinarily implies that it can be shipped (or be accepted by the consumer), even though it may contain some defective items. Arrangements between the producer and the consumer may allow for some form of credit to be given for defectives subsequently discovered by the consumer. Rejection of a lot need not mean that it is to be scrapped; a rejected lot may be subjected to closer inspection with the aim of eliminating all defective items.

Since the cost of inspection is rarely negligible (sometimes it is nearly as high as or higher than the cost of production), it is seldom desirable to inspect each item in a lot. Thus, acceptance inspection usually involves sampling; more specifically, a random sample is selected from each lot and the lot is accepted if the number of defectives found in the sample does not exceed a given *acceptance number*. This procedure is equivalent to a test of the null hypothesis that the proportion defective p in the lot equals some specified value p_0 against the alternative that it equals p_1, where $p_1 > p_0$. In acceptance sampling the value p_0 is called the *acceptable quality level* or *AQL*, and p_1 is called the *lot tolerance percent defective* or *LTPD*. The probability of a Type I error, α, can be interpreted as an upper limit to the proportion of "good" lots (lots with $p \leq p_0$) that will be rejected, and in this context it is called the *producer's risk*. The probability of a Type II error, β, gives an upper bound to the proportion of "bad" lots (lots with $p \geq p_1$) that will be accepted, and it is called the *consumer's risk*.

A *single-sampling plan* is simply a specification of the sample size and the acceptance number to be used, and its choice is usually based on a specified *AQL* and (or) *LTPD* in association with given producer's and (or) consumer's risks. A given sampling plan is best described by its operating-characteristic or *OC* curve, which gives the probability of acceptance for each value that can be assumed by the lot proportion defective p. Thus, the *OC* curve describes the degree of protection offered by the sampling plan against incoming lots of various qualities. If a sample of size n is taken from a lot containing N units, and if the acceptance number is c, the probability of accepting a lot containing the proportion of defectives p (the lot contains Np defectives) can be calculated by using the hypergeometric distribution, as follows:

$$L(p) = \sum_{x=0}^{c} h(x; n, Np, N)$$

Example. If the lot size is $N = 100$, the sample size is $n = 10$, and the acceptance number is $c = 1$, we have

$$L(p) = \frac{\binom{100p}{0}\binom{100(1 - p)}{10} + \binom{100p}{1}\binom{100(1 - p)}{9}}{\binom{100}{10}}$$

Since calculations involving the hypergeometric distribution are fairly tedious, especially when n and N are large, it is customary in acceptance sampling to approximate the hypergeometric distribution with the binomial distribution, as on page 62. For instance, for $p = 0.10$ the exact value is $L(0.10) = 0.739$, as the reader will be asked to verify in Exercise 5 on page 447, while Table 1 with $n = 10$ and $p = 0.10$ yields the binomial approximation

$$L(0.10) \simeq B(1; 10, 0.10) = 0.736$$

A sketch of the OC curve for the sampling plan $n = 10$, $c = 1$ can be made rapidly with the aid of Table 1, and the result is shown in Figure 14.7. From this curve it can be seen that the producer's risk is approximately 0.05 when the AQL is 0.04, and the consumer's risk is approximately 0.10 when the $LTPD$ is 0.34.

A sampling plan can also be described by means of its *average outgoing quality* or *AOQ* curve. This curve describes the degree of protection offered by the sampling plan by showing the average quality of outgoing lots corresponding to each quality level of incoming lots (that is, lots prior to inspection). If incoming lots are of good quality, that is, if their proportion defective is smaller than the AQL, very few lots will be rejected and the average outgoing quality or AOQ will be good. If the incoming lots are of poor quality, that is, if their proportion defective is larger than the $LTPD$, most of them will be rejected. If all rejected lots are inspected 100 percent

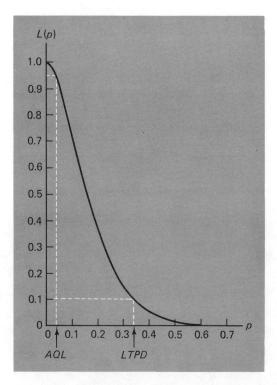

Figure 14.7. OC curve.

and all defective units are replaced by good units prior to acceptance of the lot, then the average outgoing quality will be good even though the average incoming quality is poor. It is when the average incoming quality lies between the AQL and the $LTPD$ that the poorest quality of lots will be shipped. In general, there will be a maximum AOQ over all values of incoming quality p, and this value is called the *average outgoing quality limit*, or $AOQL$.

It is not difficult to derive a formula for finding the AOQ corresponding to a given incoming quality p under the assumption that all defectives in rejected lots are replaced by acceptable items prior to their final acceptance. If the incoming quality is p, the probability that a lot will be accepted is $L(p)$, and each such lot contains the proportion p of defectives. The proportion $1 - L(p)$ of lots which are eventually rejected contain no defectives, and it follows that the AOQ is given by $p \cdot L(p) + 0 \cdot [1 - L(p)]$, or

$$AOQ = p \cdot L(p)$$

More common practice is to remove or replace defectives found in accepted as well as rejected lots, but the modification thus required in the AOQ is usually minor and it is customary to use the above formula regardless of the inspection procedure.

Continuation of Example. The AOQ curve for the sampling plan $n = 10$, $c = 1$ is shown in Figure 14.8, and it is apparent from this figure that the $AOQL$ is approximately 0.081.

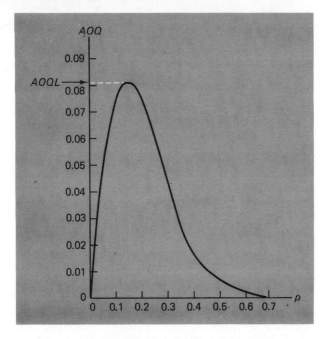

Figure 14.8. AOQ curve.

Sometimes, smaller samples (and hence, reductions in cost) can be achieved without sacrifice in the degree of protection by the use of so-called *double* or *multiple sampling*. A *double sampling plan* involves the selection of a random sample of size n_1 from a lot; if the sample contains c_1 or fewer defectives, the lot is accepted; if it contains c'_1 or more defectives $(c'_1 > c_1)$ the lot is rejected; otherwise, a second sample of size n_2

is taken from the lot, and the lot is accepted unless the total number of defectives in the combined sample of size $n_1 + n_2$ exceeds c_2. A *multiple sampling plan* is similar in nature to a double sampling plan, but it involves more than two stages.

Example. An example of a multiple sampling plan is shown in the following table:

		Combined samples		
Sample	Sample size	Size	Acceptance number	Rejection number
First	20	20		3
Second	20	40	1	4
Third	20	60	3	5
Fourth	20	80	3	6
Fifth	20	100	5	7
Sixth	20	120	6	8
Seventh	20	140	7	8

In the first step, the lot is rejected if there are 3 or more defectives, otherwise sampling continues. In the second step, the lot is accepted if the combined sample contains at most 1 defective, it is rejected if there are 4 or more defectives, otherwise sampling continues. This goes on, if necessary, until in the final step the lot is accepted if there are at most 7 defectives in the combined sample of size 140, and otherwise it is rejected.

By an appropriate choice of the sample sizes and the acceptance and rejection numbers it is possible to match the *OC* curve of a double or multiple sampling plan closely to that of an equivalent single sampling plan. Thus, the degree of protection offered by a double or multiple sampling plan can be made essentially the same as that offered by an equivalent single sampling plan.

The advantage of double or multiple sampling is that there is a high probability that a very good lot will be accepted or a very poor lot will be rejected on the basis of the first sample (or, at least, an early sample), thus reducing the required amount of inspection. On the other hand, if the lot quality is "intermediate," the total sample size required may actually be larger than that of the equivalent single sampling plan.

Example. Consider the following double sampling plan:

Sample	Sample size	Combined samples		
		Size	Acceptance number	Rejection number
First	15	15	1	5
Second	30	45	5	6

If the incoming lot quality is $p = 0.05$, the probability that a second sample will be required is the same as the probability that there are 2, 3, or 4 defectives in a sample of size 15. According to Table 1, this probability is equal to

$$B(4; 15, 0.05) - B(1; 15, 0.05) = 0.170$$

Thus, on the average, it will require a sample of size $15 + (0.170)(30) = 20.1$ to decide whether to accept or reject an incoming lot of quality $p = 0.05$. Similar calculations enable us to find the average sample size required to inspect a lot having any given incoming quality p. A graph showing the relation between the average sample size (also called the *average sample number*) and the incoming lot quality is called an *ASN* curve; the *ASN* curve for the double sampling plan described above is shown in Figure 14.9.

Several standard sampling plans have been published to facilitate the use of acceptance sampling (see the book by I. W. Burr listed in the Bibliography). Among the most widely used standard plans are those contained in the *Military Standard 105 D Tables* (see Bibliography). These plans stress the maintenance of a specified *AQL*, and they are designed to encourage the producer to offer only good products to the consumer. To accomplish this there are three general levels of inspection corresponding to different consumer's risks. (Inspection level II is normally chosen; level I uses smaller sample sizes and level III uses larger sample sizes than level II.) There are also three types of inspection: normal, tightened, and reduced. The type of inspection depends on whether the average proportion defectives for prior samples has been above or below the *AQL*, and it may be changed during the course of inspection. Under tightened inspection the producer's risk is increased and the consumer's risk is (slightly) decreased; under reduced inspection the consumer's risk is increased and the producer's risk is (slightly) decreased. Tables are available for single, double,

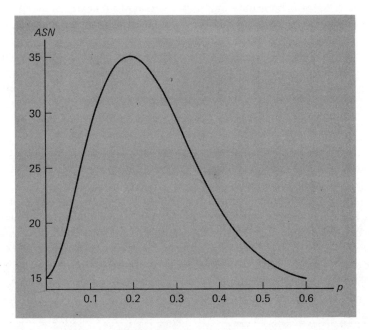

Figure 14.9. *ASN* curve.

and multiple sampling, and a brief portion of these tables is included in this book in Tables 15 and 16.

The procedure in using MIL-STD-105D for single sampling is first to find the sample size code letter corresponding to the lot size and the desired inspection level and type. Then, using the sample size code letter thus obtained and the appropriate *AQL*, one finds the sample size and the acceptance number from the master table. A portion of the table for finding sample size code letters is given in Table 15, and a portion of the master table for normal inspection is included in Table 16.

Example. To illustrate the use of MIL-STD-105D, suppose that incoming lots contain 2,000 items, and inspection level II is to be used in conjunction with normal inspection and an *AQL* of 0.025, or 2.5 percent. From Table 15 we find that the sample size code letter is *K*. Then, entering Table 16 in the row labeled *K*, we find that the sample size to be used is 125. Using the column labeled 2.5, we find that the acceptance number is 7 and the rejection number is 8. Thus, if a single sample of size 125, selected at random from a lot of 2,000 items, contains 7 or fewer defectives the lot is to be accepted; if the sample contains 8 or more defectives the lot is to be rejected.

The concept of multiple sampling is carried to its extreme in so-called *sequential sampling*. A sampling procedure is said to be *sequential* if, after each observation, one of the following decisions is made: accept whatever hypothesis is being tested, reject the hypothesis, or take another observation. Although sequential procedures are used also in connection with other kinds of problems, we shall discuss this kind of sampling only in connection with acceptance sampling, where we shall thus decide after the inspection of each successive item whether to accept a lot, reject it, or continue sampling.

The construction of a sequential sampling plan consists of finding two sequences of numbers a_n and r_n, where n is the number of observations, so that the lot is accepted as soon as the number of defectives is less than or equal to a_n for some n, the lot is rejected as soon as the number of defectives is greater than or equal to r_n for some n, and sampling continues so long as the number of defectives in a sample of size n falls between a_n and r_n. If an acceptance plan is to have p_0 and p_1 as its AQL and $LTPD$, the producer's risk α, and the consumer's risk β, it can be shown (see the book by A. Wald in the Bibliography) that the required values of a_n and r_n can be computed by means of the formulas

$$a_n = \frac{\log \dfrac{\beta}{1-\alpha} + n \cdot \log \dfrac{1-p_0}{1-p_1}}{\log \dfrac{p_1}{p_0} - \log \dfrac{1-p_1}{1-p_0}}$$

$$r_n = \frac{\log \dfrac{1-\beta}{\alpha} + n \cdot \log \dfrac{1-p_0}{1-p_1}}{\log \dfrac{p_1}{p_0} - \log \dfrac{1-p_1}{1-p_0}}$$

If a_n is not an integer, it is replaced by the largest integer less than a_n; if r_n is not an integer, it is replaced by the smallest integer greater than r_n.

Example. To illustrate this procedure, let $p_0 = 0.05$, $p_1 = 0.20$, $\alpha = 0.05$, and $\beta = 0.10$. Substituting these values into the above formulas for a_n and r_n, we obtain

$$a_n = -1.45 + 0.11n$$
$$r_n = 1.86 + 0.11n$$

and, letting $n = 1, 2, 3, \ldots$, and 25, we get the acceptance and rejection numbers shown in the second and fourth columns of the following table:

Number of items inspected n	Acceptance number a_n	Number of defectives d_n	Rejection number r_n
1	–	0	–
2	–	0	–
3	–	0	3
4	–	0	3
5	–	0	3
6	–	0	3
7	–	0	3
8	–	1	3
9	–	1	3
10	–	1	3
11	–	1	4
12	–	1	4
13	–	1	4
14	0	1	4
15	0	1	4
16	0	1	4
17	0	2	4
18	0	2	4
19	0	3	4
20	0	3	5
21	0	3	5
22	0	4	5
23	1	5	5
24	1		5
25	1		5

In the third column we have indicated the results obtained in an inspection where the 8th, 17th, 19th, 22nd, and 23rd items are defective, and where the inspection terminates with rejection of the lot after inspection of the 23rd item.

The tabular procedure illustrated can be replaced by an equivalent graphical procedure for carrying out sequential sampling inspection. Plotting the values of a_n and r_n obtained from the equations on page 444 (without rounding), we obtain two *straight lines* like those of Figure 14.10. Sampling terminates with rejection or acceptance if the number of defectives observed falls above the line for r_n or below the line for a_n, respectively.

The main advantage of sequential sampling is that it can materially reduce the required amount of inspection. Studies have shown that the

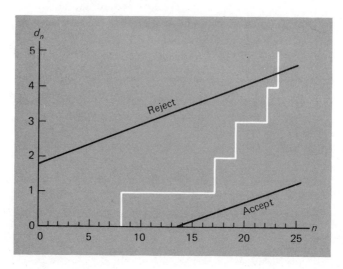

Figure 14.10. Graphical procedure for sequential sampling.

average decrease in sample size is often near 50 percent when compared with the sample size of equivalent single sampling plans. The major disadvantage is that, in a sequential sampling plan, there is no upper limit to the number of items that might have to be inspected in order to reach a decision concerning the acceptance or rejection of a lot. In fact, the sample size is a random variable and its value will occasionally be very large. For this reason it is customary to *truncate* sequential sampling procedures by selecting a number N such that a decision must be made to accept or reject a lot prior to or at $n = N$.

EXERCISES

1. A sampling plan, calling for a sample of size $n = 50$, has the acceptance number $c = 3$. Assuming that the lot size is very large, calculate the probability of accepting a lot of incoming quality 15 percent defective and the probability of rejecting a lot of incoming quality 4 percent defective:
   (a) by calculating the corresponding binomial probabilities;
   (b) by using the Poisson approximation to the binomial distribution.

2. A single sampling plan calls for a sample of size 150.
   (a) Use the normal approximation to find the acceptance number c if the AQL is to be 2.5 percent and the producer's risk is to be $\alpha = 0.05$.
   (b) Using the acceptance number obtained in part (a), determine the consumer's risk if the $LTPD$ is to be 6 percent.

3. Referring to the sampling plan of Exercise 1, use the Poisson approximation to calculate $L(p)$ for $p = 0.01, 0.02, \ldots, 0.19$, and 0.20. Sketch the OC curve for this sampling plan and read off the consumer's and producer's risks corresponding to an AQL of 4 percent and an $LTPD$ of 14 percent.

4. A single sampling plan has $n = 100$ and $c = 5$.
   (a) Find the AQL if the producer's risk is to be 0.025.
   (b) Find the $LTPD$ if the consumer's risk is to be 0.05. [*Hint*: Use the normal approximation and set up equations leading to quadratic equations in p_0 and p_1, respectively.]

5. With reference to the example of page 438, verify that
   (a) $L(0.10) = 0.739$, using the exact (hypergeometric) formula;
   (b) $L(0.10) \simeq 0.736$, using the approximate (binomial) formula;
   (c) the OC curve is as given in Figure 14.7;
   (d) the AOQ curve is as given in Figure 14.8.

6. Calculate $L(p)$ for selected values of p and sketch the OC curve for the single sampling plan $n = 120$, $c = 3$. [*Hint*: Assume a large lot size, and use the normal approximation to the binomial distribution.]

7. Sketch the AOQ curve for the sampling plan of Exercise 6, and estimate the $AOQL$.

8. Using the values of $L(p)$ obtained in Exercise 3, sketch the AOQ curve for the given sampling plan and estimate the $AOQL$.

9. Referring to the double sampling plan on page 442, use Table 1 and the normal approximation (when necessary) to calculate the producer's risk α when the AQL is $p_0 = 0.10$.

10. Calculate the average sample size required for selected values of p and sketch the ASN curve for the following double sampling plan:

		Combined samples		
Sample	Sample size	Size	Acceptance number	Rejection number
First	10	10	0	3
Second	35	45	3	4

11. A lot of 200 items is to be inspected at an AQL of 6.5 percent. If MIL-STD-105D is to be used, with normal inspection at general inspection level II, what single sampling plan is required?

12. An incoming lot of 5,000 items is to be inspected using MIL-STD-105D, with normal inspection at general inspection level II and an AQL of 2.5 percent. What single sampling plan should be used?

13. Find formulas for the acceptance and rejection numbers for the sequential

sampling plan having an AQL of 0.10, an $LTPD$ of 0.30, a producer's risk of 0.05, and a consumer's risk of 0.10. If a sample should contain defectives on the third, fifth, seventh, and eighth trials, would this plan accept or reject the corresponding lot prior to the tenth trial? If so, on which trial?

14. A sequential sampling plan is to have $p_0 = 0.01$, $p_1 = 0.10$, $\alpha = 0.05$, and $\beta = 0.20$.
    (a) Determine the acceptance and rejection numbers for $n = 1, 2, \ldots$, and 50.
    (b) Use random numbers and the acceptance and rejection numbers obtained in part (a) to simulate the inspection of a very large lot containing 20 percent defectives.

15

Applications to Reliability and Life Testing

15.1 INTRODUCTION

The task of designing and supervising the manufacture of a product has been made increasingly difficult by rapid strides in the sophistication of modern products and the severity of the environmental conditions under which they must perform. No longer can an engineer be satisfied if the operation of a product is technically feasible, or if it can be made to "work" under optimum conditions. In addition to such considerations as cost and ease of manufacture, increasing attention must now be paid to size and weight, ease of maintenance, and reliability. The magnitude of the problem of maintainability and reliability is illustrated by surveys which have uncovered the fact that frequently a high percentage of space-age electronic equipment has been in inoperative condition. Military surveys have further shown that maintenance and repair expenses for electronic equipment often exceed the original cost of procurement, even during the first year of operation.

The problem of assuring and maintaining reliability has many facets, including original equipment design, control of quality during production, acceptance inspection, field trials, life testing, and design modifications. To complicate matters further, reliability competes directly or indirectly with a host of other engineering considerations, chiefly cost, complexity, size and weight, and maintainability. In spite of its complicated engineering aspects, it is possible to give a relatively simple mathematical definition of reliability. To motivate this definition, we call the reader's attention to the fact that a product may function satisfactorily under one set of conditions but not under other conditions, and that satisfactory performance for one purpose does not assure adequate performance for another purpose. For example, a vacuum tube which is perfectly satisfactory for use in a home radio may be entirely unsatisfactory for use in the airborne guidance system of a missile. Accordingly, we shall define the reliability of a product as *the probability that it will function within specified limits for at least a specified period of time under specified environmental conditions.* Thus, the reliability of a "standard equipment" automobile tire is close to unity for 10,000 miles of normal operation on a passenger car, but it is virtually zero for use at the Indianapolis "500."

Since reliability has been defined as a probability, the theoretical treatment of this subject is based essentially on the material introduced in the early chapters of this book. Thus, the rules of probability introduced in Chapter 2 can be applied directly to the calculation of the reliability of a complex system, if the reliabilities of the individual components are known. (Estimates of the reliabilities of the individual components are usually obtained from statistical life tests, such as those discussed in Sections 15.4 and 15.5.)

Many systems can be considered to be series or parallel systems, or a combination of both. A *series system* is one in which all components are so interrelated that the entire system will fail if any one of its components fails; a *parallel system* is one that will fail only if all of its components fail.

Let us first discuss a system of n components connected in series, and let us suppose that the components are *independent*, namely, that the performance of any one part does not affect the reliability of the others. Under these conditions, the probability that the system will function is given by the special rule of multiplication for probabilities, and we have

$$R_s = \prod_{i=1}^{n} R_i$$

where R_i is the reliability of the ith component and R_s is the reliability of the series system. This simple *product law of reliabilities*, applicable to series systems of independent components, vividly demonstrates the effect of increased complexity on reliability.

Example. Suppose a system consists of five independent components in series, each having a reliability of 0.970; then the reliability of the whole series system is $(0.970)^5 = 0.859$. If the system complexity were increased so that it contained 10 similar components, its reliability would be reduced to $(0.970)^{10} = 0.738$. Looking at the effect of increasing complexity in another way, we find that each of the components in the ten-component system would require a reliability of 0.985, instead of 0.970, for the ten-component system to have a reliability equal to that of the original five-component system.

One way to increase the reliability of a system is to replace certain components by several similar components connected in parallel. If a system consists of n independent components connected in parallel, it will fail to function only if all n components fail. Thus, if $F_i = 1 - R_i$ is the "unreliability" of the ith component, we can again apply the special rule of multiplication for probabilities to obtain

$$F_p = \prod_{i=1}^{n} F_i$$

where F_p is the unreliability of the parallel system, and $R_p = 1 - F_p$ is the reliability of the parallel system. Thus, for parallel systems we have a *product law of unreliabilities* analogous to the product law of reliabilities for series systems. Writing this law in another way, we get

$$R_p = 1 - \prod_{i=1}^{n} (1 - R_i)$$

for the reliability of a parallel system.

Example. The two basic formulas for the reliability of series and parallel systems can be used in combination to calculate the reliability of a system having both series and parallel parts. To illustrate such a calculation, consider the system diagrammed in Figure 15.1, which consists of eight components having the reliabilities shown in that figure. The parallel assembly C, D, E can be replaced by an equivalent

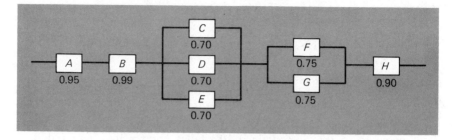

Figure 15.1. Systems reliability.

component C' having the reliability $1 - (1 - 0.70)^3 = 0.973$, without affecting the overall reliability of the system. Similarly, the parallel assembly F, G can be replaced by a single component F' having the reliability $1 - (1 - 0.75)^2 = 0.9375$. The resulting series system A, B, C', F', H, equivalent to the original system, has the reliability

$$(0.95)(0.99)(0.973)(0.9375)(0.90) = 0.772.$$

15.2 FAILURE-TIME DISTRIBUTIONS

According to the definition of reliability given in the preceding section, the reliability of a system or a component will often depend on the length of time it has been in service. Thus, of fundamental importance in reliability studies is the *distribution of the time to failure* of a component under given environmental conditions. A useful way to characterize this distribution is by means of its associated *instantaneous failure rate*, defined as follows: If $f(t)$ is the probability density of the time to failure of a given component, that is, the probability that the component will fail between times t and $t + \Delta t$ is given by $f(t) \cdot \Delta t$, then the probability that the component will fail on the interval from 0 to t is given by

$$F(t) = \int_0^t f(x)\, dx$$

and the *reliability function*, expressing the probability that it survives to time t, is given by

$$R(t) = 1 - F(t)$$

Thus, the probability that the component will fail in the interval from t to $t + \Delta t$ is $F(t + \Delta t) - F(t)$, and the conditional probability of failure in this interval, *given that the component survived to time t*, is expressed by

$$\frac{F(t + \Delta t) - F(t)}{R(t)}$$

Dividing by Δt, we obtain the average rate of failure in the interval from t to $t + \Delta t$, given that the component survived to time t:

$$\frac{F(t + \Delta t) - F(t)}{\Delta t} \cdot \frac{1}{R(t)}$$

Taking the limit as $\Delta t \to 0$, we get the *instantaneous failure rate*, or simply the *failure rate*

$$Z(t) = \frac{F'(t)}{R(t)}$$

where $F'(t)$ is the derivative of $F(t)$ with respect to t. Finally, observing that $f(t) = F'(t)$ (see page 100) we get the relation

$$Z(t) = \frac{f(t)}{R(t)} = \frac{f(t)}{1 - F(t)}$$

which expresses the failure rate in terms of the distribution of failure times.

Example. A failure-rate curve that is typical of many manufactured items is shown in Figure 15.2. The curve is conveniently divided into three parts. The first part is characterized by a decreasing failure rate

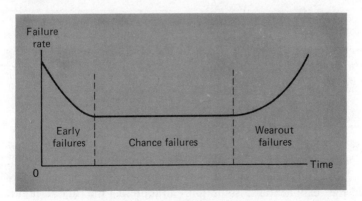

Figure 15.2. Typical failure-rate curve.

and it represents the period during which poorly manufactured items are weeded out. (It is common in the electronics industry to "burn in" components prior to actual use in order to eliminate any early failures.) The second part, which is often characterized by a constant failure rate, is normally regarded as the period of useful life during which only chance failures occur. The third part is characterized by an increasing failure rate, and it is the period during which components fail primarily because they are worn out. Note that the same general failure-rate curve is typical of human mortality, where the first part represents infant mortality, and the third part corresponds to old-age mortality.

Let us now derive an important relationship expressing the failure-time density in terms of the failure-rate function. Making use of the fact that $R(t) = 1 - F(t)$ and, hence, that $F'(t) = -R'(t)$, we can write

$$Z(t) = -\frac{R'(t)}{R(t)}$$

$$= -\frac{d[\ln R(t)]}{dt}$$

Solving this differential equation for $R(t)$, we obtain

$$R(t) = e^{-\int_0^t Z(x)\,dx}$$

and, making use of the relation $f(t) = Z(t) \cdot R(t)$, we finally get

$$f(t) = Z(t) \cdot e^{-\int_0^t Z(x)\,dx}$$

As illustrated in Figure 15.2, it is often assumed that the failure rate is constant during the period of useful life of a component. Denoting this constant failure rate by α, where $\alpha > 0$, and substituting α for $Z(t)$ in the formula for $f(t)$, we obtain

$$f(t) = \alpha \cdot e^{-\alpha t} \qquad t > 0$$

Thus, we observe that the distribution of failure times is an *exponential distribution* when it can be assumed that the failure rate is constant. For this reason, the assumption of constant failure rates is sometimes also called the "exponential assumption." Interpreting the time to failure as a *waiting time*, we can use the results of Section 4.7 to conclude that the occurrence of failures is a Poisson process, if a component which fails is immediately replaced with a new one having the same constant failure rate α. As we observed on page 119, the mean waiting time between successive failures is $1/\alpha$, or the reciprocal of the failure rate. Thus, the constant $1/\alpha$ is often referred to as the *mean time between failures*, and it is abbreviated *MTBF*.

There are situations in which the assumption of a constant failure rate is not realistic, and in many of these situations one assumes instead that the failure rate function increases or decreases "smoothly" with time. In other words, it is assumed that there are no discontinuities or turning points. This assumption would be consistent with the initial and last stages of the failure-rate curve shown in Figure 15.2.

A useful function that is often used to approximate such failure-rate curves is given by

$$Z(t) = \alpha \beta t^{\beta-1} \qquad t > 0$$

where α and β are positive constants. Note the generality of this function: If $\beta < 1$ the failure rate *decreases* with time, if $\beta > 1$ it *increases* with time, and if $\beta = 1$ the failure rate equals α. Note that the assumption of a constant failure rate, the exponential assumption, is thus included as a special case.

If we substitute the above expression for $Z(t)$ into the formula for $f(t)$ on page 454, we obtain

$$f(t) = \alpha \beta t^{\beta-1} e^{-\alpha t^{\beta}} \qquad t > 0$$

where α and β are positive constants. This density, or distribution, is the *Weibull distribution*, introduced in Section 4.9, and we shall discuss its application to problems of life testing in Section 15.5.

15.3 THE EXPONENTIAL MODEL IN RELIABILITY

If we make the exponential assumption about the distribution of failure times, some very useful results can be derived concerning the *MTBF*, the mean time between failure, of series and parallel systems. In order to use the product laws of Section 15.1, we shall first have to obtain a relation expressing the reliability of a component in terms of its service time t. Making use of the fact that

$$R(t) = 1 - F(t) = 1 - \int_0^t f(t)\,dt$$

we obtain

$$R(t) = 1 - \int_0^t \alpha e^{-\alpha x}\, dx = e^{-\alpha t}$$

for the reliability function of the exponential model. Thus, if a component has a failure rate of 0.05 per thousand hours, the probability that it will survive at least 10,000 hours of operation is $e^{-(0.05)10} = 0.607$.

Suppose now that a system consists of n components connected in *series*, and that these components have the respective failure rates α_1, $\alpha_2, \ldots$, and α_n. The product law of reliabilities can be written as

$$R_s(t) = \prod_{i=1}^{n} e^{-\alpha_i t} = e^{-\left(\sum_{i=1}^{n} \alpha_i\right)t}$$

and it can be seen that the reliability function of the series system also satisfies the exponential assumption. The failure rate of the entire series system is readily identified as $\sum_{i=1}^{n} \alpha_i$, the *sum* of the failure rates of its components. Since the *MTBF* is the reciprocal of the failure rate when each component which fails is replaced immediately with another having the identical failure rate, we obtain the formula

$$\mu_s = \frac{1}{\dfrac{1}{\mu_1} + \dfrac{1}{\mu_2} + \cdots + \dfrac{1}{\mu_n}}$$

expressing the *MTBF* μ_s of a series system in terms of the *MTBF*'s μ_i of its components. In the special case where all n components have the same failure rate α and hence the same *MTBF* μ, the system failure rate is $n\alpha$, and the system *MTBF* is $1/n\alpha = \mu/n$.

For parallel systems the results are not quite so simple. If a system consists of n components in parallel, having the respective failure rates $\alpha_1, \alpha_2, \ldots, \alpha_n$, the system "unreliability" to time t is given by

$$F_p(t) = \prod_{i=1}^{n} (1 - e^{-\alpha_i t})$$

Thus, the failure-time distribution of a parallel system is not exponential even when each of its components satisfies the exponential assumption.

The system failure-rate function can be obtained by means of the formula $Z_p(t) = F'_p(t)/R_p(t)$, but the result is fairly complicated. Note, however, that the system failure rate is not constant, but depends on t, the "age" of the system.

The mean time to failure of a parallel system is also difficult to obtain in general, but in the special case where all components have the same failure rate α, an interesting and useful result can be obtained. In this special case the system reliability function becomes

$$R_p(t) = 1 - (1 - e^{-\alpha t})^n$$

$$= \binom{n}{1}e^{-\alpha t} - \binom{n}{2}e^{-2\alpha t} + \ldots + (-1)^{n-1}e^{-n\alpha t}$$

after using the binomial theorem to expand $(1 - e^{-\alpha t})^n$. Then, making use of the fact that $f_p(t) = -R'_p(t)$, we obtain

$$f_p(t) = \alpha\binom{n}{1}e^{-\alpha t} - 2\alpha\binom{n}{2}e^{-2\alpha t} + \ldots + (-1)^{n-1}n\alpha e^{-n\alpha t}$$

and the mean of the failure time distribution is given by

$$\mu_p = \int_0^\infty t \cdot f_p(t)\, dt$$

$$= \alpha\binom{n}{1}\int_0^\infty te^{-\alpha t}\, dt - 2\alpha\binom{n}{2}\int_0^\infty te^{-2\alpha t}\, dt + \ldots$$

$$+ (-1)^{n-1}n\alpha\int_0^\infty te^{-n\alpha t}\, dt$$

$$= \frac{1}{\alpha}\binom{n}{1} - \frac{1}{2\alpha}\binom{n}{2} + \ldots + (-1)^{n-1}\frac{1}{n\alpha}$$

It can be proved by induction that this expression is equivalent to

$$\mu_p = \frac{1}{\alpha}\left(1 + \frac{1}{2} + \ldots + \frac{1}{n}\right)$$

Thus, if a parallel system consists of n components having the identical failure rate α, the mean time between failures of the system equals $\left(1 + \frac{1}{2} + \ldots + \frac{1}{n}\right)$ times the common *MTBF* of its components, provided each defective component is replaced whenever the whole parallel system fails. Thus, if we use two parallel components rather than one, the mean time to failure of the pair exceeds that of the single component by 50 percent, rather than doubling it. In general, the above formula for μ_p expresses a rather severe law of diminishing returns for parallel redundancy.

Example. To illustrate how the formulas derived in this section can be used in system design, let us again consider the system diagrammed in Figure 15.1 on page 452. Assuming the exponential model and that the reliabilities are given for 10 hours of operation, we can calculate the failure rate of component A by solving the equation $0.95 = e^{-10\alpha}$ for α, and we obtain $\alpha = (5.1)10^{-3}$ failures per hour, or 5.1 failures per thousand hours. The failure rates of all eight components (in failures per thousand hours) are as shown in the following table:

Component	A	B	C	D	E	F	G	H
Failure rate	5.1	1.0	35.7	35.7	35.7	28.8	28.8	10.5

To compute the mean time to failure for the entire system, we first obtain the mean failure times for the parallel assemblies C, D, E, and F, G, respectively. For C, D, E we have $\mu_{CDE} = \frac{1}{35.7}\left(1 + \frac{1}{2} + \frac{1}{3}\right) = 0.051$ thousand hours, or 51 hours; for F, G we have $\mu_{FG} = \frac{1}{28.8}\left(1 + \frac{1}{2}\right) = 0.052$ thousand hours, or 52 hours. Although the two parallel assemblies do not have constant failure rates, we shall approximate their respective failure rates by $1/0.051 = 19.6$ and $1/0.052 = 19.2$ failures per thousand hours, and treat the entire system as a series system. Thus, the system failure rate is given approximately by $5.1 + 1.0 + 19.6 + 19.2 + 10.5 = 55.4$ failures per thousand hours, and the mean time to failure of the system is approximately $1/55.4 = 0.018$ or 18 hours.

EXERCISES

1. An old-fashioned string of Christmas-tree lights has 8 bulbs connected in series. What would have to be the reliability of each bulb if there is to be a 95 percent chance of the string's lighting after a year's storage?

2. A system consists of 7 identical components connected in parallel. What must be the reliability of each component if the overall reliability of the system is to be 0.90?

3. A system consists of 6 components connected as in Figure 15.3. Find the overall reliability of the system, given that the reliabilities of A, B, C, D, E, and F are, respectively, 0.95, 0.80, 0.90, 0.99, 0.90, and 0.85.

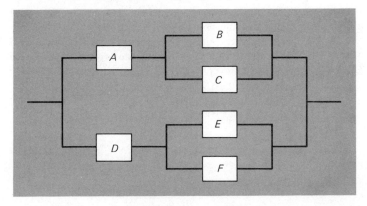

Figure 15.3. Exercise 3.

4. Suppose the flight of an aircraft is regarded as a system having the three main components A (aircraft), B (pilot), and C (airport). Suppose, furthermore, that component B can be regarded as a parallel subsystem consisting of B_1 (captain), B_2 (first officer), and B_3 (flight engineer); and C is a parallel subsystem consisting of C_1 (scheduled airport) and C_2 (alternate airport). Under given flight conditions, the reliabilities of components A, B_1, B_2, B_3, C_1, and C_2 (defined as the probabilities that they can contribute to the successful completion of the scheduled flight) are, respectively, 0.9999, 0.9995, 0.999, 0.20, 0.95, and 0.85.
  (a) What is the reliability of the system?
  (b) What is the effect on system reliability of having a flight engineer who is also a trained pilot, so that the reliability of B_3 is increased from 0.20 to 0.99?
  (c) If the flight crew did not have a first officer, what then would be the effect of increasing the reliability of B_3 from 0.20 to 0.99?
  (d) What is the effect of adding a second alternate landing point, C_3, with reliability 0.80?

5. In some reliability problems we are concerned only with initial failures, treating a component as if (for all practical purposes) it never failed, once it has survived past a certain time $t = \alpha$. In a problem like this, it may be reasonable to use the failure rate

$$Z(t) = \begin{cases} \beta\left(1 - \dfrac{t}{\alpha}\right) & \text{for } 0 < t < \alpha \\ 0 & \text{elsewhere} \end{cases}$$

  (a) Find expressions for $f(t)$ and $F(t)$.

(b) Show that the probability of an initial failure is given by

$$1 - e^{-\alpha\beta/2}$$

6. As has been indicated in the text, one often distinguishes between initial failures, random failures during the useful life of the product, and wear-out failures. Thus, suppose that for a given product the probability of an initial failure (a failure prior to time $t = \alpha$) is θ_1, the probability of a wear-out failure (a failure beyond time $t = \beta$) is θ_2, and that for the interval $\alpha \leq t \leq \beta$ the failure-time density is given by

$$f(t) = \frac{(1 - \theta_1 - \theta_2)}{\beta - \alpha}$$

(a) Find an expression for $F(t)$ for the interval $\alpha \leq t \leq \beta$.
(b) Show that for the interval $\alpha \leq t \leq \beta$ the failure rate is given by

$$Z(t) = \frac{1 - \theta_1 - \theta_2}{(\beta - \alpha)(1 - \theta_1) - (1 - \theta_1 - \theta_2)(t - \alpha)}$$

(c) Suppose that the failure of a color television set is considered to be an initial failure if it occurs during the first 100 hours of usage, and a wear-out failure if it occurs after 15,000 hours. Assuming that the model given in this exercise holds and that θ_1 and θ_2 equal 0.05 and 0.75, respectively, sketch the graph of the failure-rate function from $t = 100$ to $t = 15,000$ hours.

7. An integrated circuit chip has a constant failure rate of 0.02 per thousand hours.
(a) What is the probability that it will operate satisfactorily for at least 20,000 hours?
(b) What is the 5,000-hour reliability of a component consisting of four such chips connected in series?

8. A certain component has an exponential life distribution with a failure rate of $\alpha = 0.0045$ failures per hour.
(a) What is the probability that the component will fail during the first 250 hours it is in operation?
(b) What is the probability that two such components will both survive the first 100 hours of operation?

9. A system containing several identical components in parallel is to have a failure rate of at most $5 \cdot 10^{-5}$ per hour. What is the least number of components that must be used if each has a constant failure rate of 2.0×10^{-4} per hour?

10. A system consists of six different components connected in series. Find the MTBF of the system if the six components have exponential time-to-failure distributions with failure rates of 1.8, 2.4, 2.0, 1.3, 3.0, and 1.5 per 1,000 hours, respectively.

11. A certain part has an exponential life distribution with a mean life, that is, an *MTBF*, of 1,000 hours.
    (a) What is the probability that such a part will last at least 500 hours?
    (b) What is the probability that among three such parts at least one will fail during the first 1,000 hours?
    (c) What is the probability that among four such parts exactly two will fail during the first 600 hours?

12. If a component has the Weibull failure-time distribution with the parameters $\alpha = 0.005$ and $\beta = 0.80$, find the probability that it will operate successfully for at least 5,000 hours.

13. Throughout Sections 15.2 and 15.3 we assumed that the products with which we were concerned were in continuous operation. Consequently, the models discussed in these sections do not apply if we want to investigate the ability of light bulbs to withstand successive voltage overloads, the performance of switches which are repeatedly turned on and off, etc. In each of these cases failure can occur on the xth trial ($x = 1, 2, 3, \ldots$), and it is often assumed that the probability of failure on the xth trial equals some constant p, provided the item has not failed prior to that trial.
    (a) Show that the probability of failure on the xth trial is given by

    $$f(x) = p(1 - p)^{x-1}$$

    for $x = 1, 2, 3, \ldots$. This probability distribution is the *geometric distribution* introduced in Section 3.8.
    (b) Find $F(x)$ for the probability distribution obtained in part (a).
    (c) What is the probability that a switch will survive 2,000 cycles of operation, if the above model holds and the constant probability of failure as the result of any one of the switching cycles is $p = 6 \times 10^{-4}$?

15.4 THE EXPONENTIAL MODEL IN LIFE TESTING

An effective and widely used method of handling problems of reliability is that of life testing. For the purpose of such tests, a random sample of n components is selected from a lot, put on test under specified environmental conditions, and the times to failure of the individual components are observed. If each component that fails is immediately replaced by a new one, the resulting life test is called a *replacement test;* otherwise, the life test is called a *nonreplacement test*. Whenever the mean lifetime of the components is so large that it is not practical, or economically feasible, to test each component to failure, the life test may be *truncated*, that is, it may be terminated after the first r failures have occurred ($r \leq n$), or after a fixed period of time has been accumulated.

A special method that is often used when early results are required in connection with very-high-reliability components is that of *accelerated life testing*. In an accelerated life test the components are put on test under environmental conditions that are far more severe than those normally encountered in practice. This causes the components to fail more quickly, and it can drastically reduce both the time required for the test and the number of components that must be placed on test. Accelerated life testing can be used to compare two or more types of components for the purpose of obtaining a rapid assessment of which is the most reliable. Sometimes, preliminary experimentation is carried out to determine the relationship between the proportion of failures that can be expected under nominal conditions and under various levels of accelerated environmental conditions. The methods of Sections 11.4 and 13.2 can be applied in this connection to determine "derating curves," relating the reliability of the component to the severity of the environmental conditions under which it is to operate.

In the remainder of this section we shall assume that the exponential model holds, namely, that the failure-time distribution of each component is given by

$$f(t) = \alpha \cdot e^{-\alpha t} \qquad t > 0, \alpha > 0$$

In what follows, we shall assume that n components are put on test, life testing is discontinued after a fixed number, r ($r \leq n$), of components have failed, and that the observed failure times are $t_1 \leq t_2 \leq \ldots \leq t_r$. We shall be concerned with estimating and testing hypotheses about the mean life of the component, namely, $\mu = 1/\alpha$.

Using theory developed in the article by B. Epstein mentioned in the Bibliography, it can be shown that unbiased estimates of the mean life of the component are given by

$$\hat{\mu} = \frac{T_r}{r}$$

where T_r is the accumulated life on test until the rth failure occurs, and hence

$$T_r = \sum_{i=1}^{r} t_i + (n - r)t_r$$

for *nonreplacement tests* and

$$T_r = nt_r$$

if the test is *with replacement*. Note that if the test is without replacement and $r = n$, $\hat{\mu}$ is simply the mean of the observed times to failure.

To make inferences concerning the mean life μ of the component, we use the fact that $2T_r/\mu$ is a value of a random variable having the chi-square distribution with $2r$ degrees of freedom (see reference to B. Epstein in the Bibliography). With the appropriate expression substituted for T_r, this is true regardless of whether the test is conducted with or without replacement. Thus, in either case a two-sided $1 - \alpha$ confidence interval for μ is given by

$$\frac{2T_r}{\chi_2^2} < \mu < \frac{2T_r}{\chi_1^2}$$

where χ_1^2 and χ_2^2 cut off left- and right-hand tails of area $\alpha/2$ under the chi-square distribution with $2r$ degrees of freedom. (See Exercise 6 on page 471.)

Tests of the null hypothesis that $\mu = \mu_0$ can also be based on the sampling distribution of $2T_r/\mu$, using the appropriate expression for T_r depending on whether the test is with or without replacement. Thus, if the alternative hypothesis is $\mu > \mu_0$, we reject the null hypothesis at the level of significance α when $2T_r/\mu_0$ exceeds χ_α^2, or

$$T_r > \tfrac{1}{2}\mu_0 \chi_\alpha^2$$

where χ_α^2, to be determined for $2r$ degrees of freedom, is as defined on page 176. In Exercises 2 and 4 on page 470 the reader will be asked to construct and perform similar tests corresponding to the alternative hypotheses $\mu < \mu_0$ and $\mu \neq \mu_0$.

An alternate life-testing procedure consists of discontinuing the test after a fixed accumulated amount of lifetime T has elapsed, and treating the observed number of failures k as the value of a random variable. (In the important special case where n items are tested with replacement for a length of time t^*, we have $T = nt^*$.) Regardless of whether the test is with or without replacement, an *approximate* $1 - \alpha$ confidence interval for the mean life of the component is given by

$$\frac{2T}{\chi_4^2} < \mu < \frac{2T}{\chi_3^2}$$

Here χ_4^2 cuts off a right-hand tail of area $\alpha/2$ under the chi-square distribution with $2k + 2$ degrees of freedom, while χ_3^2 cuts off a left-hand tail of area $\alpha/2$ under the chi-square distribution with $2k$ degrees of freedom.

Example. To illustrate some of the methods presented in this section, let us consider the following example. Suppose that 50 units are placed on life test (without replacement) and that the test is to be truncated after $r = 10$ of them have failed. We shall suppose, furthermore, that the first 10 failure times are 65, 110, 380, 420, 505, 580, 650, 840, 910, and 950 hours. Thus, $n = 50$, $r = 10$,

$$T_{10} = (65 + 110 + \ldots + 950) + (50 - 10)950$$
$$= 43{,}410 \text{ hours}$$

and we estimate the mean life of the component as $\hat{\mu} = \dfrac{43{,}410}{10} = 4{,}341$ hours. The failure rate α is estimated by $1/\hat{\mu} = 0.00023$ failures per hour, or 0.23 failures per thousand hours. Also, a 0.90 confidence interval for μ is given by

$$\frac{2(43{,}410)}{31.410} < \mu < \frac{2(43{,}410)}{10.851}$$

or

$$2{,}764 < \mu < 8{,}001$$

Suppose it were also desired to use the above sample to test whether the failure rate is 0.40 failures per thousand hours against the alternative that the failure rate is less. This is equivalent to a test of the null hypothesis $\mu = 1{,}000/0.40 = 2{,}500$ hours against the alternative that $\mu > 2{,}500$ hours. Using a level of significance of 0.05, we find that the critical value for T_{10} for this test is given by $\frac{1}{2}(2{,}500)(31.410) = 39{,}263$ hours, and since this is exceeded by the observed value $T_{10} = 43{,}410$, the null hypothesis must be rejected. We conclude that the mean lifetime exceeds 2,500 hours, or equivalently, that the failure rate is less than 0.40 failures per thousand hours.

15.5 THE WEIBULL MODEL IN LIFE TESTING

Although life testing of components during the period of useful life is generally based on the exponential model, we have already pointed out that the failure rate of a component may not be constant throughout a period under investigation. In some instances the period of initial failure may be so long that the component's main use is during this period, and in other instances the main purpose of life testing may be that of determining the time to wear-out failure rather than chance failure. In such cases the exponential model generally does not apply, and it is necessary to substitute a more general assumption for that of a constant failure rate.

As we observed on page 455, the Weibull distribution adequately describes the failure times of components when their failure rate either increases or decreases with time. It has the parameters α and β, its formula is given by

$$f(t) = \alpha \beta t^{\beta-1} e^{-\alpha t^\beta} \qquad t > 0, \alpha > 0, \beta > 0$$

and it follows (see Exercise 11 on page 471) that the reliability function associated with the Weibull distribution is given by

$$R(t) = e^{-\alpha t^\beta}$$

We already showed on page 455 that the failure rate leading to the Weibull distribution is given by

$$Z(t) = \alpha \beta t^{\beta-1}$$

The range of shapes a graph of the Weibull density can take on is very broad, depending primarily on the value of the parameter β. As illustrated in Figure 15.4, the Weibull curve is asymptotic to both axes and highly skewed to the right for values of β less than 1; it is identical to that of the exponential density for $\beta = 1$, and it is "bell-shaped" but skewed for values of β greater than 1.

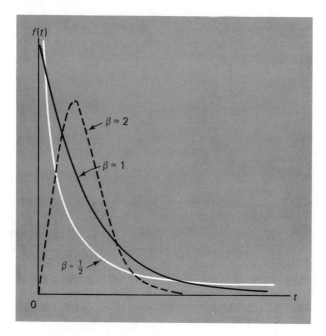

Figure 15.4. Weibull density functions ($\alpha = 1$).

The mean of the Weibull distribution having the parameters α and β may be obtained by evaluating the integral

$$\mu = \int_0^\infty t \cdot \alpha \beta t^{\beta-1} e^{-\alpha t^\beta} \, dt$$

Making the change of variable $u = \alpha t^\beta$, we get

$$\mu = \alpha^{-1/\beta} \int_0^\infty u^{1/\beta} e^{-u} \, du$$

Recognizing the integral as $\Gamma\left(1 + \dfrac{1}{\beta}\right)$, we find that the mean time to failure for the Weibull model is

$$\mu = \alpha^{-1/\beta} \Gamma\left(1 + \frac{1}{\beta}\right)$$

The reader will be asked to show in Exercise 12 on page 471 that the variance of this distribution is given by

$$\sigma^2 = \alpha^{-2/\beta}\left\{\Gamma\left(1 + \frac{2}{\beta}\right) - \left[\Gamma\left(1 + \frac{1}{\beta}\right)\right]^2\right\}$$

Estimates of the parameters α and β of the Weibull distribution are somewhat difficult to obtain. Although there exist analytical methods for estimating these parameters, they involve the solution of a system of transcendental equations, and they will not be presented here. Instead, a more rapid and commonly used method, based on a graphical technique, will be described. This method is based on the fact that the reliability function of the Weibull distribution can be transformed into a linear function of $\ln t$ by means of a double-logarithmic transformation. Taking the natural logarithm of $R(t)$, we obtain

$$\ln R(t) = -\alpha t^\beta \quad \text{or} \quad \ln\frac{1}{R(t)} = \alpha t^\beta$$

Again taking logarithms, we have

$$\ln \ln \frac{1}{R(t)} = \ln \alpha + \beta \cdot \ln t$$

and it can be seen that the right-hand side is linear in $\ln t$.

To estimate α and β we require estimates of $R(t)$ for various values of t, and the usual procedure is to place n units on life test and observe their failure times. If the ith unit fails at time t_i, we estimate $F(t_i) = 1 - R(t_i)$ by the same method as on page 143, namely, we use the estimator

$$\widehat{F(t_i)} = \frac{i - 1/2}{n}$$

Before going any further, it is customary to check whether it is actually reasonable to use the Weibull model. To this end, we plot points having the coordinates t_i and $\widehat{F(t_i)}$ on special graph paper having its scales transformed so that the divisions on the horizontal axis are proportional to $\ln t$ and those on the vertical scale are proportional to $\ln \ln \dfrac{1}{1 - F(t)}$. If the points fall reasonably close to a straight line, it can be assumed that the underlying failure-time distribution is of the Weibull type. The parameters α and β of this distribution can then be estimated by applying the linear regression methods of Chapter 11 to fit a straight line to the transformed data.

Example. To illustrate this procedure, let us consider the following numerical example. Suppose that a sample of 100 components is put on life test for 500 hours and that the times to failure of the 12 components that failed during the test are as follows: 6, 21, 50, 84, 95, 130, 205, 260, 270, 370, 440, and 480 hours. These points are plotted in Figure 15.5 on the transformed scales previously described, and it can be seen that they fall fairly close to a straight line.

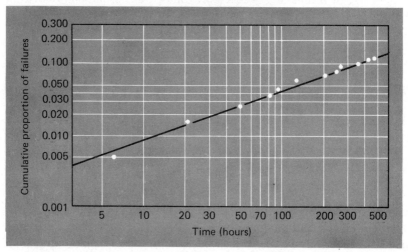

Figure 15.5. Weibull failure-time distribution.

The parameters α and β are estimated by applying the method of least squares to the transformed points (x_i, y_i), where

$$x_i = \ln t_i$$

$$y_i = \ln \ln \frac{1}{1 - \widehat{F(t_i)}}$$

Thus, in our numerical example we obtain

$\widehat{F(t_i)}$	t_i	x_i	y_i
0.005	6	1.79	−5.30
0.015	21	3.04	−4.20
0.025	50	3.91	−3.68
0.035	84	4.43	−3.33
0.045	95	4.55	−3.08
0.055	130	4.87	−2.87
0.065	205	5.32	−2.70
0.075	260	5.56	−2.55
0.085	270	5.60	−2.42
0.095	370	5.91	−2.30
0.105	440	6.09	−2.20
0.115	480	6.17	−2.10

and, by the methods of Section 11.1, the regression line becomes

$$y = -6.44 + 0.71x$$

Note that in calculating the values of y_i for the above table, it is convenient to use the approximation $\ln \ln \dfrac{1}{1-z} \simeq \ln z$ (see Exercise 13 on page 471) for small values of $\widehat{F(t_i)}$.

Thus, the parameter β of the underlying Weibull distribution is estimated as $\hat{\beta} = 0.71$, and α is estimated as $\hat{\alpha} = e^{-6.44} = 0.0016$. It follows that the mean time to failure is estimated as

$$\hat{\mu} = (0.0016)^{-1/0.71}\Gamma\left(1 + \frac{1}{0.71}\right)$$

which equals approximately 11,000 hours. Also, values of the failure-rate function may be obtained by substituting for t into

$$\widehat{Z(t)} = (0.0016)(0.71)t^{-0.29}$$

Since $\hat{\beta} < 1$ the failure rate is decreasing with time. After one hour $(t = 1)$, units are failing at the rate of $(0.0016)(0.71) = 0.00114$ units per hour, and after 1,000 hours the failure rate has decreased to $(0.00114)(1000)^{-0.29} = 0.00015$ units per hour.

EXERCISES

1. Suppose that 50 units are put on life test, each unit that fails is immediately replaced, and the test is discontinued after 8 units have failed. If the eighth failure occurred at 760 hours,

(a) construct a 0.95 confidence interval for the mean life of the units;

(b) test at the 0.05 level of significance whether or not the mean life is less than 10,000 hours.

2. In a life test with replacement, 35 space heaters were put into continuous operation, and the first five failures occurred after 250, 380, 610, 980, and 1,250 hours.

(a) Assuming the exponential model, construct a 0.99 confidence interval for the mean life of this kind of space heater.

(b) To check the manufacturer's claim that the mean life of these heaters is at least 5,000 hours, test the null hypothesis $\mu = 5,000$ against an appropriate alternative, so that the burden of proof is put on the manufacturer. Use $\alpha = 0.05$.

3. Fifteen assemblies are put on accelerated life test without replacement, and the test is truncated after four failures. If the first four failures occurred at 16.5, 19.2, 20.8, and 37.3 hours,

(a) find a 0.90 confidence interval for the failure rate of these assemblies under these accelerated conditions;

(b) test the null hypothesis that the failure rate is 0.004 failures per hour against the alternative that it is less than 0.004, using the 0.01 level of significance.

4. To investigate the average time to failure of a certain weld subjected to continuous vibration, seven welded pieces were subjected to specified frequencies and amplitudes of vibration and their times to failure were 211, 350, 384, 510, 539, 620, and 715 thousand cycles.

(a) Assuming the exponential model, construct a 0.95 confidence interval for the mean life (in thousands of cycles) of the weld under the given vibration conditions.

(b) Assuming the exponential model, test the null hypothesis that the mean life of the weld under the given vibration conditions is 500,000 cycles against the two-sided alternative. Use a level of significance of 0.10.

5. In life testing we are sometimes interested in establishing *tolerance limits* for the life of a component (see Section 14.5); in particular, we may be interested in a one-sided tolerance limit t^*, for which we can assert with a degree of confidence $1 - \alpha$ that at least $100 \cdot P$ percent of the components have a life exceeding t^*. Using the exponential model, it can be shown that a good approximation is given by

$$t^* = \frac{-2T_r(\ln P)}{\chi_\alpha^2}$$

where T_r is as defined on pages 462 and 463 and the value of χ_α^2 is to be obtained from Table 5 with $2r$ degrees of freedom.

(a) Using the data of Exercise 2, establish a lower tolerance limit for which one can assert with a degree of confidence of 0.95 that it is exceeded by at least 80 percent of the lifetimes of the heaters.

(b) Using the data of Exercise 4, establish a lower tolerance limit for which one can assert with a degree of confidence of 0.99 that it is exceeded by at least 90 percent of the lifetimes of the given welds.

6. Using the fact that $2T_r/\mu$ is a value of a random variable having the chi-square distribution with $2r$ degrees of freedom, derive the confidence interval for μ given on page 463.

7. One hundred devices are put on life test and the times to failure (in hours) of the first 10 that fail are as follows: 7.0, 14.1, 18.9, 31.6, 52.8, 80.0, 164.5, 355.4, 451.0, and 795.1. Assuming a Weibull failure-time distribution, estimate the parameters α and β as well as the failure rate at 1,000 hours. How does this value of the failure rate compare with the value we would obtain if we assumed the exponential model?

8. A sample of 300 high-reliability capacitors was placed on life test for 2,000 hours and there were no failures; the test was then terminated. Find a 0.95 *lower* confidence limit for the mean life of the capacitors.

9. To investigate the performance of a logic circuit for a small electronic calculator, a laboratory puts 75 of the circuits on life test (without replacement) under specified environmental conditions, and the first 10 failures are observed after 28, 46, 50, 63, 81, 101, 116, 137, 159, and 175 hours. Using the Weibull model, estimate the mean life of the circuit. How does this value compare with the mean life that would have been obtained under the exponential assumption?

10. Using the estimates of the parameters of the Weibull model obtained in Exercise 9, estimate the probability that this kind of circuit will perform satisfactorily for at least 100 hours.

11. Show that the reliability function associated with the Weibull failure-time distribution is given by

$$R(t) = e^{-\alpha t^\beta}$$

12. Derive the formula for the variance of the Weibull distribution given on page 467.

13. Show that $\ln \ln \dfrac{1}{1 - z}$ can be approximated by $\ln z$ for small values of z.

[*Hint*: Note that $\dfrac{1}{1 - z} = 1 + z + z^2 + z^3 \ldots$ for $|z| < 1$, and then use the Maclaurin's series for $\ln (1 + x)$.]

Bibliography

1. ENGINEERING STATISTICS

Bowker, A. H., and Lieberman, G. J., *Engineering Statistics*, 2nd ed. Englewood Cliffs, N.J.: Prentice-Hall, Inc., 1972.

Brownlee, K. A., *Statistical Theory and Methodology in Science and Engineering*, 2nd. New York: John Wiley & Sons, Inc., 1965.

Gibra, I. N., *Probability and Statistical Inference for Scientists and Engineers*. Englewood Cliffs, N.J.: Prentice-Hall, Inc., 1973.

Guttman, I., and Wilks, S. S., *Introductory Engineering Statistics*. New York: John Wiley & Sons, Inc., 1965.

Hald, A., *Statistical Theory with Engineering Applications*. New York: John Wiley & Sons, Inc., 1952.

Johnson, N. L., and Leone, F. C., *Statistics and Experimental Design: In Engineering and the Physical Sciences*, Vols. I and II. New York: John Wiley & Sons, Inc., 1964.

472

Kirkpatrick, E. G., *Introductory Statistics and Probability for Engineering, Science, and Technology.* Englewood Cliffs, N.J.: Prentice-Hall, Inc., 1974.

Walpole, R. E., and Meyers, R. H., *Probability and Statistics for Engineers and Scientists.* New York: The Macmillan Co., 1972.

2. THEORETICAL STATISTICS

Brunk, H. D., *An Introduction to Mathematical Statistics,* 3rd ed. New York: John Wiley & Sons, Inc., 1975.

Freund, J. E., *Mathematical Statistics,* 2nd ed. Englewood Cliffs, N.J.: Prentice-Hall, Inc., 1971.

Fry, T. C., *Probability and its Engineering Uses,* 2nd ed. Princeton, N.J.: D. van Nostrand Co., 1964.

Hoel, P., *Introduction to Mathematical Statistics,* 3rd ed. New York: John Wiley & Sons, Inc., 1965.

Hogg, R. V., and Craig, A. T., *Introduction to Mathematical Statistics,* 3rd ed. New York: The Macmillan Co., 1970.

Kendall, M. G., and Stuart, A., *The Advanced Theory of Statistics,* Vol. 1, 3rd ed., Vol. 2, 3rd ed., Vol. 3, 2nd ed. New York: Hafner Publishing Co., 1969, 1973, and 1968.

Mann, H. B., *Analysis and Design of Experiments.* New York: Dover Press, 1949.

Mood, A. M., and Graybill, F. A., *Introduction to the Theory of Statistics,* 2nd ed. New York: McGraw-Hill Book Co., 1963.

Wilks, S. S., *Mathematical Statistics.* New York: John Wiley & Sons, Inc., 1962.

3. EXPERIMENTAL DESIGN AND ANALYSIS OF VARIANCE

Cochran, W. G., and Cox, G. M., *Experimental Designs,* 2nd ed. New York: John Wiley & Sons, Inc., 1957.

Davies, O. L., *The Design and Analysis of Industrial Experiments.* New York: Hafner Publishing Co., 1963.

Federer, W. T. *Experimental Designs, Theory and Application.* New York: The Macmillan Co., 1955.

Fisher, R. A., *The Design of Experiments,* 4th ed. New York: Hafner Publishing Co., 1957.

Hicks, C. R., *Fundamental Concepts in the Design of Experiments.* New York: Holt, Rinehart & Winston, Inc., 1965.

Lipson, C., and Sheth, N. J., *Statistical Design and Analysis of Engineering Experiments*. New York: McGraw-Hill Book Co., 1973.

Scheffe, H., *The Analysis of Variance*. New York: John Wiley & Sons, Inc., 1959.

Snedecor, G. W., and Cochran, W. G., *Statistical Methods*, 6th ed. Ames, Iowa: Iowa University Press, 1967.

4. SPECIAL TOPICS

Barlow, R. E., and Proschan, F., *Mathematical Theory of Reliability*. New York: John Wiley & Sons, Inc., 1965.

Barlow, R. E., and Proschan, F., *Statistical Theory of Reliability and Life Testing: Probability Models*. New York: Holt, Rinehart, and Winston, 1975.

Burr, I. W., *Engineering Statistics and Quality Control*. New York: McGraw-Hill Book Co., 1953.

Cowden, D. J., *Statistical Methods in Quality Control*. Englewood Cliffs, N.J.: Prentice-Hall, Inc., 1957.

Duncan, A. J., *Quality Control and Industrial Statistics*, 3rd ed. Homewood, Ill.: Richard D. Irwin, Inc., 1965.

Grant, E. L., and Leavenworth, R., *Statistical Quality Control*, 4th ed. New York: McGraw-Hill Book Co., 1972.

Noether, G. E., *Elements of Nonparametric Statistics*. New York: John Wiley & Sons, Inc., 1967.

Wald, A., *Sequential Analysis*. New York: John Wiley & Sons, Inc., 1947.

Zelen, M., *Statistical Theory of Reliability*. Madison, Wisc.: University of Wisconsin Press, 1963.

5. GENERAL REFERENCE WORKS AND TABLES

Kendall, M. G., and Buckland, W. R., *A Dictionary of Statistical Terms*, 3rd ed. New York: Hafner Publishing Co., 1971.

Military Standard 105D. Washington, D.C.: U. S. Government Printing Office, 1963.

National Bureau of Standards Handbook 91: Experimental Statistics. Washington, D. C.: U. S. Government Printing Office, 1963.

National Bureau of Standards: Tables of the Binomial Distribution. Washington, D. C.: U. S. Government Printing Office, 1950.

Owen, D. B., *Handbook of Statistical Tables*. Reading, Mass.: Addison-Wesley Publishing Co., 1962.

Pearson, E. S., and Hartley, H. O., *Biometrika Tables for Statisticians*, 3rd ed. Cambridge: Cambridge University Press, 1966.

RAND Corporation, *A Million Random Digits with 100,000 Normal Deviates*. New York: The Macmillan Co., third printing 1966.

Romig, H. G., *50–100 Binomial Tables*. New York: John Wiley & Sons, Inc., 1953.

Statistical
Tables

Table 1

BINOMIAL DISTRIBUTION FUNCTION

$$B(x;n,p) = \sum_{k=0}^{x} \binom{n}{k} p^k (1-p)^{n-k}$$

						p					
n	x	0.05	0.10	0.15	0.20	0.25	0.30	0.35	0.40	0.45	0.50
2	2	0.9025	0.8100	0.7225	0.6400	0.5625	0.4900	0.4225	0.3600	0.3025	0.2500
	1	0.9975	0.9900	0.9775	0.9600	0.9375	0.9100	0.8755	0.8400	0.7975	0.7500
3	0	0.8574	0.7290	0.6141	0.5120	0.4219	0.3430	0.2746	0.2160	0.1664	0.1250
	1	0.9928	0.9720	0.9392	0.8960	0.8438	0.7840	0.7182	0.6480	0.5748	0.5000
	2	0.9999	0.9990	0.9966	0.9920	0.9844	0.9730	0.9571	0.9360	0.9089	0.8750
4	0	0.8145	0.6561	0.5220	0.4096	0.3164	0.2401	0.1785	0.1296	0.0915	0.0625
	1	0.9860	0.9477	0.8905	0.8192	0.7383	0.6517	0.5630	0.4752	0.3910	0.3125
	2	0.9995	0.9963	0.9880	0.9728	0.9492	0.9163	0.8735	0.8208	0.7585	0.6875
	3	1.0000	0.9999	0.9995	0.9984	0.9961	0.9919	0.9850	0.9744	0.9590	0.9375
5	0	0.7738	0.5905	0.4437	0.3277	0.2373	0.1681	0.1160	0.0778	0.0503	0.0312
	1	0.9774	0.9185	0.8352	0.7373	0.6328	0.5282	0.4284	0.3370	0.2562	0.1875
	2	0.9988	0.9914	0.9734	0.9421	0.8965	0.8369	0.7648	0.6826	0.5931	0.5000
	3	1.0000	0.9995	0.0078	0.9933	0.9844	0.9692	0.9460	0.9130	0.8688	0.8125
	4	1.0000	1.0000	0.9999	0.9997	0.9990	0.9976	0.9947	0.9898	0.9815	0.9688
6	0	0.7351	0.5314	0.3771	0.2621	0.1780	0.1176	0.0754	0.0467	0.0277	0.0156
	1	0.9672	0.8857	0.7765	0.6554	0.5339	0.4202	0.3191	0.2333	0.1636	0.1094
	2	0.9978	0.9842	0.9527	0.9011	0.8306	0.7443	0.6471	0.5443	0.4415	0.3438
	3	0.9999	0.9987	0.9941	0.9830	0.9624	0.9295	0.8826	0.8208	0.7447	0.6562
	4	1.0000	0.9999	0.9996	0.9984	0.9954	0.9891	0.9777	0.9590	0.9308	0.8906
	5	1.0000	1.0000	1.0000	0.9999	0.9998	0.9993	0.9982	0.9959	0.9917	0.9844
7	0	0.6983	0.4783	0.3206	0.2097	0.1335	0.0824	0.0490	0.0280	0.0152	0.0078
	1	0.9556	0.8503	0.7166	0.5767	0.4449	0.3294	0.2338	0.1586	0.1024	0.0625
	2	0.9962	0.9743	0.9262	0.8520	0.7564	0.6471	0.5323	0.4199	0.3164	0.2266
	3	0.9998	0.9973	0.9879	0.9667	0.9294	0.8740	0.8002	0.7102	0.6083	0.5000
	4	1.0000	0.9998	0.9988	0.9953	0.9871	0.9712	0.9444	0.9037	0.8471	0.7734
	5	1.0000	1.0000	0.9999	0.9996	0.9987	0.9962	0.9910	0.9812	0.9643	0.9375
	6	1.0000	1.0000	1.0000	1.0000	0.9999	0.9998	0.9994	0.9984	0.9963	0.9922
8	0	0.6634	0.4305	0.2725	0.1678	0.1001	0.0576	0.0319	0.0168	0.0084	0.0039
	1	0.9428	0.8131	0.6572	0.5033	0.3671	0.2553	0.1691	0.1064	0.0632	0.0352
	2	0.9942	0.9619	0.8948	0.7969	0.6785	0.5518	0.4278	0.3154	0.2201	0.1445
	3	0.9996	0.9950	0.9786	0.9437	0.8862	0.8059	0.7064	0.5941	0.4770	0.3633
	4	1.0000	0.9996	0.9971	0.9896	0.9727	0.9420	0.8939	0.8263	0.7396	0.6367
	5	1.0000	1.0000	0.9998	0.9988	0.9958	0.9887	0.9747	0.9502	0.9115	0.8555
	6	1.0000	1.0000	1.0000	0.9999	0.9996	0.9987	0.9964	0.9915	0.9819	0.9648
	7	1.0000	1.0000	1.0000	1.0000	1.0000	0.9999	0.9998	0.9993	0.9983	0.9961
9	0	0.6302	0.3874	0.2316	0.1342	0.0751	0.0404	0.0207	0.0101	0.0046	0.0020
	1	0.9288	0.7748	0.5995	0.4362	0.3003	0.1960	0.1211	0.0705	0.0385	0.0195
	2	0.9916	0.9470	0.8591	0.7382	0.6007	0.4628	0.3373	0.2318	0.1495	0.0898
	3	0.9994	0.9917	0.9661	0.9144	0.8343	0.7297	0.6089	0.4826	0.3614	0.2539
	4	1.0000	0.9991	0.9944	0.9804	0.9511	0.9012	0.8283	0.7334	0.6214	0.5000
	5	1.0000	0.9999	0.9994	0.9969	0.9900	0.9747	0.9464	0.9006	0.8342	0.7461
	6	1.0000	1.0000	1.0000	0.9997	0.9987	0.9957	0.9888	0.9750	0.9502	0.9102
	7	1.0000	1.0000	1.0000	1.0000	0.9999	0.9996	0.9986	0.9962	0.9909	0.9805
	8	1.0000	1.0000	1.0000	1.0000	1.0000	1.0000	0.9999	0.9997	0.9992	0.9980

Table 1

BINOMIAL DISTRIBUTION FUNCTION (*Continued*)

						p					
n	x	0.05	0.10	0.15	0.20	0.25	0.30	0.35	0.40	0.45	0.50
10	0	0.5987	0.3487	0.1969	0.1074	0.0563	0.0282	0.0135	0.0060	0.0025	0.0010
	1	0.9139	0.7361	0.5443	0.3758	0.2440	0.1493	0.0860	0.0464	0.0232	0.0107
	2	0.9885	0.9298	0.8202	0.6778	0.5256	0.3828	0.2616	0.1673	0.0996	0.0547
	3	0.9990	0.9872	0.9500	0.8791	0.7759	0.6496	0.5138	0.3823	0.2660	0.1719
	4	0.9999	0.9984	0.9901	0.9672	0.9219	0.8497	0.7515	0.6331	0.5044	0.3770
	5	1.0000	0.9999	0.9986	0.9936	0.9803	0.9527	0.9051	0.8338	0.7384	0.6230
	6	1.0000	1.0000	0.9999	0.9991	0.9965	0.9894	0.9740	0.9452	0.8980	0.8281
	7	1.0000	1.0000	1.0000	0.9999	0.9996	0.9984	0.9952	0.9877	0.9726	0.9453
	8	1.0000	1.0000	1.0000	1.0000	1.0000	0.9999	0.9995	0.9983	0.9955	0.9893
	9	1.0000	1.0000	1.0000	1.0000	1.0000	1.0000	1.0000	0.9999	0.9997	0.9990
11	0	0.5688	0.3138	0.1673	0.0859	0.0422	0.0198	0.0088	0.0036	0.0014	0.0005
	1	0.8981	0.6974	0.4922	0.3221	0.1971	0.1130	0.0606	0.0302	0.0139	0.0059
	2	0.9848	0.9104	0.7788	0.6174	0.4552	0.3127	0.2001	0.1189	0.0652	0.0327
	3	0.9984	0.9815	0.9306	0.8389	0.7133	0.5696	0.4256	0.2963	0.1911	0.1133
	4	0.9999	0.9972	0.9841	0.9496	0.8854	0.7897	0.6683	0.5328	0.3971	0.2744
	5	1.0000	0.9997	0.9973	0.9883	0.9657	0.9218	0.8513	0.7535	0.6331	0.5000
	6	1.0000	1.0000	0.9997	0.9980	0.9924	0.9784	0.9499	0.9006	0.8262	0.7256
	7	1.0000	1.0000	1.0000	0.9998	0.9988	0.9957	0.9878	0.9707	0.9390	0.8867
	8	1.0000	1.0000	1.0000	1.0000	0.9999	0.9994	0.9980	0.9941	0.9852	0.9673
	9	1.0000	1.0000	1.0000	1.0000	1.0000	1.0000	0.9998	0.9993	0.9978	0.9941
	10	1.0000	1.0000	1.0000	1.0000	1.0000	1.0000	1.0000	1.0000	0.9998	0.9995
12	0	0.5404	0.2824	0.1422	0.0687	0.0317	0.0138	0.0057	0.0022	0.0008	0.0002
	1	0.8816	0.6590	0.4435	0.2749	0.1584	0.0850	0.0424	0.0196	0.0083	0.0032
	2	0.9804	0.8891	0.7358	0.5583	0.3907	0.2528	0.1513	0.0834	0.0421	0.0193
	3	0.9978	0.9744	0.9078	0.7946	0.6488	0.4925	0.3467	0.2253	0.1345	0.0730
	4	0.9998	0.9957	0.9761	0.9274	0.8424	0.7237	0.5833	0.4382	0.3044	0.1938
	5	1.0000	0.9995	0.9954	0.9806	0.9456	0.8822	0.7873	0.6652	0.5269	0.3872
	6	1.0000	0.9999	0.9993	0.9961	0.9857	0.9614	0.9154	0.8418	0.7393	0.6128
	7	1.0000	1.0000	0.9999	0.9994	0.9972	0.9905	0.9745	0.9427	0.8883	0.8062
	8	1.0000	1.0000	1.0000	0.9999	0.9996	0.9983	0.9944	0.9847	0.9644	0.9270
	9	1.0000	1.0000	1.0000	1.0000	1.0000	0.9998	0.9992	0.9972	0.9921	0.9807
	10	1.0000	1.0000	1.0000	1.0000	1.0000	1.0000	0.9999	0.9997	0.9989	0.9968
	11	1.0000	1.0000	1.0000	1.0000	1.0000	1.0000	1.0000	1.0000	0.9999	0.9998
13	0	0.5133	0.2542	0.1209	0.0550	0.0238	0.0097	0.0037	0.0013	0.0004	0.0001
	1	0.8646	0.6213	0.3983	0.2336	0.1267	0.0637	0.0296	0.0126	0.0049	0.0017
	2	0.9755	0.8661	0.6920	0.5017	0.3326	0.2025	0.1132	0.0579	0.0269	0.0112
	3	0.9969	0.9658	0.8820	0.7473	0.5843	0.4206	0.2783	0.1686	0.0929	0.0461
	4	0.9997	0.9935	0.9658	0.9009	0.7940	0.6543	0.5005	0.3530	0.2279	0.1334
	5	1.0000	0.9991	0.9925	0.9700	0.9198	0.8346	0.7159	0.5744	0.4268	0.2905
	6	1.0000	0.9999	0.9987	0.9930	0.9757	0.9376	0.8705	0.7712	0.6437	0.5000
	7	1.0000	1.0000	0.9998	0.9988	0.9944	0.9818	0.9538	0.9023	0.8212	0.7095
	8	1.0000	1.0000	1.0000	0.9998	0.9990	0.9960	0.9874	0.9679	0.9302	0.8666
	9	1.0000	1.0000	1.0000	1.0000	0.9999	0.9993	0.9975	0.9922	0.9797	0.9539
	10	1.0000	1.0000	1.0000	1.0000	1.0000	0.9999	0.9997	0.9987	0.9959	0.9888
	11	1.0000	1.0000	1.0000	1.0000	1.0000	1.0000	1.0000	0.9999	0.9995	0.9983
	12	1.0000	1.0000	1.0000	1.0000	1.0000	1.0000	1.0000	1.0000	1.0000	0.9999
14	0	0.4877	0.2288	0.1028	0.0440	0.0178	0.0068	0.0024	0.0008	0.0002	0.0001
	1	0.8470	0.5846	0.3567	0.1979	0.1010	0.0475	0.0205	0.0081	0.0029	0.0009

Table 1

BINOMIAL DISTRIBUTION FUNCTION (*Continued*)

n	x	0.05	0.10	0.15	0.20	0.25	p 0.30	0.35	0.40	0.45	0.50
14	2	0.9699	0.8416	0.6479	0.4481	0.2811	0.1608	0.0839	0.0398	0.0170	0.0065
	3	0.9958	0.9559	0.8535	0.6982	0.5213	0.3552	0.2205	0.1243	0.0632	0.0287
	4	0.9996	0.9908	0.9533	0.8702	0.7415	0.5842	0.4227	0.2793	0.1672	0.0898
	5	1.0000	0.9985	0.9885	0.9561	0.8883	0.7805	0.6405	0.4859	0.3373	0.2120
	6	1.0000	0.9998	0.9978	0.9884	0.9617	0.9067	0.8164	0.6925	0.5461	0.3953
	7	1.0000	1.0000	0.9997	0.9976	0.9897	0.9685	0.9247	0.8499	0.7414	0.6074
	8	1.0000	1.0000	1.0000	0.9996	0.9978	0.9917	0.9757	0.9417	0.8811	0.7880
	9	1.0000	1.0000	1.0000	1.0000	0.9997	0.9983	0.9940	0.9825	0.9574	0.9102
	10	1.0000	1.0000	1.0000	1.0000	1.0000	0.9989	0.9989	0.9961	0.9886	0.9713
	11	1.0000	1.0000	1.0000	1.0000	1.0000	1.0000	0.9999	0.9994	0.9978	0.9935
	12	1.0000	1.0000	1.0000	1.0000	1.0000	1.0000	1.0000	0.9999	0.9997	0.9991
	13	1.0000	1.0000	1.0000	1.0000	1.0000	1.0000	1.0000	1.0000	1.0000	0.9999
15	0	0.4633	0.2059	0.0874	0.0352	0.0134	0.0047	0.0016	0.0005	0.0001	0.0000
	1	0.8290	0.5490	0.3186	0.1671	0.0802	0.0353	0.0142	0.0052	0.0017	0.0005
	2	0.9638	0.8159	0.6042	0.3980	0.2361	0.1268	0.0617	0.0271	0.0107	0.0037
	3	0.9945	0.9444	0.8227	0.6482	0.4613	0.2969	0.1727	0.0905	0.0424	0.0176
	4	0.9994	0.9873	0.9383	0.8358	0.6865	0.5155	0.3519	0.2173	0.1204	0.0592
	5	0.9999	0.9978	0.9832	0.9389	0.8516	0.7216	0.5643	0.4032	0.2608	0.1509
	6	1.0000	0.9997	0.9964	0.9819	0.9434	0.8689	0.7548	0.6098	0.4522	0.3036
	7	1.0000	1.0000	0.9996	0.9958	0.9827	0.9500	0.8868	0.7869	0.6535	0.5000
	8	1.0000	1.0000	0.9999	0.9992	0.9958	0.9848	0.9578	0.9050	0.8182	0.6964
	9	1.0000	1.0000	1.0000	0.9999	0.9992	0.9963	0.9876	0.9662	0.9231	0.8491
	10	1.0000	1.0000	1.0000	1.0000	0.9999	0.9993	0.9972	0.9907	0.9745	0.9408
	11	1.0000	1.0000	1.0000	1.0000	1.0000	0.9999	0.9995	0.9981	0.9937	0.9824
	12	1.0000	1.0000	1.0000	1.0000	1.0000	1.0000	0.9999	0.9997	0.9989	0.9963
	13	1.0000	1.0000	1.0000	1.0000	1.0000	1.0000	1.0000	1.0000	0.9999	0.9995
	14	1.0000	1.0000	1.0000	1.0000	1.0000	1.0000	1.0000	1.0000	1.0000	1.0000
16	0	0.4401	0.1853	0.0743	0.0281	0.0100	0.0033	0.0010	0.0003	0.0001	0.0000
	1	0.8108	0.5147	0.2839	0.1407	0.0635	0.0261	0.0098	0.0033	0.0010	0.0003
	2	0.9571	0.7892	0.5614	0.3518	0.1971	0.0994	0.0451	0.0183	0.0066	0.0021
	3	0.9930	0.9316	0.7899	0.5981	0.4050	0.2459	0.1339	0.0651	0.0281	0.0106
	4	0.9991	0.9830	0.9209	0.7982	0.6302	0.4499	0.2892	0.1666	0.0853	0.0384
	5	0.9999	0.9967	0.9765	0.9183	0.8103	0.6598	0.4900	0.3288	0.1976	0.1051
	6	1.0000	0.9995	0.9944	0.9733	0.9204	0.8247	0.6881	0.5272	0.3660	0.2272
	7	1.0000	0.9999	0.9989	0.9930	0.9729	0.9256	0.8406	0.7161	0.5629	0.4018
	8	1.0000	1.0000	0.9998	0.9985	0.9925	0.9743	0.9329	0.8577	0.7441	0.5982
	9	1.0000	1.0000	1.0000	0.9998	0.9984	0.9929	0.9771	0.9417	0.8759	0.7728
	10	1.0000	1.0000	1.0000	1.0000	0.9997	0.9984	0.9938	0.9809	0.9514	0.8949
	11	1.0000	1.0000	1.0000	1.0000	1.0000	0.9997	0.9987	0.9951	0.9851	0.9616
	12	1.0000	1.0000	1.0000	1.0000	1.0000	1.0000	0.9998	0.9991	0.9965	0.9894
	13	1.0000	1.0000	1.0000	1.0000	1.0000	1.0000	1.0000	0.9999	0.9994	0.9979
	14	1.0000	1.0000	1.0000	1.0000	1.0000	1.0000	1.0000	1.0000	1.0000	0.9997
	15	1.0000	1.0000	1.0000	1.0000	1.0000	1.0000	1.0000	1.0000	1.0000	1.0000
17	0	0.4181	0.1668	0.0631	0.0225	0.0075	0.0023	0.0007	0.0002	0.0000	0.0000
	1	0.7922	0.4818	0.2525	0.1182	0.0501	0.0193	0.0067	0.0021	0.0006	0.0001
	2	0.9497	0.7618	0.5198	0.3096	0.1637	0.0774	0.0327	0.0123	0.0041	0.0012
	3	0.9912	0.9174	0.7556	0.5489	0.3530	0.2019	0.1028	0.0464	0.0184	0.0063
	4	0.9988	0.9779	0.9013	0.7582	0.5739	0.3887	0.2348	0.1260	0.0596	0.0245

Table 1

BINOMIAL DISTRIBUTION FUNCTION (*Continued*)

						p					
n	x	0.05	0.10	0.15	0.20	0.25	0.30	0.35	0.40	0.45	0.50
17	5	0.9999	0.9953	0.9681	0.8943	0.7653	0.5968	0.4197	0.2639	0.1471	0.0717
	6	1.0000	0.9992	0.9917	0.9623	0.8929	0.7752	0.6188	0.4478	0.2902	0.1662
	7	1.0000	0.9999	0.9983	0.9891	0.9598	0.8954	0.7872	0.6405	0.4743	0.3145
	8	1.0000	1.0000	0.9997	0.9974	0.9876	0.9597	0.9006	0.8011	0.6626	0.5000
	9	1.0000	1.0000	1.0000	0.9995	0.9969	0.9873	0.9617	0.9081	0.8166	0.6855
	10	1.0000	1.0000	1.0000	0.9999	0.9994	0.9968	0.9880	0.9652	0.9174	0.8338
	11	1.0000	1.0000	1.0000	1.0000	0.9999	0.9993	0.9970	0.9894	0.9699	0.9283
	12	1.0000	1.0000	1.0000	1.0000	1.0000	0.9999	0.9994	0.9975	0.9914	0.9755
	13	1.0000	1.0000	1.0000	1.0000	1.0000	1.0000	0.9999	0.9995	0.9981	0.9936
	14	1.0000	1.0000	1.0000	1.0000	1.0000	1.0000	1.0000	0.9999	0.9997	0.9988
	15	1.0000	1.0000	1.0000	1.0000	1.0000	1.0000	1.0000	1.0000	1.0000	0.9999
	16	1.0000	1.0000	1.0000	1.0000	1.0000	1.0000	1.0000	1.0000	1.0000	1.0000
18	0	0.3972	0.1501	0.0536	0.0180	0.0056	0.0016	0.0004	0.0001	0.0000	0.0000
	1	0.7735	0.4503	0.2241	0.0991	0.0395	0.0142	0.0046	0.0013	0.0003	0.0001
	2	0.9419	0.7338	0.4797	0.2713	0.1353	0.0600	0.0236	0.0082	0.0025	0.0007
	3	0.9891	0.9018	0.7202	0.5010	0.3057	0.1646	0.0783	0.0328	0.0120	0.0038
	4	0.9985	0.9718	0.8794	0.7164	0.5187	0.3327	0.1886	0.0942	0.0411	0.0154
	5	0.9998	0.9936	0.9581	0.8671	0.7175	0.5344	0.3550	0.2088	0.1077	0.0481
	6	1.0000	0.9988	0.9882	0.9487	0.8610	0.7217	0.5491	0.3743	0.2258	0.1189
	7	1.0000	0.9998	0.9973	0.9837	0.9431	0.8593	0.7283	0.5634	0.3915	0.2403
	8	1.0000	1.0000	0.9995	0.9957	0.9807	0.9404	0.8609	0.7368	0.5778	0.4073
	9	1.0000	1.0000	0.9999	0.9991	0.9946	0.9790	0.9403	0.8653	0.7473	0.5927
	10	1.0000	1.0000	1.0000	0.9998	0.9988	0.9939	0.9788	0.9424	0.8720	0.7597
	11	1.0000	1.0000	1.0000	1.0000	0.9998	0.9986	0.9938	0.9797	0.9463	0.8811
	12	1.0000	1.0000	1.0000	1.0000	1.0000	0.9997	0.9986	0.9942	0.9817	0.9519
	13	1.0000	1.0000	1.0000	1.0000	1.0000	1.0000	0.9997	0.9987	0.9951	0.9846
	14	1.0000	1.0000	1.0000	1.0000	1.0000	1.0000	1.0000	0.9998	0.9990	0.9962
	15	1.0000	1.0000	1.0000	1.0000	1.0000	1.0000	1.0000	1.0000	0.9999	0.9993
	16	1.0000	1.0000	1.0000	1.0000	1.0000	1.0000	1.0000	1.0000	1.0000	0.9999
19	0	0.3774	0.1351	0.0456	0.0144	0.0042	0.0011	0.0003	0.0001	0.0000	0.0000
	1	0.7547	0.4203	0.1985	0.0829	0.0310	0.0104	0.0031	0.0008	0.0002	0.0000
	2	0.9335	0.7054	0.4413	0.2369	0.1113	0.0462	0.0170	0.0055	0.0015	0.0004
	3	0.9868	0.8850	0.6841	0.4551	0.2630	0.1332	0.0591	0.0230	0.0077	0.0022
	4	0.9980	0.9648	0.8556	0.6733	0.4654	0.2822	0.1500	0.0696	0.0280	0.0096
	5	0.9998	0.9914	0.9463	0.8369	0.6678	0.4739	0.2968	0.1629	0.0777	0.0318
	6	1.0000	0.9983	0.9837	0.9324	0.8251	0.6655	0.4812	0.3081	0.1727	0.0835
	7	1.0000	0.9997	0.9959	0.9767	0.9225	0.8180	0.6656	0.4878	0.3169	0.1796
	8	1.0000	1.0000	0.9992	0.9933	0.9713	0.9161	0.8145	0.6675	0.4940	0.3238
	9	1.0000	1.0000	0.9999	0.9984	0.9911	0.9674	0.9125	0.8139	0.6710	0.5000
	10	1.0000	1.0000	1.0000	0.9997	0.9977	0.9895	0.9653	0.9115	0.8159	0.6762
	11	1.0000	1.0000	1.0000	1.0000	0.9995	0.9972	0.9886	0.9648	0.9129	0.8204
	12	1.0000	1.0000	1.0000	1.0000	0.9999	0.9994	0.9969	0.9884	0.9658	0.9165
	13	1.0000	1.0000	1.0000	1.0000	1.0000	0.9999	0.9993	0.9969	0.9891	0.9682
	14	1.0000	1.0000	1.0000	1.0000	1.0000	1.0000	0.9999	0.9994	0.9972	0.9904
	15	1.0000	1.0000	1.0000	1.0000	1.0000	1.0000	1.0000	0.9999	0.9995	0.9978
	16	1.0000	1.0000	1.0000	1.0000	1.0000	1.0000	1.0000	1.0000	0.9999	0.9996
	17	1.0000	1.0000	1.0000	1.0000	1.0000	1.0000	1.0000	1.0000	1.0000	1.0000

Table 1

BINOMIAL DISTRIBUTION FUNCTION (*Continued*)

						p					
n	x	0.05	0.10	0.15	0.20	0.25	0.30	0.35	0.40	0.45	0.50
20	0	0.3585	0.1216	0.0388	0.0115	0.0032	0.0008	0.0002	0.0000	0.0000	0.0000
	1	0.7358	0.3917	0.1756	0.0692	0.0243	0.0076	0.0021	0.0005	0.0001	0.0000
	2	0.9245	0.6769	0.4049	0.2061	0.0913	0.0355	0.0121	0.0036	0.0009	0.0002
	3	0.9841	0.8670	0.6477	0.4114	0.2252	0.1071	0.0444	0.0160	0.0049	0.0013
	4	0.9974	0.9568	0.8298	0.6296	0.4148	0.2375	0.1182	0.0510	0.0189	0.0059
	5	0.9997	0.9887	0.9327	0.8042	0.6172	0.4164	0.2454	0.1256	0.0553	0.0207
	6	1.0000	0.9976	0.9781	0.9133	0.7858	0.6080	0.4166	0.2500	0.1299	0.0577
	7	1.0000	0.9996	0.9941	0.9679	0.8982	0.7723	0.6010	0.4159	0.2520	0.1316
	8	1.0000	0.9999	0.9987	0.9900	0.9591	0.8867	0.7624	0.5956	0.4143	0.2517
	9	1.0000	1.0000	0.9998	0.9974	0.9861	0.9520	0.8782	0.7553	0.5914	0.4119
	10	1.0000	1.0000	1.0000	0.9994	0.9961	0.9829	0.9468	0.8725	0.7507	0.5881
	11	1.0000	1.0000	1.0000	0.9999	0.9991	0.9949	0.9804	0.9435	0.8692	0.7483
	12	1.0000	1.0000	1.0000	1.0000	0.9998	0.9987	0.9940	0.9790	0.9420	0.8684
	13	1.0000	1.0000	1.0000	1.0000	1.0000	0.9997	0.9985	0.9935	0.9786	0.9423
	14	1.0000	1.0000	1.0000	1.0000	1.0000	1.0000	0.9997	0.9984	0.9936	0.9793
	15	1.0000	1.0000	1.0000	1.0000	1.0000	1.0000	1.0000	0.9997	0.9985	0.9941
	16	1.0000	1.0000	1.0000	1.0000	1.0000	1.0000	1.0000	1.0000	0.9997	0.9987
	17	1.0000	1.0000	1.0000	1.0000	1.0000	1.0000	1.0000	1.0000	1.0000	0.9998
	18	1.0000	1.0000	1.0000	1.0000	1.0000	1.0000	1.0000	1.0000	1.0000	1.0000

Table 2

POISSON DISTRIBUTION FUNCTION*

$$F(x;\lambda) = \sum_{k=0}^{x} e^{-\lambda} \frac{\lambda^k}{k!}$$

λ＼x	0	1	2	3	4	5	6	7	8	9
0.02	0.980	1.000								
0.04	0.961	0.999	1.000							
0.06	0.942	0.998	1.000							
0.08	0.923	0.997	1.000							
0.10	0.905	0.995	1.000							
0.15	0.861	0.990	0.999	1.000						
0.20	0.819	0.982	0.999	1.000						
0.25	0.779	0.974	0.998	1.000						
0.30	0.741	0.963	0.996	1.000						
0.35	0.705	0.951	0.994	1.000						
0.40	0.670	0.938	0.992	0.999	1.000					
0.45	0.638	0.925	0.989	0.999	1.000					
0.50	0.607	0.910	0.986	0.998	1.000					
0.55	0.577	0.894	0.982	0.998	1.000					
0.60	0.549	0.878	0.977	0.997	1.000					
0.65	0.522	0.861	0.972	0.996	0.999	1.000				
0.70	0.497	0.844	0.966	0.994	0.999	1.000				
0.75	0.472	0.827	0.959	0.993	0.999	1.000				
0.80	0.449	0.809	0.953	0.991	0.999	1.000				
0.85	0.427	0.791	0.945	0.989	0.998	1.000				
0.90	0.407	0.772	0.937	0.987	0.998	1.000				
0.95	0.387	0.754	0.929	0.984	0.997	1.000				
1.00	0.368	0.736	0.920	0.981	0.996	0.999	1.000			
1.1	0.333	0.699	0.900	0.974	0.995	0.999	1.000			
1.2	0.301	0.663	0.879	0.966	0.992	0.998	1.000			
1.3	0.273	0.627	0.857	0.957	0.989	0.998	1.000			
1.4	0.247	0.592	0.833	0.946	0.986	0.997	0.999	1.000		
1.5	0.223	0.558	0.809	0.934	0.981	0.996	0.999	1.000		
1.6	0.202	0.525	0.783	0.921	0.976	0.994	0.999	1.000		
1.7	0.183	0.493	0.757	0.907	0.970	0.992	0.998	1.000		
1.8	0.165	0.463	0.731	0.891	0.964	0.990	0.997	0.999	1.000	
1.9	0.150	0.434	0.704	0.875	0.956	0.987	0.997	0.999	1.000	
2.0	0.135	0.406	0.677	0.857	0.947	0.983	0.995	0.999	1.000	

*Reprinted by kind permission from E. C. Molina, *Poisson's Exponential Binomial Limit*, D. Van Nostrand Company, Inc., Princeton, N.J., 1947.

Table 2

POISSON DISTRIBUTION FUNCTION (*Continued*)

λ \ x	0	1	2	3	4	5	6	7	8	9
2.2	0.111	0.355	0.623	0.819	0.928	0.975	0.993	0.998	1.000	
2.4	0.091	0.308	0.570	0.779	0.904	0.964	0.988	0.997	0.999	1.000
2.6	0.074	0.267	0.518	0.736	0.877	0.951	0.983	0.995	0.999	1.000
2.8	0.061	0.231	0.469	0.692	0.848	0.935	0.976	0.992	0.998	0.999
3.0	0.050	0.199	0.423	0.647	0.815	0.916	0.966	0.988	0.996	0.999
3.2	0.041	0.171	0.380	0.603	0.781	0.895	0.955	0.983	0.994	0.998
3.4	0.033	0.147	0.340	0.558	0.744	0.871	0.942	0.977	0.992	0.997
3.6	0.027	0.126	0.303	0.515	0.706	0.844	0.927	0.969	0.988	0.996
3.8	0.022	0.107	0.269	0.473	0.668	0.816	0.909	0.960	0.984	0.994
4.0	0.018	0.092	0.238	0.433	0.629	0.785	0.889	0.949	0.979	0.992
4.2	0.015	0.078	0.210	0.395	0.590	0.753	0.867	0.936	0.972	0.989
4.4	0.012	0.066	0.185	0.359	0.551	0.720	0.844	0.921	0.964	0.985
4.6	0.010	0.056	0.163	0.326	0.513	0.686	0.818	0.905	0.955	0.980
4.8	0.008	0.048	0.143	0.294	0.476	0.651	0.791	0.887	0.944	0.975
5.0	0.007	0.040	0.125	0.265	0.440	0.616	0.762	0.867	0.932	0.968
5.2	0.006	0.034	0.109	0.238	0.406	0.581	0.732	0.845	0.918	0.960
5.4	0.005	0.029	0.095	0.213	0.373	0.546	0.702	0.822	0.903	0.951
5.6	0.004	0.024	0.082	0.191	0.342	0.512	0.670	0.797	0.886	0.941
5.8	0.003	0.021	0.072	0.170	0.313	0.478	0.638	0.771	0.867	0.929
6.0	0.002	0.017	0.062	0.151	0.285	0.446	0.606	0.744	0.847	0.916

λ \ x	10	11	12	13	14	15	16
2.8	1.000						
3.0	1.000						
3.2	1.000						
3.4	0.999	1.000					
3.6	0.999	1.000					
3.8	0.998	0.999	1.000				
4.0	0.997	0.999	1.000				
4.2	0.996	0.999	1.000				
4.4	0.994	0.998	0.999	1.000			
4.6	0.992	0.997	0.999	1.000			
4.8	0.990	0.996	0.999	1.000			
5.0	0.986	0.995	0.998	0.999	1.000		
5.2	0.982	0.993	0.997	0.999	1.000		
5.4	0.977	0.990	0.996	0.999	1.000		
5.6	0.972	0.988	0.995	0.998	0.999	1.000	
5.8	0.965	0.984	0.993	0.997	0.999	1.000	
6.0	0.957	0.980	0.991	0.996	0.999	0.999	1.000

Table 2

POISSON DISTRIBUTION FUNCTION (*Continued*)

λ \ x	0	1	2	3	4	5	6	7	8	9
6.2	0.002	0.015	0.054	0.134	0.259	0.414	0.574	0.716	0.826	0.902
6.4	0.002	0.012	0.046	0.119	0.235	0.384	0.542	0.687	0.803	0.886
6.6	0.001	0.010	0.040	0.105	0.213	0.355	0.511	0.658	0.780	0.869
6.8	0.001	0.009	0.034	0.093	0.192	0.327	0.480	0.628	0.755	0.850
7.0	0.001	0.007	0.030	0.082	0.173	0.301	0.450	0.599	0.729	0.830
7.2	0.001	0.006	0.025	0.072	0.156	0.276	0.420	0.569	0.703	0.810
7.4	0.001	0.005	0.022	0.063	0.140	0.253	0.392	0.539	0.676	0.788
7.6	0.001	0.004	0.019	0.055	0.125	0.231	0.365	0.510	0.648	0.765
7.8	0.000	0.004	0.016	0.048	0.112	0.210	0.338	0.481	0.620	0.741
8.0	0.000	0.003	0.014	0.042	0.100	0.191	0.313	0.453	0.593	0.717
8.5	0.000	0.002	0.009	0.030	0.074	0.150	0.256	0.386	0.523	0.653
9.0	0.000	0.001	0.006	0.021	0.055	0.116	0.207	0.324	0.456	0.587
9.5	0.000	0.001	0.004	0.015	0.040	0.089	0.165	0.269	0.392	0.522
10.0	0.000	0.000	0.003	0.010	0.029	0.067	0.130	0.220	0.333	0.458

λ	10	11	12	13	14	15	16	17	18	19
6.2	0.949	0.975	0.989	0.995	0.998	0.999	1.000			
6.4	0.939	0.969	0.986	0.994	0.997	0.999	1.000			
6.6	0.927	0.963	0.982	0.992	0.997	0.999	0.999	1.000		
6.8	0.915	0.955	0.978	0.990	0.996	0.998	0.999	1.000		
7.0	0.901	0.947	0.973	0.987	0.994	0.998	0.999	1.000		
7.2	0.887	0.937	0.967	0.984	0.993	0.997	0.999	0.999	1.000	
7.4	0.871	0.926	0.961	0.980	0.991	0.996	0.998	0.999	1.000	
7.6	0.854	0.915	0.954	0.976	0.989	0.995	0.998	0.999	1.000	
7.8	0.835	0.902	0.945	0.971	0.986	0.993	0.997	0.999	1.000	
8.0	0.816	0.888	0.936	0.966	0.983	0.992	0.996	0.998	0.999	1.000
8.5	0.763	0.849	0.909	0.949	0.973	0.986	0.993	0.997	0.999	0.999
9.0	0.706	0.803	0.876	0.926	0.959	0.978	0.989	0.995	0.998	0.999
9.5	0.645	0.752	0.836	0.898	0.940	0.967	0.982	0.991	0.996	0.998
10.0	0.583	0.697	0.792	0.864	0.917	0.951	0.973	0.986	0.993	0.997

λ	20	21	22
8.5	1.000		
9.0	1.000		
9.5	0.999	1.000	
10.0	0.998	0.999	1.000

Table 2

POISSON DISTRIBUTION FUNCTION (*Continued*)

λ \ x	0	1	2	3	4	5	6	7	8	9
10.5	0.000	0.000	0.002	0.007	0.021	0.050	0.102	0.179	0.279	0.397
11.0	0.000	0.000	0.001	0.005	0.015	0.038	0.079	0.143	0.232	0.341
11.5	0.000	0.000	0.001	0.003	0.011	0.028	0.060	0.114	0.191	0.289
12.0	0.000	0.000	0.001	0.002	0.008	0.020	0.046	0.090	0.155	0.242
12.5	0.000	0.000	0.000	0.002	0.005	0.015	0.035	0.070	0.125	0.201
13.0	0.000	0.000	0.000	0.001	0.004	0.011	0.026	0.054	0.100	0.166
13.5	0.000	0.000	0.000	0.001	0.003	0.008	0.019	0.041	0.079	0.135
14.0	0.000	0.000	0.000	0.000	0.002	0.006	0.014	0.032	0.062	0.109
14.5	0.000	0.000	0.000	0.000	0.001	0.004	0.010	0.024	0.048	0.088
15.0	0.000	0.000	0.000	0.000	0.001	0.003	0.008	0.018	0.037	0.070

λ	10	11	12	13	14	15	16	17	18	19
10.5	0.521	0.639	0.742	0.825	0.888	0.932	0.960	0.978	0.988	0.994
11.0	0.460	0.579	0.689	0.781	0.854	0.907	0.944	0.968	0.982	0.991
11.5	0.402	0.520	0.633	0.733	0.815	0.878	0.924	0.954	0.974	0.986
12.0	0.347	0.462	0.576	0.682	0.772	0.844	0.899	0.937	0.963	0.979
12.5	0.297	0.406	0.519	0.628	0.725	0.806	0.869	0.916	0.948	0.969
13.0	0.252	0.353	0.463	0.573	0.675	0.764	0.835	0.890	0.930	0.957
13.5	0.211	0.304	0.409	0.518	0.623	0.718	0.798	0.861	0.908	0.942
14.0	0.176	0.260	0.358	0.464	0.570	0.669	0.756	0.827	0.883	0.923
14.5	0.145	0.220	0.311	0.413	0.518	0.619	0.711	0.790	0.853	0.901
15.0	0.118	0.185	0.268	0.363	0.466	0.568	0.664	0.749	0.819	0.875

λ	20	21	22	23	24	25	26	27	28	29
10.5	0.997	0.999	0.999	1.000						
11.0	0.995	0.998	0.999	1.000						
11.5	0.992	0.996	0.998	0.999	1.000					
12.0	0.988	0.994	0.997	0.999	0.999	1.000				
12.5	0.983	0.991	0.995	0.998	0.999	0.999	1.000			
13.0	0.975	0.986	0.992	0.996	0.998	0.999	1.000			
13.5	0.965	0.980	0.989	0.994	0.997	0.998	0.999	1.000		
14.0	0.952	0.971	0.983	0.991	0.995	0.997	0.999	0.999	1.000	
14.5	0.936	0.960	0.976	0.986	0.992	0.996	0.998	0.999	0.999	1.000
15.0	0.917	0.947	0.967	0.981	0.989	0.994	0.997	0.998	0.999	1.000

Table 2

POISSON DISTRIBUTION FUNCTION (*Continued*)

λ \ x	4	5	6	7	8	9	10	11	12	13
16	0.000	0.001	0.004	0.010	0.022	0.043	0.077	0.127	0.193	0.275
17	0.000	0.001	0.002	0.005	0.013	0.026	0.049	0.085	0.135	0.201
18	0.000	0.000	0.001	0.003	0.007	0.015	0.030	0.055	0.092	0.143
19	0.000	0.000	0.001	0.002	0.004	0.009	0.018	0.035	0.061	0.098
20	0.000	0.000	0.000	0.001	0.002	0.005	0.011	0.021	0.039	0.066
21	0.000	0.000	0.000	0.000	0.001	0.003	0.006	0.013	0.025	0.043
22	0.000	0.000	0.000	0.000	0.001	0.002	0.004	0.008	0.015	0.028
23	0.000	0.000	0.000	0.000	0.000	0.001	0.002	0.004	0.009	0.017
24	0.000	0.000	0.000	0.000	0.000	0.000	0.001	0.003	0.005	0.011
25	0.000	0.000	0.000	0.000	0.000	0.000	0.001	0.001	0.003	0.006

λ \ x	14	15	16	17	18	19	20	21	22	23
16	0.368	0.467	0.566	0.659	0.742	0.812	0.868	0.911	0.942	0.963
17	0.281	0.371	0.468	0.564	0.655	0.736	0.805	0.861	0.905	0.937
18	0.208	0.287	0.375	0.469	0.562	0.651	0.731	0.799	0.855	0.899
19	0.150	0.215	0.292	0.378	0.469	0.561	0.647	0.725	0.793	0.849
20	0.105	0.157	0.221	0.297	0.381	0.470	0.559	0.644	0.721	0.787
21	0.072	0.111	0.163	0.227	0.302	0.384	0.471	0.558	0.640	0.716
22	0.048	0.077	0.117	0.169	0.232	0.306	0.387	0.472	0.556	0.637
23	0.031	0.052	0.082	0.123	0.175	0.238	0.310	0.389	0.472	0.555
24	0.020	0.034	0.056	0.087	0.128	0.180	0.243	0.314	0.392	0.473
25	0.012	0.022	0.038	0.060	0.092	0.134	0.185	0.247	0.318	0.394

λ \ x	24	25	26	27	28	29	30	31	32	33
16	0.978	0.987	0.993	0.996	0.998	0.999	0.999	1.000		
17	0.959	0.975	0.985	0.991	0.995	0.997	0.999	0.999	1.000	
18	0.932	0.955	0.972	0.983	0.990	0.994	0.997	0.998	0.999	1.000
19	0.893	0.927	0.951	0.969	0.980	0.988	0.993	0.996	0.998	0.999
20	0.843	0.888	0.922	0.948	0.966	0.978	0.987	0.992	0.995	0.997
21	0.782	0.838	0.883	0.917	0.944	0.963	0.976	0.985	0.991	0.994
22	0.712	0.777	0.832	0.877	0.913	0.940	0.959	0.973	0.983	0.989
23	0.635	0.708	0.772	0.827	0.873	0.908	0.936	0.956	0.971	0.981
24	0.554	0.632	0.704	0.768	0.823	0.868	0.904	0.932	0.953	0.969
25	0.473	0.553	0.629	0.700	0.763	0.818	0.863	0.900	0.929	0.950

λ \ x	34	35	36	37	38	39	40	41	42	43
19	0.999	1.000								
20	0.999	0.999	1.000							
21	0.997	0.998	0.999	0.999	1.000					
22	0.994	0.996	0.998	0.999	0.999	1.000				
23	0.998	0.993	0.996	0.997	0.999	0.999	1.000			
24	0.979	0.987	0.992	0.995	0.997	0.998	0.999	0.999		
25	0.966	0.978	0.985	0.991	0.994	0.997	0.998	0.999	1.000	

Table 3

NORMAL DISTRIBUTION FUNCTION

$$F(z) = \frac{1}{\sqrt{2\pi}} \int_{-\infty}^{z} e^{-\frac{1}{2}t^2} dt$$

z	0.00	0.01	0.02	0.03	0.04	0.05	0.06	0.07	0.08	0.09
0.0	0.5000	0.5040	0.5080	0.5120	0.5160	0.5199	0.5239	0.5279	0.5319	0.5359
0.1	0.5398	0.5438	0.5478	0.5517	0.5557	0.5596	0.5636	0.5675	0.5714	0.5753
0.2	0.5793	0.5832	0.5871	0.5910	0.5948	0.5987	0.6026	0.6064	0.6103	0.6141
0.3	0.6179	0.6217	0.6255	0.6293	0.6331	0.6368	0.6406	0.6443	0.6480	0.6517
0.4	0.6554	0.6591	0.6628	0.6664	0.6700	0.6736	0.6772	0.6808	0.6844	0.6879
0.5	0.6915	0.6950	0.6985	0.7019	0.7054	0.7088	0.7123	0.7157	0.7190	0.7224
0.6	0.7257	0.7291	0.7324	0.7357	0.7389	0.7422	0.7454	0.7486	0.7517	0.7549
0.7	0.7580	0.7611	0.7642	0.7673	0.7704	0.7734	0.7764	0.7794	0.7823	0.7852
0.8	0.7881	0.7910	0.7939	0.7967	0.7995	0.8023	0.8051	0.8078	0.8106	0.8133
0.9	0.8159	0.8186	0.8212	0.8238	0.8264	0.8289	0.8315	0.8340	0.8365	0.8389
1.0	0.8413	0.8438	0.8461	0.8485	0.8508	0.8531	0.8554	0.8577	0.8599	0.8621
1.1	0.8643	0.8665	0.8686	0.8708	0.8729	0.8749	0.8770	0.8790	0.8810	0.8830
1.2	0.8849	0.8869	0.8888	0.8907	0.8925	0.8944	0.8962	0.8980	0.8997	0.9015
1.3	0.9032	0.9049	0.9066	0.9082	0.9099	0.9115	0.9131	0.9147	0.9162	0.9177
1.4	0.9192	0.9207	0.9222	0.9236	0.9251	0.9265	0.9279	0.9292	0.9306	0.9319
1.5	0.9332	0.9345	0.9357	0.9370	0.9382	0.9394	0.9406	0.9418	0.9429	0.9441
1.6	0.9452	0.9463	0.9474	0.9484	0.9495	0.9505	0.9515	0.9525	0.9535	0.9545
1.7	0.9554	0.9564	0.9573	0.9582	0.9591	0.9599	0.9608	0.9616	0.9625	0.9633
1.8	0.9641	0.9649	0.9656	0.9664	0.9671	0.9678	0.9686	0.9693	0.9699	0.9706
1.9	0.9713	0.9719	0.9726	0.9732	0.9738	0.9744	0.9750	0.9756	0.9761	0.9767
2.0	0.9772	0.9778	0.9783	0.9788	0.9793	0.9798	0.9803	0.9808	0.9812	0.9817
2.1	0.9821	0.9826	0.9830	0.9834	0.9838	0.9842	0.9846	0.9850	0.9854	0.9857
2.2	0.9861	0.9864	0.9868	0.9871	0.9875	0.9878	0.9881	0.9884	0.9887	0.9890
2.3	0.9893	0.9896	0.9898	0.9901	0.9904	0.9906	0.9909	0.9911	0.9913	0.9916
2.4	0.9918	0.9920	0.9922	0.9925	0.9927	0.9929	0.9931	0.9932	0.9934	0.9936
2.5	0.9938	0.9940	0.9941	0.9943	0.9945	0.9946	0.9948	0.9949	0.9951	0.9952
2.6	0.9953	0.9955	9.9956	0.9957	0.9959	0.9960	0.9961	0.9962	0.9963	0.9964
2.7	0.9965	0.9966	0.9967	0.9968	0.9969	0.9970	0.9971	0.9972	0.9973	0.9974
2.8	0.9974	0.9975	0.9976	0.9977	0.9977	0.9978	0.9979	0.9979	0.9980	0.9981
2.9	0.9981	0.9982	0.9982	0.9983	0.9984	0.9984	0.9985	0.9985	0.9986	0.9986
3.0	0.9987	0.9987	0.9987	0.9988	0.9988	0.9989	0.9989	0.9989	0.9990	0.9990
3.1	0.9990	0.9991	0.9991	0.9991	0.9992	0.9992	0.9992	0.9992	0.9993	0.9993
3.2	0.9993	0.9993	0.9994	0.9994	0.9994	0.9994	0.9994	0.9995	0.9995	0.9995
3.3	0.9995	0.9995	0.9995	0.9996	0.9996	0.9996	0.9996	0.9996	0.9996	0.9997
3.4	0.9997	0.9997	0.9997	0.9997	0.9997	0.9997	0.9997	0.9997	0.9997	0.9998

Table 4

Values of t_α*

ν	$\alpha = 0.10$	$\alpha = 0.05$	$\alpha = 0.025$	$\alpha = 0.01$	$\alpha = 0.005$	ν
1	3.078	6.314	12.706	31.821	63.657	1
2	1.886	2.920	4.303	6.965	9.925	2
3	1.638	2.353	3.182	4.541	5.841	3
4	1.533	2.132	2.776	3.474	4.604	4
5	1.476	2.015	2.571	3.365	4.032	5
6	1.440	1.943	2.447	3.143	3.707	6
7	1.415	1.895	2.365	2.998	3.499	7
8	1.397	1.860	2.306	2.896	3.355	8
9	1.383	1.833	2.262	2.821	3.250	9
10	1.372	1.812	2.228	2.764	3.169	10
11	1.363	1.796	2.201	2.718	3.106	11
12	1.356	1.782	2.179	2.681	3.055	12
13	1.350	1.771	2.160	2.650	3.012	13
14	1.345	1.761	2.145	2.624	2.977	14
15	1.341	1.753	2.131	2.602	2.947	15
16	1.337	1.746	2.120	2.583	2.921	16
17	1.333	1.740	2.110	2.567	2.898	17
18	1.330	1.734	2.101	2.552	2.878	18
19	1.328	1.729	2.093	2.539	2.861	19
20	1.325	1.725	2.086	2.528	2.845	20
21	1.323	1.721	2.080	2.518	2.831	21
22	1.321	1.717	2.074	2.508	2.819	22
23	1.319	1.714	2.069	2.500	2.807	23
24	1.318	1.711	2.064	2.492	2.797	24
25	1.316	1.708	2.060	2.485	2.787	25
26	1.315	1.706	2.056	2.479	2.779	26
27	1.314	1.703	2.052	2.473	2.771	27
28	1.313	1.701	2.048	2.467	2.763	28
29	1.311	1.699	2.045	2.462	2.756	29
inf	1.282	1.645	1.960	2.326	2.576	inf.

*This table is abridged from Table IV of R. A. Fisher, *Statistical Methods for Research Workers*, published by Oliver and Boyd, Ltd., Edinburgh, by permission of the author and publishers.

Table 5

VALUES OF χ_α^2*

ν	$\alpha = 0.995$	$\alpha = 0.99$	$\alpha = 0.975$	$\alpha = 0.95$	$\alpha = 0.05$	$\alpha = 0.025$	$\alpha = 0.01$	$\alpha = 0.005$	ν
1	0.0000393	0.000157	0.000982	0.00393	3.841	5.024	6.635	7.879	1
2	0.0100	0.0201	0.0506	0.103	5.991	7.378	9.210	10.597	2
3	0.0717	0.115	0.216	0.352	7.815	9.348	11.345	12.838	3
4	0.207	0.297	0.484	0.711	9.488	11.143	13.277	14.860	4
5	0.412	0.554	0.831	1.145	11.070	12.832	13.086	16.750	5
6	0.676	0.872	1.237	1.635	12.592	14.449	16.812	18.548	6
7	0.989	1.239	1.690	2.167	14.067	16.013	18.475	20.278	7
8	1.344	1.646	2.180	2.733	15.507	17.535	20.090	21.955	8
9	1.735	2.088	2.700	3.325	16.919	19.023	21.666	23.589	9
10	2.156	2.558	3.247	3.940	18.307	20.483	23.209	25.188	10
11	2.603	3.053	3.816	4.575	19.675	21.920	24.725	26.757	11
12	3.074	3.571	4.404	5.226	21.026	23.337	26.217	28.300	12
13	3.565	4.107	5.009	5.892	22.362	24.736	27.688	29.819	13
14	4.075	4.660	5.629	6.571	23.685	26.119	29.141	31.319	14
15	4.601	5.229	6.262	7.261	24.996	27.488	30.578	32.801	15
16	5.142	5.812	6.908	7.962	26.296	28.845	32.000	34.267	16
17	5.697	6.408	7.564	8.672	27.587	30.191	33.409	35.718	17
18	6.265	7.015	8.231	9.390	28.869	31.526	34.805	37.156	18
19	6.844	7.633	8.907	10.117	30.144	32.852	36.191	38.582	19
20	7.434	8.260	9.591	10.851	31.410	34.170	37.566	39.997	20
21	8.034	8.897	10.283	11.591	32.671	35.479	38.932	41.401	21
22	8.643	9.542	10.982	12.338	33.924	36.781	40.289	42.796	22
23	9.260	10.196	11.689	13.091	35.172	38.076	41.638	44.181	23
24	9.886	10.856	12.401	13.484	36.415	39.364	42.980	45.558	24
25	10.520	11.524	13.120	14.611	37.652	40.646	44.314	46.928	25
26	11.160	12.198	13.844	15.379	38.885	41.923	45.642	48.290	26
27	11.808	12.879	14.573	16.151	40.113	43.194	46.963	49.645	27
28	12.461	13.565	15.308	16.928	41.337	44.461	48.278	50.993	28
29	13.121	14.256	16.047	17.708	42.557	45.772	49.588	52.336	29
30	13.787	14.953	16.791	18.493	43.773	46.979	50.892	53.672	30

*This table is based on Table 8 of *Biometrika Tables for Statisticians*, Vol. I, by permission of the *Biometrika* trustees.

Table 6 (a)

VALUES OF $F_{.05}$ *

v_1 = Degrees of freedom for numerator

v_2 = Degrees of freedom for denominator	1	2	3	4	5	6	7	8	9	10	12	15	20	24	30	40	60	120	∞
1	161	200	216	225	230	234	237	239	241	242	244	246	248	249	250	251	252	253	254
2	18.50	19.00	19.20	19.20	19.30	19.30	19.40	19.40	19.40	19.40	19.40	19.40	19.40	19.50	19.50	19.50	19.50	19.50	19.50
3	10.10	9.55	9.28	9.12	9.01	8.94	8.89	8.85	8.81	8.79	8.74	8.70	8.66	8.64	8.62	8.59	8.57	8.55	8.53
4	7.71	6.94	6.59	6.39	6.26	6.16	6.09	6.04	6.00	5.96	5.91	5.86	5.80	5.77	5.75	5.72	5.69	5.66	5.63
5	6.61	5.79	5.41	5.19	5.05	4.95	4.88	4.82	4.77	4.74	4.68	4.62	4.56	4.53	4.50	4.46	4.43	4.40	4.37
6	5.99	5.14	4.76	4.53	4.39	4.28	4.21	4.15	4.10	4.06	4.00	3.94	3.87	3.84	3.81	3.77	3.74	3.70	3.67
7	5.59	4.74	4.35	4.12	3.97	3.87	3.79	3.73	3.68	3.64	3.57	3.51	3.44	3.41	3.38	3.34	3.30	3.27	3.23
8	5.32	4.46	4.07	3.84	3.69	3.58	3.50	3.44	3.39	3.35	3.28	3.22	3.15	3.12	3.08	3.04	3.01	2.97	2.93
9	5.12	4.26	3.86	3.63	3.48	3.37	3.29	3.23	3.18	3.14	3.07	3.01	2.94	2.90	2.86	2.83	2.79	2.75	2.71
10	4.96	4.10	3.71	3.48	3.33	3.22	3.14	3.07	3.02	2.98	2.91	2.85	2.77	2.74	2.70	2.66	2.62	2.58	2.54
11	4.84	3.98	3.59	3.36	3.20	3.09	3.01	2.95	2.90	2.85	2.79	2.72	2.65	2.61	2.57	2.53	2.49	2.45	2.40
12	4.75	3.89	3.49	3.26	3.11	3.00	2.91	2.85	2.80	2.75	2.69	2.62	2.54	2.51	2.47	2.38	2.38	2.30	2.30
13	4.67	3.81	3.41	3.18	3.03	2.92	2.83	2.77	2.71	2.67	2.60	2.53	2.46	2.42	2.38	2.34	2.30	2.25	2.21
14	4.60	3.74	3.34	3.11	2.96	2.85	2.76	2.70	2.65	2.60	2.53	2.46	2.39	2.35	2.31	2.27	2.22	2.18	2.13
15	4.54	3.68	3.29	3.06	2.90	2.79	2.71	2.64	2.59	2.54	2.48	2.40	2.33	2.29	2.25	2.20	2.16	2.11	2.07
16	4.49	3.63	3.24	3.01	2.85	2.74	2.66	2.59	2.54	2.49	2.42	2.35	2.28	2.24	2.19	2.15	2.11	2.06	2.01
17	4.45	3.59	3.20	2.96	2.81	2.70	2.61	2.55	2.49	2.45	2.38	2.31	2.23	2.19	2.15	2.10	2.06	2.01	1.96
18	4.41	3.55	3.16	2.93	2.77	2.66	2.58	2.51	2.46	2.41	2.34	2.27	2.19	2.15	2.11	2.06	2.02	1.97	1.93
19	4.38	3.52	3.13	2.90	2.74	2.63	2.54	2.48	2.42	2.38	2.31	2.23	2.16	2.11	2.07	2.03	1.98	1.93	1.88
20	4.35	3.49	3.10	2.87	2.71	2.60	2.51	2.45	2.39	2.35	2.28	2.20	2.12	2.08	2.04	1.99	1.95	1.90	1.84
21	4.32	3.47	3.07	2.84	2.68	2.57	2.49	2.42	2.37	2.32	2.25	2.18	2.10	2.05	2.01	1.96	1.92	1.87	1.81
22	4.30	3.44	3.05	2.82	2.66	2.55	2.46	2.40	2.34	2.30	2.23	2.15	2.07	2.03	1.98	1.94	1.89	1.84	1.78
23	4.28	3.42	3.03	2.80	2.64	2.53	2.44	2.37	2.32	2.27	2.20	2.13	2.05	2.01	1.96	1.91	1.86	1.81	1.76
24	4.26	3.40	3.01	2.78	2.62	2.51	2.42	2.36	2.30	2.25	2.18	2.11	2.03	1.98	1.94	1.89	1.84	1.79	1.73
25	4.24	3.39	2.99	2.76	2.60	2.49	2.40	2.34	2.28	2.24	2.16	2.09	2.01	1.96	1.92	1.87	1.82	1.77	1.71
30	4.17	3.32	2.92	2.69	2.53	2.42	2.33	2.27	2.21	2.16	2.09	2.01	1.93	1.89	1.84	1.79	1.74	1.68	1.62
40	4.08	3.23	2.84	2.61	2.45	2.34	2.25	2.18	2.12	2.08	2.00	1.92	1.84	1.79	1.74	1.69	1.64	1.58	1.51
60	4.00	3.15	2.76	2.53	2.37	2.25	2.17	2.10	2.04	1.99	1.92	1.84	1.75	1.70	1.65	1.59	1.53	1.47	1.39
120	3.92	3.07	2.68	2.45	2.29	2.18	2.09	2.02	1.96	1.91	1.83	1.75	1.66	1.61	1.55	1.50	1.43	1.35	1.25
∞	3.84	3.00	2.60	2.37	2.21	2.10	2.01	1.94	1.88	1.83	1.75	1.67	1.57	1.52	1.46	1.39	1.32	1.22	1.00

*This table is reproduced from M. Merrington and C. M. Thompson, "Tables of percentage points of the inverted beta (F) distribution," *Biometrika*, Vol. 33 (1943), by permission of the *Biometrika* trustees.

Table 6 (b)

VALUES OF $F_{.01}$ *

v_1 = Degrees of freedom for numerator

v_2 = Degrees of freedom for denominator	1	2	3	4	5	6	7	8	9	10	12	15	20	24	30	40	60	120	∞
1	4,052	5,000	5,403	5,625	5,764	5,859	5,928	5,982	6,023	6,056	6,106	6,157	6,209	6,235	6,261	6,287	6,313	6,339	6,366
2	98.50	99.00	99.20	99.20	99.30	99.30	99.40	99.40	99.40	99.40	99.40	99.40	99.40	99.50	99.50	99.50	99.50	99.50	99.50
3	34.10	30.80	29.50	28.70	28.20	27.90	27.70	27.50	27.30	27.20	27.10	26.90	26.70	26.60	26.50	26.40	26.30	26.20	26.10
4	21.20	18.00	16.70	16.00	15.50	15.20	15.00	14.80	14.70	14.50	14.40	14.20	14.00	13.90	13.80	13.70	13.70	13.60	13.50
5	16.30	13.30	12.10	11.40	11.00	10.70	10.50	10.30	10.20	10.10	9.89	9.72	9.55	9.47	9.38	9.29	9.20	9.11	9.02
6	13.70	10.90	9.78	9.15	8.75	8.47	8.26	8.10	7.98	7.87	7.72	7.56	7.40	7.31	7.23	7.14	7.06	6.97	6.88
7	12.20	9.55	8.45	7.85	7.46	7.19	6.99	6.84	6.72	6.62	6.47	6.31	6.16	6.07	5.99	5.91	5.82	5.74	5.65
8	11.30	8.65	7.59	7.01	6.63	6.37	6.18	6.03	5.91	5.81	5.67	5.52	5.36	5.28	5.20	5.12	5.03	4.95	4.83
9	10.60	8.02	6.99	6.42	6.06	5.80	5.61	5.47	5.35	5.26	5.11	4.96	4.81	4.73	4.65	4.57	4.48	4.40	4.31
10	10.00	7.56	6.55	5.99	5.64	5.39	5.20	5.06	4.94	4.85	4.71	4.56	4.41	4.33	4.25	4.17	4.08	4.00	3.91
11	9.65	7.21	6.22	5.67	5.32	5.07	4.89	4.74	4.63	4.54	4.40	4.25	4.10	4.02	3.94	3.86	3.78	3.69	3.60
12	9.33	6.93	5.95	5.41	5.06	4.82	4.64	4.50	4.39	4.30	4.16	4.01	3.86	3.78	3.70	3.62	3.54	3.45	3.36
13	9.07	6.70	5.74	5.21	4.86	4.62	4.44	4.30	4.19	4.10	3.96	3.82	3.66	3.59	3.51	3.43	3.34	3.25	3.17
14	8.86	6.51	5.56	5.04	4.70	4.46	4.28	4.14	4.03	3.94	3.80	3.66	3.51	3.43	3.35	3.27	3.18	3.09	3.00
15	8.68	6.36	5.42	4.89	4.56	4.32	4.14	4.00	3.89	3.80	3.67	3.52	3.37	3.29	3.21	3.13	3.05	2.96	2.87
16	8.53	6.23	5.29	4.77	4.44	4.20	4.03	3.89	3.78	3.69	3.55	3.41	3.26	3.18	3.10	3.02	2.93	2.84	2.75
17	8.40	6.11	5.19	4.67	4.34	4.10	3.93	3.79	3.68	3.59	3.46	3.31	3.16	3.08	3.00	2.92	2.83	2.75	2.65
18	8.29	6.01	5.09	4.58	4.25	4.01	3.84	3.71	3.60	3.51	3.37	3.23	3.08	3.00	2.92	2.84	2.75	2.66	2.57
19	8.19	5.93	5.01	4.50	4.17	3.94	3.77	3.63	3.52	3.43	3.30	3.15	3.00	2.92	2.84	2.76	2.67	2.58	2.49
20	8.10	5.85	4.94	4.43	4.10	3.87	3.70	3.56	3.46	3.37	3.23	3.09	2.94	2.86	2.78	2.69	2.61	2.52	2.42
21	8.02	5.78	4.87	4.37	4.04	3.81	3.64	3.51	3.40	3.31	3.17	3.03	2.88	2.80	2.72	2.64	2.55	2.46	2.36
22	7.95	5.72	4.82	4.31	3.99	3.76	3.59	3.45	3.35	3.26	3.12	2.98	2.83	2.75	2.67	2.58	2.50	2.40	2.31
23	7.88	5.66	4.76	4.26	3.94	3.71	3.54	3.41	3.30	3.21	3.07	2.93	2.78	2.70	2.62	2.54	2.45	2.35	2.26
24	7.82	5.61	4.72	4.22	3.90	3.67	3.50	3.36	3.26	3.17	3.03	2.89	2.74	2.66	2.58	2.49	2.40	2.31	2.21
25	7.77	5.57	4.68	4.18	3.86	3.63	3.46	3.32	3.22	3.13	2.99	2.85	2.70	2.62	2.53	2.45	2.36	2.27	2.17
30	7.56	5.39	4.51	4.02	3.70	3.47	3.30	3.17	3.07	2.98	2.84	2.70	2.55	2.47	2.39	2.30	2.21	2.11	2.01
40	7.31	5.18	4.31	3.83	3.51	3.29	3.12	2.99	2.89	2.80	2.66	2.52	2.37	2.29	2.20	2.11	2.02	1.92	1.80
60	7.08	4.98	4.13	3.65	3.34	3.12	2.95	2.82	2.72	2.63	2.50	2.35	2.20	2.12	2.03	1.94	1.84	1.73	1.60
120	6.85	4.79	3.95	3.48	3.17	2.96	2.79	2.66	2.56	2.47	2.34	2.19	2.03	1.95	1.86	1.76	1.66	1.53	1.38
∞	6.63	4.61	3.78	3.32	3.02	2.80	2.64	2.51	2.41	2.32	2.18	2.04	1.88	1.79	1.70	1.59	1.47	1.32	1.00

*This table is reproduced from M. Merrington and C. M. Thompson, "Tables of percentage points of the inverted beta (F) distribution," *Biometrika*. Vol. 33 (1943), by permission of the *Biometrika* trustees.

Table 7

RANDOM DIGITS*

1306	1189	5731	3968	5606	5084	8947	3897	1636	7810
0422	2431	0649	8085	5053	4722	6598	5044	9040	5121
6597	2022	6168	5060	8656	6733	6364	7649	1871	4328
7965	6541	5645	6243	7658	6903	9911	5740	7824	8520
7695	6937	0406	8894	0441	8135	9797	7285	5905	9539
5160	7851	8464	6789	3938	4197	6511	0407	9239	2232
2961	0551	0539	8288	7478	7565	5581	5771	5442	8761
1428	4183	4312	5445	4854	9157	9158	5218	1464	3634
3666	5642	4539	1561	7849	7520	2547	0756	1206	2033
6543	6799	7454	9052	6689	1946	2574	9386	0304	7945
9975	6080	7423	3175	9377	6951	6519	8287	8994	5532
4866	0956	7545	7723	8085	4948	2228	9583	4415	7065
8239	7068	6694	5168	3117	1568	0237	6160	9585	1133
8722	9191	3386	3443	0434	4586	4150	1224	6204	0937
1330	9120	8785	8382	2929	7089	3109	6742	2468	7025
2296	2952	4764	9070	6356	9192	4012	0618	2219	1109
3582	7052	3132	4519	9250	2486	0830	8472	2160	7046
5872	9207	7222	6494	8973	3545	6967	8490	5264	9821
1134	6324	6201	3792	5651	0538	4676	2064	0584	7996
1403	4497	7390	8503	8239	4236	8022	2914	4368	4529
3393	7025	3381	3553	2128	1021	8353	6413	5161	8583
1137	7896	3602	0060	7850	7626	0854	6565	4260	6220
7437	5198	8772	6927	8527	6851	2709	5992	7383	1071
8414	8820	3917	7238	9821	6073	6658	1280	9643	7761
8398	5224	2749	7311	5740	9771	7826	9533	3800	4553
0995	8935	2939	3092	2496	0359	0318	4697	7181	4035
6657	0755	9685	4017	6581	7292	5643	5064	1142	1297
8875	8369	7868	0190	9278	1709	4253	9346	4335	3769
8399	6702	0586	6428	7985	2979	4513	1970	1989	3105
6703	1024	2064	0393	6815	8502	1375	4171	6970	1201
4730	1653	9032	9855	0957	7366	0325	5178	7959	5371
8400	6834	3187	8688	1079	1480	6776	9888	7585	9998
3647	8002	6726	0877	4552	3238	7542	7804	3933	9475
6789	5197	8037	2354	9262	5497	0005	3986	1767	7981
2630	2721	2810	2185	6323	5679	4931	8336	6662	3566
1374	8625	1644	3342	1587	0762	6057	8011	2666	3759
1519	7625	9110	4409	0239	7059	3415	5537	2250	7292
9678	2877	7579	4935	0449	8119	6969	5383	1717	6719
0882	6781	3538	4090	3092	2365	6001	3446	9985	6007
0006	4205	2389	4365	1981	8158	7784	6256	3842	5603
4611	9861	7916	9305	2074	9462	0254	4827	9198	3974
1093	3784	4190	6332	1175	8599	9735	8584	6581	7194
3374	3545	6865	8819	3342	1676	2264	6014	5012	2458
3650	9676	1436	4374	4716	5548	8276	6235	6742	2154
7292	5749	7977	7602	9205	3599	3880	9537	4423	2330
2353	8319	2850	4026	3027	1708	3518	7034	7132	6903
1094	2009	8919	5676	7283	4982	9642	9235	8167	3366
0568	4002	0587	7165	1094	2006	7471	0940	4366	9554
5606	4070	5233	4339	6543	6695	5799	58?'	3953	9458
8285	7537	1181	2300	5294	6892	1627	3372	1952	3028

*From Donald B. Owen, *Handbook of Statistical Tables*. Reading, Mass.: Addison Wesley, 1962.

Table 7

RANDOM DIGITS (*Continued*)

2444	9039	4803	8568	1590	2420	2547	2470	8179	4617
5748	7767	2800	6289	2814	8281	1549	9519	3341	1192
7761	8583	0852	5619	6864	8506	9643	7763	9611	1289
6838	9280	2654	0812	3988	2146	5095	0150	8043	9079
6440	2631	3033	9167	4998	7036	0133	7428	9702	1376
8829	0094	2887	3802	5497	0318	5168	6377	9216	2802
9845	4796	2951	4449	1999	2691	5328	7674	7004	6212
5072	9000	3887	5739	7920	6074	4715	3681	2721	2701
9035	0553	1272	2600	3828	8197	8852	9092	8027	6144
5562	1080	2222	0336	1411	0303	7424	3713	9278	1918
2757	2650	8727	3953	9579	2442	8041	9869	2887	3933
6397	1848	1476	0787	4990	4666	1208	2769	3922	1158
9208	7641	3575	4279	1282	1840	5999	1806	7809	5885
2418	9289	6120	8141	3908	5577	3590	2317	8975	4593
7300	9006	5659	8258	3662	0332	5369	3640	0563	7939
6870	2535	8916	3245	2256	4350	6064	2438	2002	1272
2914	7309	4045	7513	3195	4166	0878	5184	6680	2655
0868	8657	8118	6340	9452	7460	3291	5778	1167	0312
7994	6579	6461	2292	9554	8309	5036	0974	9517	8293
8587	0764	6687	9150	1642	2050	4934	0027	1376	5040
8016	8345	2257	5084	8004	7949	3205	3972	7640	3478
5581	5775	7517	9076	4699	8313	8401	7147	9416	7184
2015	3364	6688	2631	2152	2220	1637	8333	4838	5699
7327	8987	5741	0102	1173	7350	7080	7420	1847	0741
3589	1991	1764	8355	9684	9423	7101	1063	4151	4875
2188	6454	7319	1215	0473	6589	2355	9579	7004	6209
2924	0472	9878	7966	2491	5662	5635	2789	2564	1249
1961	1669	2219	1113	9175	0260	4046	8142	4432	2664
2393	9637	0410	7536	0972	5153	0708	1935	1143	1704
7585	4424	2648	6728	2233	3518	7267	1732	1926	3833
0197	4021	9207	7327	9212	7017	8060	6216	1942	6817
9719	5336	5532	8537	2980	8252	4971	0110	6209	1556
8866	4785	6007	8006	9043	4109	5570	9249	9905	2152
5744	3957	8786	9023	1472	7275	1014	1104	0832	7680
7149	5721	1389	6581	7196	7072	6360	3084	7009	0239
7710	8479	9345	7773	9086	1202	8845	3163	7937	6163
5246	5651	0432	8644	6341	9661	2361	8377	8673	6098
3576	0013	7381	0124	8559	9813	9080	6984	0926	2169
3026	1464	2671	4691	0353	5289	8754	2442	7799	8983
6591	4365	8717	2365	5686	8377	8675	9798	7745	6360
0402	3257	0480	5038	1998	2935	1306	1190	2406	2596
7105	7654	4745	4482	8471	1424	2031	7803	4367	6816
7181	4140	1046	0885	1264	7755	1653	8924	5822	4401
3655	3282	2178	8134	3291	7262	8229	2866	7065	4806
5121	6717	3117	1901	5184	6467	8954	3884	0279	8635
3618	3098	9208	7429	1578	1917	7927	2696	3704	0833
0166	3638	4947	1414	4799	9189	2459	5056	5982	6154
6187	9653	3658	4730	1652	8096	8288	9368	5531	7788
1234	1448	0276	7290	1667	2823	3755	5642	4854	8844
8949	8731	4875	5724	2962	1182	2930	7539	4526	7252

Table 7

RANDOM DIGITS (*Continued*)

4357	4146	8353	9952	8004	7945	1530	5207	4730	1967
5339	7325	6862	7584	8634	3485	2278	5832	0612	8118
6583	8433	0717	0606	9284	2719	1888	2889	0285	2765
6564	3526	2171	3809	3428	5523	9078	0648	7768	3326
4811	1933	3763	6265	8931	0649	8085	6177	4450	2139
6931	7236	1230	0441	4013	1352	6563	1499	7332	3068
8755	3390	6120	7825	9005	7012	1643	9934	4044	7022
6742	2260	3443	0190	9278	1816	7697	7933	0067	2906
6655	3930	9014	6032	7574	1685	5258	3100	5358	1929
8514	4806	4124	9286	0449	5051	4772	4651	0038	1580
8135	5004	7299	8981	4689	1950	2271	2201	8344	3852
4414	6855	0127	5489	5157	6386	7492	3736	7164	0498
3727	7959	5056	5983	8021	0204	7616	4325	7454	5039
5434	7342	0314	7525	0067	2800	6292	4706	3454	6881
7195	8828	9869	2785	3186	8375	7414	7232	0401	2483
2705	8245	6251	9611	1077	0641	0195	7024	6202	3899
1547	8981	4972	1280	4286	5678	0338	8096	8284	7010
3424	1435	1354	7631	7260	7361	0151	8903	9056	8684
8969	7551	3695	4915	7921	2913	3840	9031	9747	9735
5225	8720	8898	2478	3342	9200	8836	7269	2992	6284
6432	9861	1516	2849	2539	2208	4595	8616	6170	5865
3085	5903	8319	2744	0814	7318	8619	7614	3265	5999
0264	1246	3687	9759	6995	6565	3949	1012	0179	0059
8710	2419	6065	0036	9650	2027	6042	5467	1839	5577
5736	9001	3132	4521	9973	5070	8078	4150	2276	5059
7529	1339	4802	5751	3785	7125	4922	8877	9530	6499
5133	7995	8030	7408	2186	0725	5554	5664	6791	9677
3170	9915	6960	2621	6718	4059	9919	1007	6469	5410
3024	0680	1127	8088	0200	5868	0084	6362	6808	3727
4398	3121	7749	8191	2087	8270	5233	3980	6774	8522
0082	5419	7659	2061	2506	7573	1157	3979	2309	0811
4351	6516	6814	5898	3973	8103	3616	2049	7843	0568
3268	0086	7580	1337	3884	5679	4830	4509	9587	2184
4391	8487	4884	1488	2249	6661	5774	7205	2717	7030
7328	0705	0652	9424	7082	8579	5647	5571	9667	8555
3835	2938	2671	4691	0559	8382	2825	4928	5379	8635
8731	4980	8674	4506	7262	8127	2022	2178	7463	4842
2995	7868	0683	3768	0625	9887	7060	0514	0034	8600
5597	9028	5660	5006	8325	9677	2169	3196	0357	7811
3081	5876	8150	1360	1868	9265	3277	8465	7502	6458
7406	4439	5683	6877	2920	9588	3002	2869	3746	3690
5969	9442	7696	7510	1620	4973	1911	1288	6160	9797
4765	9647	4364	1037	4975	1998	1359	1346	6125	5078
3219	2532	7577	2815	8696	9248	9410	9282	6572	3940
6906	8859	5044	8826	6218	3206	9034	0843	9832	2703
7993	3141	0103	4528	7988	4635	8478	9094	9077	5306
2549	3737	7686	0723	4505	6841	1379	6460	1869	5700
3672	7033	4844	0149	7412	6370	1884	0717	5740	8477
2217	0293	3978	5933	1032	5192	1732	2137	9357	5941
3162	9968	6369	1258	0416	4326	7840	6525	2608	5255

Table 7

RANDOM DIGITS (*Continued*)

1758	1489	2774	6033	9813	1052	1816	7484	1699	7350
6430	8803	0478	4157	5626	1603	1339	4666	1207	2135
4893	8857	1717	1533	6572	8408	2173	4754	0272	1305
1516	2733	7326	8674	9233	1799	5281	0797	0885	0947
4950	3171	5756	3036	9047	8719	8498	1312	7124	4787
0549	6775	9360	6639	0990	0037	7309	4702	0812	4195
1018	7027	7569	7549	2539	2315	8030	7663	3881	8264
2241	9965	9729	7092	4891	9239	0738	1804	3025	1030
1602	0708	2201	9848	6241	1084	8142	8555	7291	5016
5840	8381	1549	9902	6935	3681	6420	0214	8489	5911
1676	0367	7484	1595	5693	3008	9816	7311	6162	1024
6048	4175	8940	9029	8306	8892	4127	1709	4043	6591
5549	9621	2563	0515	0560	9021	0632	4309	4044	7010
5317	4584	9418	4600	0640	9668	6379	6515	6310	7916
2532	7784	6469	4793	5957	4123	6555	3237	6915	6960
2300	5412	3106	4877	6936	4109	8060	1896	6881	7028
1499	8699	4534	5367	7557	2701	2587	2521	2159	6991
6201	3791	2946	2863	5684	5517	7448	2227	8991	7505
6839	9736	8312	8068	7339	5395	9559	3416	6169	5484
0092	5537	1933	3186	8482	6680	2656	1864	4535	2193
1862	3253	6515	6299	2929	2219	9145	7511	2146	4962
9886	6744	3097	8894	0446	3494	8211	1723	6138	3181
5289	9071	1231	0651	9109	7448	2228	9700	0224	4595
2685	7104	7193	5506	2993	7028	4830	3866	8698	0277
6055	7092	4786	6847	4543	7448	2017	4114	8385	3625
4092	4995	0280	9371	3375	3503	4496	8642	5388	1831
5951	4937	3670	5797	5030	6524	2265	7748	7875	6976
7687	3849	4821	2373	1157	4208	3623	9399	7349	6663
9886	3463	2055	4872	2702	1807	9056	8576	4237	8757
3193	3011	8899	6721	0086	2623	7977	7578	4024	1997
9181	7365	9135	1669	2007	7784	6363	6913	6017	6588
9459	2175	5728	4933	0111	6703	1234	2410	1620	4859
9874	5278	2849	3163	6372	2600	9887	7060	3919	1111
7729	2099	7513	2774	6030	8260	9023	1368	9513	6122
4699	8102	3001	7947	1659	6571	1969	7152	7356	1062
1872	7244	3954	7422	2688	8649	0156	1965	5012	2461
9636	0123	2438	1757	4204	1650	2486	0002	4724	7412
6403	9054	7632	7469	8973	3332	0294	5062	0303	7315
4433	3293	2314	7431	2389	4094	5062	0118	0046	6070
2361	3933	8026	0431	8012	3214	8927	7355	0585	7638
4077	8463	6580	6983	0181	3327	6812	1755	3387	4569
6678	0006	3686	8478	9187	2291	9032	9852	1450	7940
6499	2582	5207	4627	0456	4245	3583	8996	1006	5839
6663	9021	0319	7908	1241	9977	7042	6923	2539	2103
9999	4503	6105	5525	1068	6272	7036	0200	6291	2841
9048	6982	3845	6865	9029	8700	0349	3416	8236	1129
5136	9653	3654	2863	5565	8923	5596	8389	9927	9092
9906	1070	5693	3012	1218	7309	4361	3041	4327	8423
4198	7035	8182	6270	6461	2079	2998	5507	9605	6734
2030	5878	8989	6789	4359	1820	5063	9199	7751	6337

Table 8

OPERATING CHARACTERISTIC CURVES FOR TESTS ABOUT MEANS*

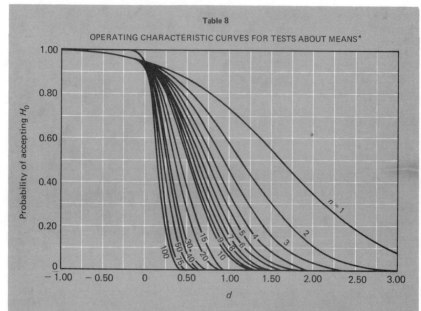

(a) One-tail tests based on normal distribution; $\alpha = 0.05$.

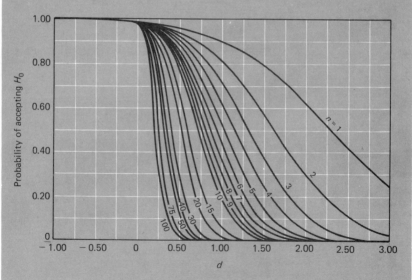

(b) One-tail tests based on normal distribution; $\alpha = 0.01$.

*Reproduced from the 2nd edition of *Engineering Statistics* by A.H. Bowker and G.J. Lieberman, Prentice-Hall, Inc., 1972, with the permission of the authors and publishers.

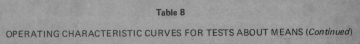

Table 8

OPERATING CHARACTERISTIC CURVES FOR TESTS ABOUT MEANS (*Continued*)

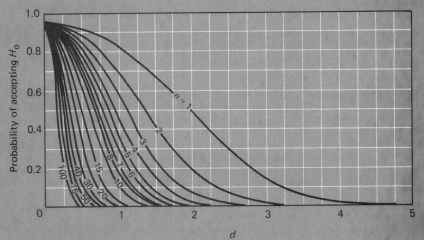

(c) Two-tail tests based on normal distribution; $\alpha = 0.05$

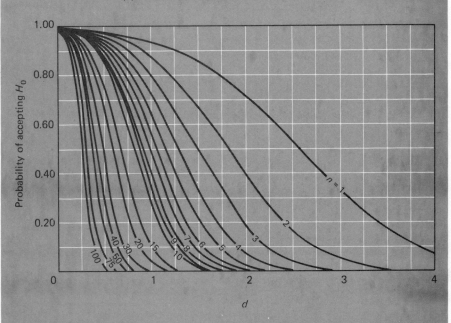

(d) Two-tail tests based on normal distribution; $\alpha = 0.01$

Table 9 (a)

0.95 CONFIDENCE INTERVALS FOR PROPORTIONS*

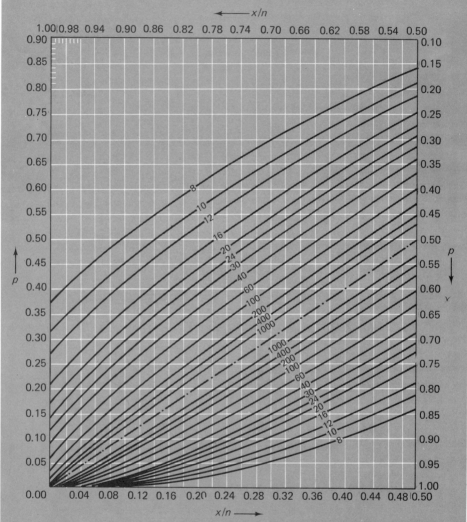

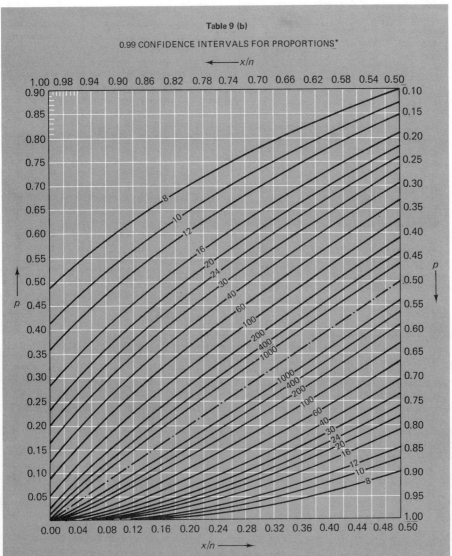

Table 9 (b)

0.99 CONFIDENCE INTERVALS FOR PROPORTIONS*

Table 10

CRITICAL VALUES OF D*

Sample size n	$D_{.10}$	$D_{.05}$	$D_{.01}$
1	0.950	0.975	0.995
2	0.776	0.842	0.929
3	0.642	0.708	0.828
4	0.564	0.624	0.733
5	0.510	0.565	0.669
6	0.470	0.521	0.618
7	0.438	0.486	0.577
8	0.411	0.457	0.543
9	0.388	0.432	0.514
10	0.368	0.410	0.490
11	0.352	0.391	0.468
12	0.338	0.375	0.450
13	0.325	0.361	0.433
14	0.314	0.349	0.418
15	0.304	0.338	0.404
16	0.295	0.328	0.392
17	0.286	0.318	0.381
18	0.278	0.309	0.371
19	0.272	0.301	0.363
20	0.264	0.294	0.356
25	0.24	0.27	0.32
30	0.22	0.24	0.29

*Adapted from F. J. Massey, Jr., "The Kolgomorov-Smirnov test for goodness of fit," *J. Amer. Statist. Ass.*, Vol. 46 (1951), p. 70, with the kind permission of the author and publisher

Table 11

$$\text{VALUES OF } Z = \frac{1}{2} \ln \frac{1+r}{1-r}$$

r	.00	.01	.02	.03	.04	.05	.06	.07	.08	.09
0.0	0.000	0.010	0.020	0.030	0.040	0.050	0.060	0.070	0.080	0.090
0.1	0.100	0.110	0.121	0.131	0.141	0.151	0.161	0.172	0.182	0.192
0.2	0.203	0.213	0.224	0.234	0.245	0.255	0.266	0.277	0.288	0.299
0.3	0.310	0.321	0.332	0.343	0.354	0.365	0.377	0.388	0.400	0.412
0.4	0.424	0.436	0.448	0.460	0.472	0.485	0.497	0.510	0.523	0.536
0.5	0.549	0.563	0.576	0.590	0.604	0.618	0.633	0.648	0.662	0.678
0.6	0.693	0.709	0.725	0.741	0.758	0.775	0.793	0.811	0.829	0.848
0.7	0.867	0.887	0.908	0.929	0.950	0.973	0.996	1.020	1.045	1.071
0.8	1.099	1.127	1.157	1.188	1.221	1.256	1.293	1.333	1.376	1.422
0.9	1.472	1.528	1.589	1.658	1.738	1.832	1.946	2.092	2.298	2.647

For negative values of r put a minus sign in front of the corresponding Z's, and vice versa.

Table 12 (a)

VALUES OF r_p FOR $\alpha = 0.05$*

$d.f.$ \ p	2	3	4	5	6	7	8	9	10
1	17.97								
2	6.09	6.09							
3	4.50	4.52	4.52						
4	3.93	4.01	4.03	4.03					
5	3.64	3.75	3.80	3.81	3.81				
6	3.46	3.59	3.65	3.68	3.69	3.70			
7	3.34	3.48	3.55	3.59	3.61	3.62	3.63		
8	3.26	3.40	3.48	3.52	3.55	3.57	3.57	3.58	
9	3.20	3.34	3.42	3.47	3.50	3.52	3.54	3.54	3.55
10	3.15	3.29	3.38	3.43	3.47	3.49	3.51	3.52	3.52
11	3.11	3.26	3.34	3.40	3.44	3.46	3.48	3.49	3.50
12	3.08	3.23	3.31	3.37	3.41	3.44	3.46	3.47	3.48
13	3.06	3.20	3.29	3.35	3.39	3.42	3.46	3.46	3.47
14	3.03	3.18	3.27	3.33	3.37	3.40	3.43	3.44	3.46
15	3.01	3.16	3.25	3.31	3.36	3.39	3.41	3.43	3.45
16	3.00	3.14	3.23	3.30	3.34	3.38	3.40	3.42	3.44
17	2.98	3.13	3.22	3.28	3.33	3.37	3.39	3.41	3.43
18	2.97	3.12	3.21	3.27	3.32	3.36	3.38	3.40	3.42
19	2.96	3.11	3.20	3.26	3.31	3.35	3.38	3.40	3.41
20	2.95	3.10	3.19	3.25	3.30	3.34	3.37	3.39	3.41
24	2.92	3.07	3.16	3.23	3.28	3.31	3.35	3.37	3.39
30	2.89	3.03	3.13	3.20	3.25	3.29	3.32	3.35	3.37
40	2.86	3.01	3.10	3.17	3.22	3.27	3.30	3.33	3.35
60	2.83	2.98	3.07	3.14	3.20	3.24	3.28	3.31	3.33
120	2.80	2.95	3.04	3.12	3.17	3.22	3.25	3.29	3.31
∞	2.77	2.92	3.02	3.09	3.15	3.19	3.23	3.27	3.29

*This table is reproduced from H. L. Harter, "Critical values for Duncan's new multiple range test." It contains some corrected values to replace those given by D. B. Duncan in "Multiple Range and Multiple F Tests," *Biometrics*, Vol. 11 (1955). The above table is reproduced with the permission of the author and the editor of *Biometrics*.

Table 12 (b)

VALUES OF r_p FOR $\alpha = 0.01$*

p / $d.f.$	2	3	4	5	6	7	8	9	10
1	90.02								
2	14.04	14.04							
3	8.26	8.32	8.32						
4	6.51	6.68	6.74	6.76					
5	5.70	5.90	5.99	6.04	6.07				
6	5.24	5.44	5.55	5.62	5.66	5.68			
7	4.95	5.15	5.26	5.33	5.38	5.42	5.44		
8	4.74	4.94	5.06	5.13	5.19	5.23	5.26	5.28	
9	4.60	4.79	4.91	4.99	5.04	5.09	5.12	5.14	5.16
10	4.48	4.67	4.79	4.88	4.93	4.98	5.01	5.04	5.06
11	4.39	4.58	4.70	4.78	4.84	4.89	4.92	4.95	4.97
12	4.32	4.50	4.62	4.71	4.77	4.81	4.85	4.88	4.91
13	4.26	4.44	4.56	4.64	4.71	4.75	4.79	4.82	4.85
14	4.21	4.39	4.51	4.59	4.66	4.70	4.74	4.77	4.80
15	4.17	4.34	4.46	4.55	4.61	4.66	4.70	4.73	4.76
16	4.13	4.31	4.43	4.51	4.57	4.62	4.66	4.70	4.72
17	4.10	4.27	4.39	4.47	4.54	4.59	4.63	4.66	4.69
18	4.07	4.25	4.36	4.45	4.51	4.56	4.60	4.64	4.66
19	4.05	4.22	4.33	4.42	4.48	4.53	4.57	4.61	4.64
20	4.02	4.20	4.31	4.40	4.46	4.51	4.55	4.59	4.62
24	3.96	4.13	4.24	4.32	4.39	4.44	4.48	4.52	4.55
30	3.89	4.06	4.17	4.25	4.31	4.36	4.41	4.45	4.48
40	3.82	3.99	4.10	4.18	4.24	4.29	4.33	4.38	4.41
60	3.76	3.92	4.03	4.11	4.18	4.23	4.37	4.31	4.34
120	3.70	3.86	3.97	4.04	4.11	4.16	4.20	4.24	4.27
∞	3.64	3.80	3.90	3.98	4.04	4.09	4.13	4.17	4.21

*This table is reproduced from H. L. Harter, "Critical values for Duncan's new multiple range test." It contains some corrected values to replace those given by D. B. Duncan in "Multiple Range and Multiple F Tests," *Biometrics*, Vol. 11 (1955). The above table is reproduced with the permission of the author and the editor of *Biometrics*.

Table 13

CONTROL CHART CONSTANTS*

Number of observations in sample, n	Chart for averages			Chart for standard deviations					Chart for ranges				
	Factors for control limits			Factor for central line	Factors for control limits				Factor for central line	Factors for control limits			
	A	A_1	A_2	c_2	B_1	B_2	B_3	B_4	d_2	D_1	D_2	D_3	D_4
2	2.121	3.760	1.880	0.5642	0	1.843	0	3.267	1.128	0	3.686	0	3.267
3	1.732	2.394	1.023	0.7236	0	1.858	0	2.568	1.693	0	4.358	0	2.575
4	1.500	1.880	0.729	0.7979	0	1.808	0	2.266	2.059	0	4.698	0	2.282
5	1.342	1.596	0.577	0.8407	0	1.756	0	2.089	2.326	0	4.918	0	2.115
6	1.225	1.410	0.483	0.8686	0.026	1.711	0.030	1.970	2.534	0	5.078	0	2.004
7	1.134	1.277	0.419	0.8882	0.105	1.672	0.118	1.882	2.704	0.205	5.203	0.076	1.924
8	1.061	1.175	0.373	0.9027	0.167	1.638	0.185	1.815	2.847	0.387	5.307	0.136	1.864
9	1.000	1.094	0.337	0.9139	0.219	1.609	0.239	1.761	2.970	0.546	5.394	0.184	1.816
10	0.949	1.028	0.308	0.9227	0.262	1.584	0.284	1.716	3.078	0.687	5.469	0.223	1.777
11	0.905	0.973	0.285	0.9300	0.299	1.561	0.321	1.679	3.173	0.812	5.534	0.256	1.744
12	0.866	0.925	0.266	0.9359	0.331	1.541	0.354	1.646	3.258	0.924	5.592	0.284	1.716
13	0.832	0.884	0.249	0.9410	0.359	1.523	0.382	1.618	3.336	1.026	5.646	0.308	1.692
14	0.802	0.848	0.235	0.9453	0.384	1.507	0.406	1.594	3.407	1.121	5.693	0.329	1.671
15	0.775	0.816	0.223	0.9490	0.406	1.492	0.428	1.572	3.472	1.207	5.737	0.348	1.652

*This table is reproduced, by permission, from the *ASTM Manual on Quality Control of Materials*, American Society for Testing and Materials, Philadelphia, Pa., 1951.

Table 14

	$1 - \alpha = 0.95$			$1 - \alpha = 0.99$		
n	$P \geq$ 0.90	0.95	0.99	0.90	0.95	0.99
2	32.019	37.674	48.430	160.193	188.491	242.300
3	8.380	9.916	12.861	18.930	22.401	29.055
4	5.369	6.370	8.299	9.398	11.150	14.527
5	4.275	5.079	6.634	6.612	7.855	10.260
6	3.712	4.414	5.775	5.337	6.345	8.301
7	3.369	4.007	5.248	4.613	5.488	7.187
8	3.136	3.732	4.891	4.147	4.936	6.468
9	2.967	3.532	4.631	3.822	4.550	5.966
10	2.839	3.379	4.433	3.582	4.265	5.594
11	2.737	3.259	4.277	3.397	4.045	5.308
12	2.655	3.162	4.150	3.250	3.870	5.079
13	2.587	3.081	4.044	3.130	3.727	4.893
14	2.529	3.012	3.955	3.029	3.608	4.737
15	2.480	2.954	3.878	2.945	3.507	4.605
16	2.437	2.903	3.812	2.872	3.421	4.492
17	2.400	2.858	3.754	2.808	3.345	4.393
18	2.366	2.819	3.702	2.753	3.279	4.307
19	2.337	2.784	3.656	2.703	3.221	4.230
20	2.310	2.752	3.615	2.659	3.168	4.161
25	2.208	2.631	3.457	2.494	2.972	3.904
30	2.140	2.549	3.350	2.385	2.841	3.733
35	2.090	2.490	3.272	2.306	2.748	3.611
40	2.052	2.445	3.213	2.247	2.677	3.518
45	2.021	2.408	3.165	2.200	2.621	3.444
50	1.996	2.379	3.126	2.162	2.576	3.385
55	1.976	2.354	3.094	2.130	2.538	3.335
60	1.958	2.333	3.066	2.103	2.506	3.293
65	1.943	2.315	3.042	2.080	2.478	3.257
70	1.929	2.299	3.021	2.060	2.454	3.225
75	1.917	2.285	3.002	2.042	2.433	3.197
80	1.907	2.272	2.986	2.026	2.414	3.173
85	1.897	2.261	2.971	2.012	2.397	3.150
90	1.889	2.251	2.958	1.999	2.382	3.130
95	1.881	2.241	2.945	1.987	2.368	3.112
100	1.874	2.233	2.934	1.977	2.355	3.096
150	1.825	2.175	2.859	1.905	2.270	2.983
200	1.798	2.143	2.816	1.865	2.222	2.921
250	1.780	2.121	2.788	1.839	2.191	2.880
300	1.767	2.106	2.767	1.820	2.169	2.850
400	1.749	2.084	2.739	1.794	2.138	2.809
500	1.737	2.070	2.721	1.777	2.117	2.783
600	1.729	2.060	2.707	1.764	2.102	2.763
700	1.722	2.052	2.697	1.755	2.091	2.748
800	1.717	2.046	2.688	1.747	2.082	2.736
900	1.712	2.040	2.682	1.741	2.075	2.726
1000	1.709	2.036	2.676	1.736	2.068	2.718
∞	1.645	1.960	2.576	1.645	1.960	2.576

*Adapted by permission from *Techniques of Statistical Analysis* by C. Eisenhart, M. W. Hastay, and W. A. Wallis. Copyright 1947, McGraw-Hill Book Company, Inc.

See p. 435

Table 15

SAMPLE SIZE CODE LETTERS FROM MIL-STD-105 D

Lot or batch size			General inspection levels		
			I	II	III
2	to	8	A	A	B
9	to	15	A	B	C
16	to	25	B	C	D
26	to	50	C	D	E
51	to	90	C	E	F
91	to	150	D	F	G
151	to	280	E	G	H
281	to	500	F	H	J
501	to	1,200	G	J	K
1,201	to	3,200	H	K	L
3,201	to	10,000	J	L	M
10,001	to	35,000	K	M	N
35,001	to	150,000	L	N	P
150,001	to	500,000	M	P	Q
500,001	and	over	N	Q	R

Table 16

MASTER TABLE FOR SINGLE SAMPLING (NORMAL INSPECTION)
MIL-STD-105 D

Acceptable quality levels (normal inspection). Values shown are **Ac Re** (Ac = Acceptance number, Re = Rejection number). ↓ = use first sampling plan below arrow; ↑ = use first sampling plan above arrow.

Sample size code letter	Sample size	0.010	0.015	0.025	0.040	0.065	0.10	0.15	0.25	0.40	0.65	1.0	1.5	2.5	4.0	6.5	10	15	25	40	65	100	150	250	400	650	1000
A	2	↓	↓	↓	↓	↓	↓	↓	↓	↓	↓	↓	↓	↓	↓	↓	↓	0 1	1 2	2 3	3 4	5 6	7 8	10 11	14 15	21 22	30 31
B	3	↓	↓	↓	↓	↓	↓	↓	↓	↓	↓	↓	↓	↓	↓	↓	0 1	1 2	2 3	3 4	5 6	7 8	10 11	14 15	21 22	30 31	44 45
C	5	↓	↓	↓	↓	↓	↓	↓	↓	↓	↓	↓	↓	↓	↓	0 1	1 2	2 3	3 4	5 6	7 8	10 11	14 15	21 22	30 31	44 45	↑
D	8	↓	↓	↓	↓	↓	↓	↓	↓	↓	↓	↓	↓	↓	0 1	1 2	2 3	3 4	5 6	7 8	10 11	14 15	21 22	30 31	44 45	↑	↑
E	13	↓	↓	↓	↓	↓	↓	↓	↓	↓	↓	↓	↓	0 1	1 2	2 3	3 4	5 6	7 8	10 11	14 15	21 22	30 31	44 45	↑	↑	↑
F	20	↓	↓	↓	↓	↓	↓	↓	↓	↓	↓	↓	0 1	1 2	2 3	3 4	5 6	7 8	10 11	14 15	21 22	30 31	44 45	↑	↑	↑	↑
G	32	↓	↓	↓	↓	↓	↓	↓	↓	↓	↓	0 1	1 2	2 3	3 4	5 6	7 8	10 11	14 15	21 22	30 31	44 45	↑	↑	↑	↑	↑
H	50	↓	↓	↓	↓	↓	↓	↓	↓	↓	0 1	1 2	2 3	3 4	5 6	7 8	10 11	14 15	21 22	30 31	44 45	↑	↑	↑	↑	↑	↑
J	80	↓	↓	↓	↓	↓	↓	↓	↓	0 1	1 2	2 3	3 4	5 6	7 8	10 11	14 15	21 22	30 31	44 45	↑	↑	↑	↑	↑	↑	↑
K	125	↓	↓	↓	↓	↓	↓	↓	0 1	1 2	2 3	3 4	5 6	7 8	10 11	14 15	21 22	30 31	44 45	↑	↑	↑	↑	↑	↑	↑	↑
L	200	↓	↓	↓	↓	↓	↓	0 1	1 2	2 3	3 4	5 6	7 8	10 11	14 15	21 22	30 31	44 45	↑	↑	↑	↑	↑	↑	↑	↑	↑
M	315	↓	↓	↓	↓	↓	0 1	1 2	2 3	3 4	5 6	7 8	10 11	14 15	21 22	30 31	44 45	↑	↑	↑	↑	↑	↑	↑	↑	↑	↑
N	500	↓	↓	↓	↓	0 1	1 2	2 3	3 4	5 6	7 8	10 11	14 15	21 22	30 31	44 45	↑	↑	↑	↑	↑	↑	↑	↑	↑	↑	↑
P	800	↓	↓	↓	0 1	1 2	2 3	3 4	5 6	7 8	10 11	14 15	21 22	30 31	44 45	↑	↑	↑	↑	↑	↑	↑	↑	↑	↑	↑	↑
Q	1250	↓	↓	0 1	1 2	2 3	3 4	5 6	7 8	10 11	14 15	21 22	30 31	44 45	↑	↑	↑	↑	↑	↑	↑	↑	↑	↑	↑	↑	↑
R	2000	↓	0 1	1 2	2 3	3 4	5 6	7 8	10 11	14 15	21 22	30 31	44 45	↑	↑	↑	↑	↑	↑	↑	↑	↑	↑	↑	↑	↑	↑

Ac = Acceptance number.
Re = Rejection number.

↓ = Use first sampling plan below arrow. If sample size equals, or exceeds, lot or batch size, do 100 percent inspection.

↕ = Use first sampling plan above arrow.

Answers to Odd-Numbered Exercises

CHAPTER 2

PAGE 13

7. 6,561.

9. (b) *B* is the event that three graduate assistants are present, *C* is the event that as many graduate assistants as professors are present, *D* is the event that three persons (professors or graduate assistants) are present;

(c) (1, 1), (1, 2), (2, 1), (2, 2); at most two graduate assistants are present;

(d) Yes.

11. (a) 2; (b) 7; (c) 1 and 4; (d) 1, 2, 3, and 5; (e) 8.

PAGE 27

1. $\varnothing(0)$, $I \cap D'(20)$, $I \cap D(10)$, $I' \cap D(5)$, $I' \cap D'(465)$, $I(30)$, $D(15)$, $I'(470)$, $D'(485)$, $(I \cap D') \cup (I' \cap D)(25)$, $(I \cap D) \cup (I' \cap D')(475)$, $I \cup D(35)$, $I' \cup D'(490)$, $D \cup I'(480)$, $I \cup D'(495)$, $S(500)$.
3. Claim should be questioned, since $N(R \cap P') + N(R \cap P) + N(R' \cap P) + N[(R \cup P)'] = 430$.
5. (a) Yes; (b) No, sum exceeds 1; (c) No, $P(C) < 0$; (d) No, sum is less than 1; (e) Yes.
7. (a) 0.50; (b) 0.40; (c) 0.50; (d) 0.22; (e) 0.25; (f) 0.20.
11. $0.75 \leq P < 0.80$.
13. Inconsistent, since $P(A \cup B) = 1/3 + 1/4 > 1/2$.
15. (a) 1/26; (b) 1/2; (c) 4/13; (d) 1/13.
17. (a) 84/150; (b) 99/150; (c) 43/150; (d) 63/150.
19. (a) 0.94; (b) 0.21; (c) 0.86; (d) 0.59.
21. (a) 0.29; (b) 0.18.
23. 0.0024.

PAGE 40

1. $P(I \mid D) = 2/3$, $P(I \mid D') = 4/97$.
3. (a) 62/85; (b) 74/84; (c) 29/51.
5. (a) 13/30; (b) 1/2; (c) 7/16.
9. (a) 1/4; (b) 25/102.
11. $P(E_1 \cap P_1 \cap C_1) = 0.07$, $P(E_1)P(P_1)P(C_1) = 0.072$, dependent events.
13. (a) 1/20; (b) 0.0768; (c) 0.025.
15. (a) 0.885; (b) 0.115.
17. 6/17.
19. (a) 0.781; (b) 0.219.
21. Industrial sabotage, $P(\text{sabotage} \mid \text{explosion}) = 0.344$.

PAGE 47

1. $0.14; $0.22.
3. (a) Yes; (b) equally profitable; (c) more profitable to test.
5. $1.20.
7. (a) $P(\text{Winning}) < 1/3$; (b) $P(\text{Winning}) > 1/3$.
9. (a) First job, since $E_1 = \$165,000$ and $E_2 = \$135,000$;
   (b) Second job so that he will at least have a chance.
11. $np = m(1 - p)$; $p = 3/4$.

13. (a) *FF* costs $6.25, *LWR* costs $10.53/*KW*/year of useful life;
     (b) *FF* costs more than 1.68 times *LWR* fuel;
     (c) *LWR* now costs $11.40/*KW*/year, *FF* fuel cost must now exceed
         1.82 times *LWR* fuel cost.

CHAPTER 3

PAGE 63

1. $f(0) = 1/20$, $f(1) = 2/20$, $f(2) = 3/20$, $f(3) = 4/20$, $f(4) = 4/20$,
    $f(5) = 3/20$, $f(6) = 2/20$, $f(7) = 1/20$.
3. (a) Yes; (b) Yes; (c) No, $f(3) < 0$; (d) Yes; (e) No, $\sum f(x) < 1$.
9. (a) 0.3169; (b) 0.1442; (c) 0.0861; (d) 0.0746; (e) 0.4862; (f) 0.7052.
11. (a) $\binom{8}{3}(.2)^3(.8)^5 = 0.1468$; (b) 0.1468.
13. (a) 0.3412; (b) 0.9804; (c) 0.1184.
15. (a) 0.5367; (b) 0.2059.
17. (a) 102/253; (b) 459/1012; (c) 145/1012.
19. (a) 11/21; (b) 44/105; (c) 2/35.
21. (a) 0.117; (b) 0.116, error = 0.85%.

PAGE 76

1. (a) 1.0; (b) 1.0.
3. $\mu = \frac{1}{2}(n + 1)$, $\sigma^2 = \frac{1}{12}(n^2 - 1)$.
5. $\mu = 3/2$, $\sigma^2 = 15/28$.
7. | k | Chebyshev | Binomial |
    |---|-----------|----------|
    | 1 | 1.0000 | 0.4544 |
    | 2 | 0.2500 | 0.0768 |
    | 3 | 0.1111 | 0.0042 |
9. 15,625.

PAGE 86

3. (a) 0.173; (b) 0.091; (c) 0.809.
5. $\lambda = 3$, $p = 0.224$.
7. (a) 0.191; (b) 0.874; (c) 0.473.
9. 0.007.
11. (a) 0.981; (b) 0.577; (c) 0.978.
13. 0.024.
15. (a) 1.5; (b) 0.9; (c) 3 min.
17. Savings are $68.56; not worthwhile.

PAGE 90

1. 0.058.

3. (a) $\dfrac{\binom{a_1}{x_1}\binom{a_2}{x_2}\cdots\binom{a_k}{x_k}}{\binom{N}{n}}$; (b) 0.195.

PAGE 94

1. (a) 00–40, 41–77, 78–93, 94–98, 99.
3. (a) 0000–2465, 2466–5917, 5918–8334, 8335–9462, 9463–9857, 9858–9968, 9969–9994, 9995–9999.

CHAPTER 4

PAGE 102

3. (a) 0.30; (b) 0.50;
   (c) $F(x) = 0 \ (x \leq 0)$, $F(x) = \frac{1}{2}x^2 \ (0 < x < 1)$,
   $F(x) = 2x - \frac{1}{2}x^2 - 1 \ (1 \leq x < 2)$, $F(x) = 1 \ (x \geq 2)$;
   (d) 0.02; (e) 0.84.
5. (a) 5/9; (b) 9/100; (c) $f(x) = 0 \ (x \leq 2)$, $f(x) = 8x^{-3} \ (x > 2)$.
7. (a) 0.393; (b) 0.148; (c) 0.223.
9. $\mu = 3/8$, $\sigma^2 = 19/320$.
11. $\mu = 4$, σ^2 does not exist.

PAGE 110

1. (a) 0.9332; (b) 0.1151; (c) 0.0154; (d) 0.9599;
   (e) 0.4987; (f) 0.1018; (g) 0.2789; (h) 0.9093.
3. (a) 0.6826; (b) 0.9544; (c) 0.9974.
5. (a) 0.3520; (b) 0.2033; (c) 0.9030; (d) 0.0502.
7. 0.1292.
9. 0.8093.
11. (a) 0.0062; (b) 0.1587; (c) 0.8664.
13. 83.14%.
15. $\mu = 2.984$.

PAGE 123

1. $F(x) = 0 \ (x \leq \alpha), F(x) = \dfrac{x - \alpha}{\beta - \alpha} \ (\alpha < x < \beta), F(x) = 1 \ (x \geq \beta).$
7. (a) $\mu = 4, \sigma = \sqrt{8}$; (b) 0.594.
9. $n = 25$, maximum profit $= \$125$.
11. (a) $e^{-\alpha t}$; (b) $e^{-\alpha t}$; (c) $f(t) = 0 \ (t \leq 0), f(t) = \alpha e^{-\alpha t} \ (t > 0)$.
13. (a) 0.362; (b) 0.067.
17. 0.264.

PAGE 130

1. (a) 1/4; (b) 1/24;
   (c) $F(x_1, x_2) = 0 \ (x_1 \leq 0, x_2 \leq 0),$
   $F(x_1, x_2) = \frac{1}{4}x_1^2 x_2^2 \ (0 < x_1 < 1, 0 < x_2 < 2),$
   $F(x_1, x_2) = 1 \ (x_1 \geq 1, x_2 \geq 1),$
   $F(x_1, x_2) = x^2, \ (0 < x_1 < 1, x_2 < 2),$
   $F(x_1, x_2) = \frac{1}{4}x_2^2 \ (x_1 > 1, 0 < x_2 < 2).$
3. (a) $f(x_1 \,|\, 0.25) = x_1 + \frac{1}{2} \ (0 < x_1 < 1);$
   (b) $f(x_2 \,|\, x_1) = \dfrac{x_1 + 2x_2}{x_1 + 1}.$
5. (a) 0.3264; (b) 0.472.
7. 2.
9. $\mu = LW, \sigma^2 = \frac{1}{12}(a^2 W^2 + b^2 L^2 + \frac{1}{12}a^2 b^2).$

CHAPTER 5

PAGE 146

15. No. The volume, not the height and width, should be doubled.

PAGE 156

3. (a) $\bar{x} = 103.7$, claim seems reasonable but is hardly proved;
   (b) $s = 11.1$, 67%.
7. (a) $\bar{x} = 459.2$; (b) $s^2 = 953.79$; (c) C.V. $= 6.73\%$.
9. (a) $\bar{x} = 12.92, s = 8.92$; (b) impossible, open class.
11. $\bar{x} = 27.7, s = 5.16$.
13. (a) 11.4; (b) 27.45.
15. (a) 73; (b) 8.02%.

CHAPTER 6

PAGE 170

7. (a) Becomes 1/3 as large; (b) becomes 2/3 as large;
(c) becomes 5 times as large.
13. 0.6284.
15. A process which is "in control" will be judged "out of control" with a probability less than 0.0013.

PAGE 178

1. $t = 3.21$; $t_{.005} = 2.797$ (24 d.f.); information does not support claim.
3. $t = 4.74$; $t_{.005} = 3.250$ (9 d.f.); process is out of control.
5. 0.94.
7. 0.02.

CHAPTER 7

PAGE 191

3. $E = 1.48$ beats per minute.
5. (a) $E = 0.84$; (b) $11.97 < \mu < 13.49$.
7. $E = 0.33$.
9. $n = 171$.
11. $E = 1.16$ minutes.
13. $E = 5.2$.
15. $135,639 < \mu < 157,745$.
19. (a) 835; (b) 727; (c) 901.

PAGE 207

1. (a) $\alpha = 0.11$; (b) $\beta = 0.11$.
7. (b) $\beta = 0.800$ for $\mu = 0.5$, $\beta = 0.255$ for $\mu = 1.0$, $\beta = 0.015$ for $\mu = 1.5$.
9. (a) 41; (b) 40.
11. (a) 0.91; (b) 0.78; (c) 0.55; (d) 0.31; (e) 0.13; (f) 0.04.
13. (a) 0.60; (b) 0.31; (c) 0.12; (d) 0.03; (e) 0.00.

PAGE 221

1. (a) $\mu_1 - \mu_2 < 0$, buy radial tires only if H_0 can be rejected;
   (b) $\mu_1 - \mu_2 > 0$, buy radial tires unless H_0 can be rejected.
3. $z = 2.07$, reject H_0.
5. $t = 1.48$, cannot reject H_0.
7. (a) $z = 1.15$, cannot reject H_0; (b) 0.04.
9. $t = 3.09$, reject the claim.
11. (a) $\mu = 0.006$, $\sigma = 0.0036$; (b) 0.0475.
13. $n = 25$.
15. $t = 1.83$, cannot reject H_0.
17. $t = 4.16$, reject H_0 (substantiates claim).
19. $\pm = 2.21$, cannot reject H_0.
23. $(\bar{x}_1 - \bar{x}_2) \pm t_{\alpha/2}\sqrt{\dfrac{[(n_1 - 1)s_1^2 + (n_2 - 1)s_2^2](n_1 + n_2)}{n_1 n_2 (n_1 + n_2 - 2)}}$,
   where $t_{\alpha/2}$ has $n_1 + n_2 - 2$ d.f.

CHAPTER 8

PAGE 231

1. (a) $s = 4.2$; (b) 4.3.
3. (a) 3.29; (b) 3.35.
5. $3.56 < \sigma^2 < 72.93$.
7. $6.56 < \sigma < 10.59$.
9. $2.30 < \sigma < 5.77$.
11. $1.31 < \sigma < 2.05$.

PAGE 238

1. $\chi^2 = 5.83$, cannot reject H_0.
3. $z = 3.51$, reject H_0.
5. $\chi^2 = 19.21$, cannot reject H_0.
7. $z = -2.10$, reject H_0 (inspector not making satisfactory measurements).
9. $F = 1.20$, cannot reject H_0 (t-test justified).
11. $F = 1.51$, cannot reject H_0 (claim not substantiated).
13. (a) No, variances are unequal; (b) base test on *logarithms* of the observations.

CHAPTER 9

PAGE 249

1. $0.35 < p < 0.49$
3. (a) $0.52 < p < 0.64$;
   (b) $0.521 < p < 0.633$.
5. (a) $0.49 < p < 0.63$;
   (b) $0.491 < p < 0.629$.
7. $E = 0.116$.
9. (a) $n = 1068$; (b) $n = 1025$.
11. $0.520 < p < 0.631$.
13. $p < 0.0232$.
15. (a) $0.40, 0.56, 0.04$; (b) $0.598, 0.133, 0.269$.

PAGE 258

1. $z = 2.19$, reject H_0 ($p = 0.30$) in favor of alternative $p > 0.30$.
3. $z = -1.83$, cannot reject H_0.
5. $B(4; 13, 0.60) = 0.0321$, reject H_0.
7. (b) 0.0345
9. $z = -0.64$, difference not significant.
11. $z = -2.91$, reject H_0; yes.
13. (a) $\chi^2 = 34.9$, significant difference; (b) $z = -5.91$.
15. $z = 1.76$, reject H_0.
17. $\chi^2 = 7.10$, differences not significant.
19. $\chi^2 = 9.39$; yes, proportions not the same.

PAGE 265

1. $\chi^2 = 3.46$, cannot reject H_0.
3. $\chi^2 = 0.20$, cannot reject H_0.
5. $\chi^2 = 22.6$, reject the claim.
7. $\text{d.f.} = 8$.
9. $\chi^2 = 9.16$, good fit.
11. (a) $0.0313, 0.1071, 0.2223, 0.2811, 0.2163, 0.1013, 0.0289$;
    (b) $2.5, 8.6, 17.8, 22.5, 17.3, 8.1, 2.3$;
    (c) $\chi^2 = 1.454$, good fit; since $\bar{x}$, s, and $\sum f_i$ were needed from the data, $\text{d.f.} = k - 3$.

CHAPTER 10

PAGE 279

1. $z = 3.91$, reject H_0.
3. P (9 or more successes in 15) $= 0.3036$, cannot reject H_0.
9. $z = -0.10$, cannot reject H_0.
11. $z = 2.92$, difference is significant.
13. $z = 0.10$, cannot reject H_0.
15. $H = 1.53$, cannot reject H_0.
17. $H = 26.1$, populations not identical.

PAGE 287

1. $z = 0.24$, cannot reject H_0.
3. $z = 2.43$, reject H_0 (signal could contain a message).
5. $z = -0.11$, cannot reject hypothesis of randomness.
7. $z = -4.00$, reject H_0.

CHAPTER 11

PAGE 302

1. (a) $a = 39.05$, $b = 0.764$; (b) $y' = 65.8$.
3. $11.86 < \beta < 17.12$.
5. $y' = 3.34 + 24.93x$; $22.32 < \beta < 27.54$.
7. $53.01 < \alpha + \beta x_0 < 61.99$.
9. (a) $b = \sum xy / \sum x^2$; (b) $b = 14.75$; (c) $b = 25.48$.
11. $y = -2.24 + 0.100x$.

PAGE 317

1. $y' = 2.00(0.988)^x$.
3. $y' = 28.3e^{0.000153x}$.
5. (a) $\hat{y} = 1.50$; (b) $1.39 < \hat{y} < 1.61$, assume $\log p_i$ are independent and normally distributed with equal variances.
7. 0.928.

9. $y' = 1.751 - 0.316x + 0.025x^2$.
11. (a) $y' = 10.48 - 0.383x$, $t = -2.27$, not significant, $\hat{\sigma}_1^2 = 1.70$;
    (b) $\hat{\sigma}_2^2 = 0.27$, $F = 5.3$, not significant.
15. $y' = 961x_1 + 2,976x_2 - 16,279$; \$34,061.

PAGE 328

1. (a) $r = 0.81$; (b) $z = 5.16$, reject H_0.
3. (a) $r = -0.58$; (b) $-0.85 < \rho < -0.07$.
9. $z = 0.627$, cannot reject H_0.
11. (a) $0.47 < \rho < 0.95$; (b) $-0.33 < \rho < 0.52$; (c) $-0.75 < \rho < 0.52$.
13. (a) $r' = 0.70$; (b) $r' = 0.83$; (c) $r' = 0.92$.

CHAPTER 12

PAGE 342

5. $F = 0.68$, not significant.
7. $F = 2.80$, not significant.
9. $F = 11.3$, significant.

PAGE 354

1. For threads, $F = 8.3$, significant; for instruments, $F = 0.06$, not significant.
3. For students, $F = 20.6$, significant; for forms, $F = 0.51$, not significant.
5. For flow, $F = 7.8$, significant; for precipitators, $F = 8.1$, significant.
13.

A	C	B
72	76	86

15.

2	4	3	5	1
25.65	23.70	23.53	20.90	20.68

PAGE 364

1. For treatments, $F = 6.55$; significant;
   for rows, $F = 1.29$; not significant;
   for columns, $F = 26.83$; significant;
   for replicates, $F = 0.02$; not significant.

3. For golf-ball designs, $F = 66.6$; significant at 0.01 level; for pros, $F = 24.3$; significant at 0.01 level; for drivers, $F = 2.73$, not significant; for fairways, $F = 73.3$; significant at 0.01 level.

5. For needles, $F = 92.5$; significant; for threads, $F = 13.0$; significant; for operators, $F = 1.3$; not significant; for machines, $F = 0.2$; not significant; for weeks, $F = 0.9$; not significant.

PAGE 372

1. $F = 10.3$; significant; $\hat{\delta} = -57.0$.
3. For analysis of covariance, $F = 14.2$, significant at 0.05 level; for one-way analysis of variance, $F = 15.0$, also significant.

CHAPTER 13

PAGE 389

1. Significant F–values for temperature ($F = 11.5$, $\alpha = 0.01$), concentration ($F = 48.0$, $\alpha = 0.01$), and their interaction ($F = 3.8$, $\alpha = 0.05$). Optimum conditions are 10 grams at 75°F, with $37.32 < \mu < 46.02$.

3. For detergents, $F = 0.05$; not significant; for machines, $F = 6.9$; significant at 0.05 level; for interaction, $F = 0.81$; not significant.

5. Hours, dosages, and the strain-dosage interaction are significant at the 0.01 level with corresponding F values of 39.4, 22.9, and 7.3; the hours-dosage interaction as well as replications are significant at the 0.05 level with corresponding F values of 3.8 and 5.1.

7. The F values for C and D are 12.7 and 86.5, respectively; both are significant at the 0.01 level.

PAGE 401

1. (a) $SST = 110.44$, $SS(Tr) = 103.94$, $SS(Rep) = 1.56$, $SSE = 4.94$;
   (b) $[A] = -1$, $[B] = -13$, $[AB] = 3$, $[C] = -25$, $[AC] = 3$, $[BC] = -29$, $[ABC] = 3$;

(c) $SSA = 0.06$, $SSB = 10.56$, $SS(AB) = 0.56$, $SSC = 39.06$, $SS(AC) = 0.56$, $SS(BC) = 52.56$, $SS(ABC) = 0.56$;

(e) B, C, and BC are significant at the 0.01 level, corresponding F values are 14.9, 55.0, and 74.0.

3. A and E are significant at the 0.01 level with corresponding F values of 13.4 and 18.2; B, C, AD, ACD, BCD, and BDE are significant at the 0.05 level with corresponding F values of 6.0, 6.3, 6.6, 5.0, 4.4, and 4.4.

5. C, B, BE, BC, BCE, A, AC, AB, ABD, and ABC are significant at the 0.01 level with corresponding F values of 408.0, 1051.9, 18.2, 18.2, 15.9, 2891.3, 11.7, 149.7, 15.9, and 46.6; E and ACE are significant at the 0.05 level with corresponding F values of 6.6 and 6.6.

9. $\hat{a}_0 = 64.8$, $\widehat{(\alpha\beta)}_{00} = -35.1$, $\hat{y}_0 = 84.0$, and $\widehat{(\alpha\beta\gamma)}_{000} = -35.0$.

PAGE 415

1. (a) *Block 1:* 1, b, ac, abc, ad, abd, cd, bcd, ae, abe, ce, bce, de, bde, acde, abcde;

   Block 2: a, ab, c bc, d, bd, acd, abcd, e, be, ace, abce, ade, abde, cde, bcde;

   (b) *Block 1:* 1, ac, abd, bcd, be, abce, ade, cde;

   Block 2: a, c, bd, abcd, abe, bce, de, acde;

   Block 3: b, abc, ad, cd, e, ace, abde, bcde;

   Block 4: ab, bc, d, acd, ae, ce, bde, abcde.

3. 32 blocks of size 2; confound on AB, AC, DE, DF, and AD.

5. The intrablock SSE equals 4.39; B, C, and BC are significant at the 0.01 level with corresponding F values of 14.5, 53.5, and 72.0

11. (a) A and BCD, B and ACD, C and ABD, D and ABC, AB and CD, AC and BD, AD and BC.

    (b) 1 and BC and ABD and ACD, A and ABC and BD and CD, B and C and AD and $ABCD$, AB and AC and D and BCD.

13. 1, bc, adf, aeg, defg, abcdf, abceg, bcdefg, bdg, bef, cef, abfg, acfg, abde, acde, cdg, ab, ac, bdf, beg, cdf, ceg, acdefg, abdefg, de, fg, adg, aef, bcde, bcfg, abcdg, abcef.

15. (a) 4920, -360, -420, -144, -84, -68, 12, -40, -374, 118, -126, 122, 22, 2, -22, -22, -84, 16, -32, 64, -80, 44, -24, 120, -138, 70, -90, 42, -54, 18, -2, 82;

    (b) $[I]$, $[A]$, $[B]$, $[AB]$, $[C]$, $[AC]$, $[BC]$, error, $[D]$, $[AD]$, $[BD]$, error, $[CD]$, error, error, $[EF]$, $[E]$, $[AE]$, $[BE]$, error, $[CE]$, error, error, $[DF]$, $[DE]$, error, error, $[CF]$, error, $[BF]$, $[AF]$, $[F]$;

(c) $MSE = 123.41$; A, B, and D are significant at the 0.01 level with F values of 32.8, 44.7, and 35.4, respectively; AB is significant at the 0.05 level with an F value of 5.2.

CHAPTER 14

PAGE 430

1. (a) central line $= 0.150$, $UCL = 0.153$, $LCL = 0.147$;
   (b) central line $= 0.005$, $UCL = 0.010$, $LCL = 0$;
   (c) $\bar{x}$: 8th, 16th, and 17th sample values outside limits;
       R: All sample values within limits.
3. (a) central line $= 48.1$, $UCL = 50.3$, $LCL = 46.0$;
   (b) central line $= 2.9$, $UCL = 6.7$, $LCL = 0$;
   (c) process mean out of control, process variability in control;
   (d) $z = -2.12$, there is a trend;
   (e) no, process is not in control.
5. (a) $\bar{x}$: central line $= 21.7$, $UCL = 25.2$, $LCL = 18.2$;
       σ: central line $= 1.05$, $UCL = 3.74$, $LCL = 0$;
   (b) yes, process is in control.
7. central line $= 0.0017$, $UCL = 0.0042$, $LCL = 0$.
9. (a) central line $= 0.037$, $UCL = 0.093$, $LCL = 0$.
11. Yes, central line for c-chart is 4.9, $UCL = 11.6$, and $LCL = 0$.

PAGE 435

1. We can assert with 95% confidence that 99% of the pieces will have yield strength between 36,897 and 68,703 *psi*.
3. (a) 0.1063 ± 0.0008; (b) 0.1063 ± 0.0001.

PAGE 446

1. (a) 0.046, 0.139; (b) 0.059, 0.143.
3. Producer's risk $= 0.143$, consumer's risk $= 0.082$.
7. $AOQL \approx 0.025$.
9. $\alpha = 0.15$.
11. Sample size is 32, acceptance number is 5, rejection number is 6.
13. $a_n = -1.67 + 0.19n$, $r_n = 2.14 + 0.19n$; reject on 8th trial.

CHAPTER 15

PAGE 458

1. 0.9936.
3. 0.9983.
5. (a) $f(t) = \beta(1 - t/\alpha)e^{-\beta t(1 - t/2\alpha)}$ for $0 < t < \alpha$,
    $f(t) = 0$ elsewhere,
    $F(t) = 1 - e^{-\beta t(1 - t/2\alpha)}$ for $0 < t < \alpha$,
    $F(t) = 1 - e^{-\alpha\beta/2}$ for $t > \alpha$.
7. (a) 0.6703; (b) 0.6703.
9. 31.
11. (a) 0.6065; (b) 0.9502; (c) 0.3679.
13. (b) $F(x) = 1 - (1 - p)^x$ for $x = 1, 2, 3, \ldots$; (c) 0.301.

PAGE 469

1. (a) $2{,}635 < \mu < 11{,}002$; (b) $T_8 = 39{,}810$, reject H_0.
3. (a) $65.0 < \mu < 368.9$; (b) $T_r < 2511.25$, cannot reject H_0.
5. (a) $t^* = 998.7$; (b) $t^* = 24.1$.
7. $\hat{\alpha} = 0.0039$, $\hat{\beta} = 0.536$; $R(1000) = 0.854$; using exponential model,
    $\hat{\alpha} = 0.126$ per thousand hours.
9. $\hat{\alpha} = 6.75 \times 10^{-5}$, $\hat{\beta} = 1.51$, $\hat{\mu} = 580\Gamma(1.66) = 523$.

Index

A